LONG NIGHT
OF THE TANKERS

BEYOND BOUNDARIES:
CANADIAN DEFENCE AND
STRATEGIC STUDIES SERIES

Rob Huebert, Series Editor

ISSN 1716-2645 (Print) ISSN 1925-2919 (Online)

Canada's role in international military and strategic studies ranges from peace-building and Arctic sovereignty to unconventional warfare and domestic security. This series provides narratives and analyses of the Canadian military from both an historical and a contemporary perspective.

DAVID J. BERCUSON
AND HOLGER H. HERWIG

LONG NIGHT
OF THE TANKERS

HITLER'S WAR AGAINST CARIBBEAN OIL

UNIVERSITY OF
CALGARY
PRESS

BEYOND BOUNDARIES: CANADIAN
DEFENCE AND STRATEGIC STUDIES SERIES

ISSN 1716-2645 (Print) ISSN 1925-2919 (Online)

University of Calgary Press
2500 University Drive NW
Calgary, Alberta
Canada T2N 1N4
www.uofcpress.com

LIBRARY AND ARCHIVES CANADA CATALOGUING IN PUBLICATION

Bercuson, David Jay, 1945-, author
        Long night of the tankers : Hitler's war against Caribbean oil / David J. Bercuson and Holger H. Herwig.

(Beyond boundaries ; 4)
Includes bibliographical references and index.
Issued in print and electronic formats.
ISBN 978-1-55238-759-7 (pbk.).—ISBN 978-1-55238-760-3 (open access pdf).—
ISBN 978-1-55238-761-0 (edist pdf).—ISBN 978-1-55238-762-7 (epub).—
ISBN 978-1-55238-763-4 (mobi)

        1. World War, 1939-1945—Naval operations—Submarine. 2. World War, 1939-1945—Naval operations, German. 3. World War, 1939-1945—Campaigns—Caribbean Area. 4. Petroleum industry and trade—Caribbean Area—History—20th century. I. Herwig, Holger H., author II. Title. III. Series: Beyond boundaries ; 4

D781.B47 2014   940.54'51   C2014-902466-5
                                                               C2014-902507-6

The University of Calgary Press acknowledges the support of the Government of Alberta through the Alberta Media Fund for our publications. We acknowledge the financial support of the Government of Canada through the Canada Book Fund for our publishing activities. We acknowledge the financial support of the Canada Council for the Arts for our publishing program.

This book has been published with support from the Centre for Military and Strategic Studies at the University of Calgary.

Printed and bound in Canada by Marquis
♻ This book is printed on FSC Enviro 100 paper

Cover design, page design, and typesetting by Melina Cusano

To Barrie and Lorraine

TABLE OF CONTENTS

ACKNOWLEDGMENTS

I am indebted to Marjan Eggermont, Senior Instructor at the Schulich School of Engineering, University of Calgary, for assisting me with the translation of Dutch materials concerning the Chinese stokers' uprising on Aruba in 1942; and Dr. Herb Emery, Department of Economics, University of Calgary, for his help in converting 1942 florins and guilder to current US dollars. The book owes much to the ongoing efforts to provide documents, maps, and photographs on the part of the staff at both the Deutsches U-Boot Museum (formerly Traditionsarchiv Unterseeboote, U-Boot Archiv) at Cuxhaven-Altenbruch, Germany; and the Federal Military Archive (Bundesarchiv-Militaerarchiv) at Freiburg, Germany. And most pleasantly, Admiral Pierre Martinez, Commandant la Marine á Lorient, provided me with a personal tour of the Villa Kerillon on July 22, 2006.

– HOLGER H. HERWIG

Thanks to Alex Heard, Marshall Horne, Stephen Randall, and Nancy Pearson Mackie.

– DAVID J. BERCUSON

PROLOGUE

The waters north of Scotland are nasty at the best of times. During the long winter months, Force 6 to 10 storms with sleet and ice are normal. Winds of 30 to 40 knots howl over the heavy gray waves, with breaking crests forming streaks of foam. But Kapitänleutnant[1] Albrecht ("Ajax") Achilles considered himself to be a lucky man – in fact, doubly lucky. In December 1941, Hans Witt, the first commander of the brand new *U-161*, had broken his leg in an accident on shore, and on the last day of the year, Achilles, until then First Watch (or Executive) Officer, at age 28 had been given command of the 1,200-ton Type IXC boat. His companion from pre-war merchant shipping days, Oberleutnant[2] Werner Bender, became the new executive officer. And now, the second piece of luck: the first week of 1942 brought only moderate Force 2 to 3 light breezes in those usually turbulent seas between the Shetland and Faeroe Islands. The sky was overcast, good protection from patrolling British aircraft. Gray boat. Gray seas. Gray skies.

U-161 was running well, covering more than 230 nautical miles per day. Ahead to the southeast lay the German-occupied French naval bases in the Bay of Biscay, the boat's first port of call. On January 7, Achilles received a garbled message from Group North that a convoy had been sighted just west of his position, but he was too far off to the south to join the hunt. Then, around 12:30 p.m.[3] on January 9, *U-161* received a terse, coded "for-officers-only" radio message from Vice Admiral Karl Dönitz, Commander U-Boats. "Proceed to Lorient at once."[4] What emergency had prompted this sudden haste, Achilles wondered? Was it a routine dispatch announcing that Lorient was to be *U-161*'s new home? Or was there a new theater of operations in the cards? Orders were orders. Achilles at once shaped course for Lorient. Within minutes, a smoke smudge appeared on the horizon. "Ajax" gave chase, but to his dismay, the target was moving too fast and was protected not only by a surface escort but also by an aircraft. *U-161* again shaped course south-southwest for Lorient.

Shortly before noon on January 10, the Old Man put his crew through the paces of an emergency dive. Within minutes, the boat became heavy by the bow and quickly plunged to depth "A," 80 meters. Chief Engineer

Klaus Ehrhardt managed to trim the boat at that depth and reported that Dive Tank IV on the starboard side had unexpectedly taken on air due to a faulty seal and that seawater was penetrating the tank. Moreover, Ehrhardt suspected that oil was leaking into the dive tank from its hydraulic pressure hoses.

Back on the surface, the Atlantic was beginning to show its true winter face: Force 5 winds, with a fresh breeze and long waves cresting into foam and spray. On the bridge, Achilles discovered that *U-161* was indeed leaving an oil slick behind. No choice but to proceed as ordered. Fortunately, heavy leaden skies kept enemy bombers away. For the next five days, *U-161* crashed through the rising seas en route to Lorient. On the afternoon of January 15, the bridge watch sighted the low, gray shape of the Île de Groix, an eight-kilometer-long rock that protected Lorient from the often tempestuous southwesterly gales.[5] The Kéroman and Scorff River U-boat pens lay further inland.

As Achilles carefully steered his boat toward the naval base, he could not help but take in the history and geography of the German command post. Off his port side lay Larmor-Plage, recently upgraded with a steel-reinforced concrete artillery post; off his starboard side was the massive stone fortress Port-Louis, originally built to protect the trade of the *Compagnie des Indes* and under Louis XIV expanded into a star-shaped citadel. The waters between the two points were about a kilometer wide, but this was deceiving since countless submerged rocks and mud banks studded the Larmor-Plage shore; in reality, the navigable channel of the Kernével Narrows was a mere 200 meters wide. Mariners since the days of the Celts and Julius Caesar had passed on the adage, "You must be crazy to moor in the Blavet River," the main tributary into the harbor.

With the heavy gray winter sun sinking off its port side, *U-161* passed through the narrow channel and entered the *Rade de Lorient*, a two-kilometer-wide bowl formed by the confluence of the Scorff, Ter, and Blavet rivers. It was a maelstrom of fresh water, tidal sea water, heavy silt, and harbor offal. Off the left bow, Achilles could make out three elegant late-nineteenth-century villas along the beach of the resort town of Kernével: Kerillon, Margaret, and Kerozen. Completed in 1899 by a wealthy Breton engaged in the sardine fishing trade, they had been confiscated by German naval commander Dönitz in mid-October 1940 and the owners given 24 hours to vacate. The middle building, the Villa Kerillon, was the headquarters of Commander U-Boats; the two flanking structures housed his staff.

Achilles was ordered to put into the narrow basin leading up to the large Kéroman U-boat pens. He made fast at the pier at precisely 6:50 p.m. The next day, Chief Engineer Ehrhardt would supervise repairs to the faulty diving tank by way of an ingenious system of wet and dry bunkers.[6] *U-161* was scheduled to be taken into an enclosed "wet" berth, on whose sloping floor rested a 45-meter-long cradle. Once secured on the cradle, water would be pumped out of the berth and cradle and the U-boat, secured by an overhead crane, would be lowered onto a wheeled trolley. The boat would then be hauled out of the water, up a sloped slipway, placed on a 48-meter-long traversing unit on eight sets of rails, and thereby aligned with and directed into any of the two sets of five "dry" Kéroman bunkers on either side of the traversing unit. The operation would take up to two hours. Amazingly, never once did Allied bombers manage to damage a single boat undergoing this transfer process.

Would there be time for shore leave, Achilles wondered? Perhaps a quick trip to Paris? No such luck. No sooner had *U-161* been safely berthed, than "Ajax" was peremptorily ordered to report to Dönitz's headquarters. As darkness set in, a staff car took him across the Ter River to the Villa Kerillon. It was a veritable fortress: an anti-tank ditch surrounded the modest château and three 5-cm anti-tank guns as well as the turret of an old French tank protected it against land attack; numerous small-caliber anti-aircraft guns mounted in concrete pillboxes and countless searchlights studded the coastline along the narrow channel guarding against hostile aircraft.[7] The wiry, athletic Achilles quickly bounced up the eight stairs of the villa and via a small foyer entered a vast space of three interconnected rooms. This was the admiral's operations nerve center. Elegance abounded: the ceilings were six-meters high, the floors had been constructed of inlaid oak planks, plate-glass windows offered splendid views of the harbor channel as well as the flood-lit openings of the "wet" Kéroman bunkers, and an exquisite spiraling wooden staircase led to the upper two floors of what Dönitz's staff had dubbed "le château des sardines."[8]

From the great window of the central room, Achilles could see that the lawn leading down to the beach had been replaced with a brownish concrete slab – the roof of three steel-reinforced concrete bunkers completed by the Organisation Todt[9] in 1941 as protection against enemy bombs, which had first fallen on Lorient on September 1 and 27, 1940, just to remind the Germans of the air dimension to the Battle of the Atlantic. The bunkers housed Dönitz's communications center, called "Berlin" by its staff. Further off toward the land approach to the Villa Kerillon was another set of massive bunkers, these for the command post's naval security detail.

The villa's three rooms were Dönitz's operations center.[10] Maps and charts studded the walls in the two "Situation Rooms." Pins and small flags marked the positions of the U-boats on patrol as well as anticipated convoys and known dispositions of Allied anti-submarine warfare (ASW) forces. Others consisted of weather charts, world time zones, ice and fog conditions in the North Atlantic, dates on which U-boats were expected back from patrol, and times when new boats were scheduled to deploy. A globe one meter in diameter gave a realistic picture of the broad sweeps of the Atlantic, allowing better distance calculations due to the curvature of the ocean's surface. The third room was the so-called "Museum," where yet more charts and graphs tracked sinkings at sea, U-boat losses, average sinkings per day at sea, and success rates against convoys.

Achilles noticed immediately that several men were already sitting around a massive oak table. All were of the same rank as he – Kapitänleutnant. He recognized the senior member of the group, the 33-year-old Werner Hartenstein, the gruff commander of *U-156*. Korvettenkapitän Viktor Schütze, 2nd Flotilla Leader, then introduced himself as well as three other skippers: Jürgen von Rosenstiel of *U-502*, Günther Müller-Stöckheim of *U-67*, and Asmus Nicolai Clausen of *U-129*. Obviously, Dönitz had chosen his skippers carefully. All were "regular navy," men who had graduated from the Naval Academy and then served on surface warships. All were senior commanders who had just turned 30 or were about to reach that milestone. Dönitz calculated that they would be up to the rigors of two- to three-month-long journeys over some 10,000 nautical miles, much of it in temperatures reaching 40 degrees Celsius inside the boat. Two civilians completed the group.[11]

After brief acknowledgments, Schütze introduced Captains Strüwing and Kregohl, both former merchant skippers of the Hamburg-Amerika Line. Both had plied the waters of the Caribbean Sea before the war. Achilles sat up at once. So, this was the reason for the terse command to head for Lorient without delay. He, Achilles, also had worked for the same shipping line as a cadet officer, mainly in the waters around Trinidad, as had his Executive Officer, Bender. For the next hour, the U-boat captains took detailed notes as Strüwing and Kregohl briefed them on currents and reefs, shipping routes and harbor installations, and the sailing patterns of numerous Caribbean steamship lines.

Late in the evening, "the Great Lion," as Dönitz was called by his U-boat crews, joined the group. His large forehead and ears and thin mouth gave his head an unbalanced look. But his chin was set and his small, steely blue eyes penetrating. His admiral's uniform sat immaculately on his lanky frame. He had not put on a pound since his days as commander of *U-68* in the Adriatic Sea during the Great War. He took his place at the head of the table and eyed each man in turn. Then he got down to business. Whereas Adolf Hitler until recently had vetoed all plans by the navy to interdict the trans-Caribbean flow of crude and refined oil or to shell the large refineries because "oil centers belong to Standard Oil, thus American corporation,"[12] now that the United States was officially in the war, there was no further impediment to action. The boats were to mount a special operation, code-named "Neuland," or New Land, an assault on the oil tankers and bauxite carriers that plied the Caribbean basin. The operations orders were precise: "Surprise, concentric attack on the traffic in the waters adjacent to the West Indies Islands. The core of the task thus consists in the surprising and synchronized appearance at the main stations of Aruba a[nd] Curaçao."[13] The group was to commence operations during the new moon period beginning on February 16, 1942. Müller-Stöckheim's *U-67* was to take up station off Curaçao; Hartenstein's *U-156* and Rosenstiel's *U-502* off Aruba; Achilles' *U-161* was to attack Port of Spain, Trinidad; and Clausen's *U-129* was to patrol the coast of the Guianas. Primary targets, apart from the oil tankers and bauxite freighters, were also the mammoth oil refineries on Aruba that produced almost 500,000 barrels of gasoline and diesel fuel per day. An ocean-going tanker with 3.5 million gallons of refined gasoline in its bunkers would be a splendid target!

Dönitz then pushed back the pile of papers on the table before him and assumed a more relaxed posture. His skippers knew well that the time had come for the customary pep talk. The admiral impressed on them the importance of the operation and its expected effect on enemy land, sea, and air operations. He informed them of the rich harvest that the six boats currently deployed in Operation Drumbeat (Paukenschlag) were taking off the United States' eastern seaboard. He expected no less from Neuland. He reminded them yet again that the Atlantic was "the decisive theater of the war." He demanded victory at all cost. "Be strong! Do not falter!" The Führer and his Wehrmacht stood at the gates of Moscow. "Faith in the Führer is a German officer's first and foremost duty," Dönitz sternly lectured the Kaleus. "Find, engage, destroy!" "Attack, attack like wolves!" The pep talk behind him, "the Great Lion" turned the briefing back over to Schütze and his staff.

"Operations Order No. 51 'West Indies'," formalized on January 17, 1942, defined specific targets. Aruba stood at the top of the list. The oil refineries, first and foremost the Standard Oil of New Jersey Lago plant at San Nicolas and secondarily the Royal Dutch Shell refinery north of Oranjestad, were the main targets. Willemstad on Curaçao was home to a much larger Royal Dutch Refinery. "The oil is brought to Aruba as well as Curaçao from the Gulf of Maracaibo [Venezuela] in shallow-draft tankers of about 12 to 1,500 tons with a draft of 2 to 3 m[eters], is refined there and loaded onto large ocean-going tankers." The Gulf of Maracaibo was protected by a large sand bank and as a result of the shelling of Maracaibo's Fort San Carlos in January 1903 by the German cruiser *Vineta*,[14] Juan Vicente Gómez, the Venezuelan dictator, had refused to dredge the sand bank for fear that other foreign warships might enter the Gulf. Thus, only small tankers could exit Maracaibo and only at high tide, "usually at day break." Trinidad offered another target-rich environment, as it not only contained oil refineries and tank farms but was also the port of destination and transshipment site from the Guianas of valuable bauxite, vital for airplane production. Furthermore, it was the departure point for traffic bound for Cape Town, South Africa. A third target was the Florida Strait characterised by elastic or elasto-plastic fracture mechanics parameters in most cases. When the effect of plasticity can be neglected, the stress and strain states around a crack tip are characterised by K, and the crack growth per cycle da/dN is often plotted against stress the intensity factor range ΔK which is defined as the difference between the maximum K_{max} and minimum K_{min} stress intensity factors applied each load cycle. ΔK is given by;

$$\Delta K = Y \Delta \sigma \sqrt{a} \qquad (136)$$

where $\Delta \sigma$ is the range of the applied stress, and Y is the non-dimensional geometry factor shown in equation (37). The crack growth rate da/dN is correlated with ΔK using the power law relation [89] in steady state fatigue as,

$$\frac{da}{dN} = C_f \Delta K^m \qquad (137)$$

where C_f and m are material constants and m is typically around 3. Typically, da/dN is sensitive to the mean stress or the load ratio R defined by;

$$R = \frac{\sigma_{min}}{\sigma_{max}} \qquad (138)$$

where σ_{min} and σ_{max} are the minimum stress and the maximum stress, respectively [89]. To include the effect of the load ratio in equation (137), the effective stress intensity factor range ΔK_{eff} is used. Several formulae to define ΔK_{eff} are proposed using ΔK and R [89-91].

6.2 Elevated temperature cyclic crack growth

In the previous section cyclic crack growth was considered where time dependent effects were not important and where cracking was controlled mainly by fatigue mechanisms. As temperature is increased, time dependent processes become more significant. Creep and environmentally assisted crack growth can take place more readily since they are aided by diffusion and rates of diffusion increase with rise in temperature. The effects of temperature, frequency, mean stress and environment will be considered in turn [91-99].

Frequency effects can be investigated in continuous cycling tests or by introducing hold periods into a cycle as illustrated in Fig 21b. However, it is found that wave-shape is relatively unimportant compared with the temperature and mean load at which the cycling is taking place. Generally the influence of frequency on crack propagation rate is more pronounced with increase in temperature and R. Figure 38 shows a schematic description of cracking rate versus ΔK for cyclic cracking at elevated temperatures showing the effects of frequency, R-ratio and temperature. A higher crack propagation rate is observed with increase in load range. In all cases, crack propagation rate increases slowly at first before accelerating rapidly as final fracture is approached. As a consequence, most of the time in these tests is spent in extending the crack a small distance and the number of cycles to failure is not significantly influenced by the crack size at which fracture occurs.

and the tankers that traversed it en route to New Orleans, Galveston, and Port Arthur.

Antisubmarine defenses, the former Hamburg-Amerika merchant captains reported, existed only at Trinidad. But it was likely, Schütze's staff allowed, that the first "wave" of attacks would in time bring antisubmarine nets, aerial reconnaissance and surface U-boat hunters to the Caribbean. Still, the lack of war experience of what was expected to be hastily dispatched and inexperienced American forces would render ASW "of little fighting value." All U-boats were to proceed to the West Indies running on one diesel engine only, to save fuel oil. Once they crossed the line 40 degrees west longitude, they were to radio in their position and fuel supply. Kernével would then give the signal to commence operations: "Neuland 186," with the first and third letters denoting the day, February 16. The initial attacks were to be driven home "five hours before day break."

Werner Hartenstein was to command the assault group. The skippers were to interpret their zones of attack liberally and independently – a departure for Dönitz, who liked to keep tight control of operations. They were free to repeat their attacks after initial strikes. "Thus, do not break off [operations] too soon!" They were to use their torpedoes first and thereafter their 10.5-cm deck guns if land targets were in the offing. Last but not least, Schütze handed the commanders commercial sea charts for Aruba, Curaçao, and Trinidad, as well as the most recent sailing plots for the West Indies.

Unbeknown to the Kaleus, a bitter dispute as to targeting had broken out behind the scenes between Dönitz and Grand Admiral Erich Raeder, Commander in Chief Kriegsmarine. While Dönitz as ever was fixated simply on "tonnage war" (sinking ships), Raeder demanded that shore installations such as refineries and tank farms be given priority. He had a point. The world's largest oil refinery was the Standard Oil "Esso" facility at San Nicolas, Aruba; and with the Royal Dutch Shell refinery at Eagle Beach, they together produced 5,000 barrels per day of critical 100-octane gasoline for aircraft alone. Raeder also knew that Pointe-à-Pierre on Trinidad was home to the largest refinery in the British Empire, Trinidad Leaseholds Ltd. The "Great Lion" chose to leave the targeting issue for further discussion.

"Ajax" Achilles was delighted. He and Bender had sailed the waters off Trinidad before the war and they knew intimately its reefs and currents as well as harbors and onshore installations. They planned to exploit this advantage. Moreover, the Caribbean was virgin territory for the U-boats. Surprise was thus assured. Surely, Knight's Crosses (*Ritterkreuze*) would be in the offing. And what a welcome relief the warm waters of the Caribbean would be from the frigid wastes of the North Atlantic. The meeting broke up precisely at 10 p.m., Dönitz's self-imposed bedtime.

* * *

Operation New Land was, of course, but one part of the greater Battle of the Atlantic, "the most prolonged naval campaign in history."[15] For six long years, German surface and subsurface raiders fought a tenacious battle for control of the North Atlantic sea lanes that connected Britain to its vital allies in North America. Most specifically, Karl Dönitz launched more than 1,000 of his "gray sharks" from their lairs in the Bay of Biscay in so-called "wolf packs" against the Allied lifelines; roughly 780 boats and 30,000 sailors never returned from the Atlantic. For the Allies, 175 warships, 2,700 merchant ships, and 30,000 merchant sailors met a similar fate. In time, an army of technical experts mounted a complex and sophisticated air and sea assault against the U-boats, while especially American industry ramped up merchant-ship production to the point where already by July 1941 the number of new vessels entering the Allied shipping pool surpassed total losses.

As the war escalated, especially after America's entry as a result of the Japanese attack on Pearl Harbor on December 7, 1941, Dönitz sent his U-boats ever further west, seeking out the Allied convoys at their North American point of egress. His most spectacular campaign was dubbed Operation Drumbeat ("Paukenschlag"), launched on January 13, 1942, with the arrival of five U-boats in the waters between the Gulf of St. Lawrence and Cape Hatteras; eight boats followed in March and April.[16] It was a stunning surprise: in what S. E. Morison, the official historian of the US Navy in World War II termed "a merry massacre," the raiders destroyed 470,000 tons of Allied shipping off the eastern seaboard of the United States in February, and 1.15 million tons to the end of April 1942.

Thereafter, sinkings declined precipitously as the US Navy finally adopted convoy, blackened its ports, and concentrated its air and sea resources against the German raiders.

The greater story of the Battle of the Atlantic is well known and well told – nearly 300 titles in the catalog of the Library of Congress and 90.9 million Google hits[17] attest to this. It is not our story. Rather, we concentrate on the post-Drumbeat period, when Dönitz redirected his "gray sharks" to the waters of the Caribbean and the Gulf of Mexico to wreak havoc with the Allied supply of vital stocks of crude oil, refined diesel and gasoline, and bauxite. For without those resources, the Battle of the Atlantic would have ground to a halt.

INTRODUCTION

No man knew more about the importance of oil for the Allied war effort than British Prime Minister Winston S. Churchill. On June 22, 1941 – the day Nazi forces invaded the Soviet Union – Churchill informed his countrymen during a BBC radio broadcast that Germany's "terrible military machine must be fed not only with flesh but with oil."[1] As First Lord of the Admiralty prior to World War I, he had been the key figure pressing the Royal Navy to change from a coal-fired to an oil-fired navy. He agreed with Sir Marcus Samuel, one of the principal owners of Royal Dutch Shell (formed in a merger of Samuel's Shell Oil Company and Royal Dutch Petroleum in 1906) that oil was a much more efficient fuel for warships than coal. Samuel had campaigned for the conversion since 1899, but the tradition-bound Admiralty had dragged its heels, even though some of the newest and most powerful British warships, such as HMS *Dreadnought* (launched in 1906), were in fact fitted with oil-fired boilers.

In 1912 Churchill established the Royal Commission on Oil Supplies, headed by First Sea Lord Sir John Fisher to examine the advantages of oil. The commission's finding was predictable – Fisher was a strong advocate of the conversion – and the result was decisive: coal was obsolete; oil would fire all Royal Navy ships in future. In Churchill's words, "oil gave a large excess of speed over coal. It enabled ... speed to be obtained with far greater rapidity. It gave 40 per cent greater radius of action for the same rate of coal. It ... made it possible in every type of vessel to have more gun-power and more speed for less size and cost."[2]

The sudden conversion of the world's principal navies – especially the British, American, French, and Japanese – to oil, combined with the rapid expansion of those navies in the decade prior to the war, made the problem of securing oil supplies a matter of utmost importance. Britain was particularly concerned since the Royal Navy was the United Kingdom's principal source of international power and the guardian of both its trade and its independence. Thus, Churchill was also in the forefront of Britain's drive to ensure that the Royal Navy had both secure and adequate supplies. The world's largest producer of oil by far was the United States, which was also self-sufficient in oil. Since it was simply inconceivable to the British government that it might rely largely on US sources, Churchill was forced to secure the UK's own oil supply.

The British Isles had virtually no oil resources. Thus, Britain encouraged UK financiers such as Samuel to secure whole or partial ownership of as many new or newly expanding fields in other parts of the world as possible. Russia, Romania, Persia (Iran), Iraq, and the Caribbean were the best choices. As a result, British-owned companies such as British Petroleum, the Anglo-Persian Oil Company (51 per cent owned by the UK government), and Royal Dutch Shell (40 per cent owned by Samuel) came to dominate the international oil scene outside the United States. In the interwar years Britain came to depend almost exclusively on supplies from the Far East (the Dutch East Indies), Iraq and Iran in the Middle East, and Venezuela, which by 1939 was the world's third largest producer and second largest exporter. The British supply system constituted a global network of oilfields, pipelines, refineries, tank farms, and oil ports, linked by some 500 British-flagged tankers capable of moving about 20 million barrels of oil at any one time – the largest tanker fleet in the world by far.

On the outbreak of World War II, both the Axis and the Anglo-French allies were fully aware of the importance of securing their oil supplies. Not only were the economies of these industrialized societies highly dependent on oil, but oil was essential for their war machines. The Allies appeared to be in a much more favorable position than the Axis since neither Germany nor Italy had natural oil reserves and both countries had very little tanker capacity. The British and French set out almost immediately to lease as many tankers as they could from countries such as Norway (which was neutral until April 1940), both to ensure their own supplies and to deny those ships to the Germans. But Germany was far from bereft of oil.

German scientists had been working on the means to produce synthetic oil from coal since before World War I. Even though the synthetic product proved to be six times as expensive as natural crude, the exorbitant cost was a price Adolf Hitler was willing to pay, at least until Germany could acquire natural crude through diplomacy or by conquest. Thus, the Nazis contracted with the chemical giant I.G. Farben to subsidize synthetic oil production. By 1939, synthetic oil accounted for just over one-third of Germany's oil needs. Immediately after the German army rolled into Poland in September 1939, special units fanned out to seize existing oil stocks and to take control of the small fields in Galicia before the Soviet Union did. This move was only partially successful; the Red Army occupied a sizeable chunk of that region as it took the territory allotted Moscow under the Nazi-Soviet Non-Aggression Pact. But Stalin was willing to be reasonable: the USSR sold oil to Germany literally until June 22, 1941, the day Hitler pounced on Russia. Romania was another major German supplier, especially after it joined the Axis in the summer of 1940. Romania soon became Germany's primary source of both crude oil and refined products from the giant and very modern refining complex at Ploesti. After the French surrender in June 1940, a small French-controlled field in Alsace was also commandeered. Although the German army seized key oilfields in the Soviet Caucasus in the summer of 1942, the Wehrmacht was expelled by the Red Army before any significant amount of Caucasus oil could be sent to Germany.[3]

Allied oil resources suffered from severe structural weaknesses which quickly became apparent. The greatest of these was the reliance on oil tankers. As well, the United States proclaimed strict neutrality at the start of the war and adopted a policy of "cash and carry" toward the belligerents. In other words, the Allies or the Axis could purchase whatever they wanted from the United States, but in cash, and they had to carry it away in their own ships. While this policy allowed both Britain and France to purchase oil or refined products from the United States, it was unclear whether the more than 400 US-flagged tankers would be allowed to carry oil or refined petroleum products to Allied ports in charter. Normally, neither country would even think of relying on American (or other neutral) tankers – except that the Battle of the Atlantic and the "tonnage war" waged by the German submarine force began to cut into Allied tanker capability from almost the very start of the war. And the more Allied tankers lost, the greater the damage to the worldwide system of supply that Britain had carefully built up since the earliest days of the oil-fired Royal Navy.

The initial blows fell early. On September 8, 1939, submarines sank the British tankers *Kennebec* (5,548 tons) and *Regent Tiger* (10,177 tons). Two more tankers were lost by the end of September, for a total deficit of

34,007 tons. By the end of May 1940, twenty-two British flagged tankers had gone down to torpedoes, gunfire, or mines, for a loss of about 150,000 tons – with another 67,000 idled by damage, mostly from mines. Other Allied tankers and neutral tankers were also lost, though not nearly as many.[4] At first glance, the British tanker loss appears minor compared to the total tanker fleet – about 4 per cent of capacity over a nine-month period. But at a time when British ship-building capacity was taken up almost exclusively by the production of warships, when tanker construction had fallen, and when Britain had also lost about 200,000 tons of other merchant shipping that had to be replaced, 4 per cent was a significant number. If current losses continued, a deep cut in British tanker capacity seemed unavoidable. But then German attacks appeared to slacken, and negotiations between Britain and Norway resulted in an increase in the number of Norwegian tankers available for charter by the Allies. Thus, by the end of March 1940, in the words of the official history of British oil policy and oil administration in World War II, "the barometer [measuring the future of British oil supplies] was set 'fair'."[5]

* * *

On April 9, 1940, Germany attacked Norway. The fighting raged over much of the country's coastal areas. The Norwegians were aided by British, French, and other Allied forces, but surrendered in early June. The Norwegian king and government fled to London. A few Norwegian tankers were in port at the time of the German victory and were seized by the Nazis, but the vast majority were at sea, in charter, and now available without restriction to the Allies. That was certainly a positive event, but it was more than cancelled out when, in May 1940, Germany attacked France and the Low Countries and Italy entered the war. The Dutch and the Belgians were both quickly defeated; their governments fled to the UK; and control of the vast majority of their tankers was assumed by London. "These summer months of 1940," in the words of the official history, "formed a unique interlude in the history of wartime oil supply; a period when tankers were in surplus."[6]

When France surrendered in late June 1940, its navy remained under the control of the ostensibly neutral but decidedly pro-Axis French Vichy government. So did all of French North Africa. Suddenly, the Mediterranean had effectively become an Axis lake – *mare nostrum* ("our sea"), as the fascist Italian regime put it. From Suez to Gibraltar, the Royal Navy's only port of call was Malta and most of the northern and southern shores of the Mediterranean were hostile or neutral. Britain's shipping in the Mediterranean was subject to air attack along most of its length. Its oil supplies from the Middle East were virtually cut off. It could still obtain oil from Iraq and Iran or from the Dutch East Indies via the Cape of Good Hope, but that route was very long and vulnerable to submarine attack. Since Britain began the war with virtually no strategic reserve, the oil supply picture suddenly grew very dark once again.[7]

The British government did everything it could to reallocate fuel from civilian to military consumption. A government-appointed Oil Control Board, consisting of both government and industry representatives, took control of all British oil companies and their operations. Strict civilian rationing was imposed, storage tanks were moved to areas less susceptible to bombing, underground tanks were hurriedly prepared, and aviation gas was carefully husbanded. All of these moves helped, but none came close to alleviating Britain's growing petroleum shortage.

In the high summer of 1940, the German Air Force began a concerted bombing campaign against the British Isles, beginning with attacks on coastal shipping and ending with the London Blitz – the ceaseless, mostly nightly raids against the capital that began in the fall and continued until late May 1941. At first the Germans' main objective was the bases and installations of the Royal Air Force. But they also attacked docks and oil terminals, oil storage facilities and refineries, rail yards, and transportation hubs, not to mention most of the industrial cities and shipbuilding and ship repair yards that were in range. The attacks on Britain's east coast ports, which were closer to German air fields, were especially damaging. In raids against Plymouth and the Clydeside, Royal Navy oil stocks suffered severe damage. If the Luftwaffe had sustained its attack against British refineries and oil storage facilities, it might have done considerable damage, but it did not. Bombing of British oil installations was sporadic and ineffective much of the time. Hence, although British stocks ran low – sometimes dangerously low – they never came close to running out.

The problem for Britain was not so much the maintenance of daily stocks as it was trying to ensure future stocks in what was certain to be a long war, especially now that Britain stood alone. That was a significant challenge because in late 1940 its tanker fleet began to deteriorate once again. France's surrender gave Germany two major advantages over the UK it had not had at the outbreak of war. First, French ports and naval bases on the Atlantic were now open for use by submarines; a substantial number of U-boats were transferred from Germany to newly built shelters and maintenance facilities along the Bay of Biscay. This significantly cut the distance that U-boats needed to travel to get to the North Atlantic and to waters south of Newfoundland. It also increased their time on station and thus their ability to find and sink ships. Second, German aircraft based at French airfields could cover much more of the UK while long-range aircraft, such as the four-engine Focke Wulf 200 Condor, could fly far out into the Atlantic to attack convoys or to vector U-boats to them.

One British response to these dangers was to curtail shipping to its east coast and the Thames River. This kept ships somewhat out of harm's way but led to massive congestion in the UK's west coast ports and the rail lines and roads that ran from and into them. Congestion led to delays, which made the ships and their cargoes more vulnerable. There was an increase in both sunk and damaged tankers. Soon, British shipyards were overwhelmed. Oil stocks began to slide again; this time it "was beginning to look catastrophic,"[8] despite everything being done to speed up tanker unloading and to ease rail and road congestion. By February 1941, more than a million tons of tanker capacity was immobilized.

It was not just tanker losses that put a major squeeze on Britain's oil reserves. As Churchill told the House of Commons in a secret session on June 25, 1941:

> The protective measures of the Admiralty – convoy, diversion, degaussing (mine-proofing of steel hulls), mine clearance, the closing of the Mediterranean, generally the lengthening of the voyages in time and distance, to all of which must be added delays at the ports through enemy action and the blackout – have reduced the operative fertility of our shipping to an extent even more serious than the actual loss.[9]

Put simply, the problem was the friction caused by war added to the normal business of conveying oil. Convoying, as Churchill mentioned, was a particular difficulty. Even those tankers which were not sunk, or damaged, were greatly impeded in their passage by measures that Britain was forced to take to protect wartime shipping from the U-boats. Prior to the war, for example, Caribbean crude or refined products (which made up a large part of Britain's domestic oil supply) were shipped directly to the UK from the refineries on Trinidad, Aruba, and Curaçao. But very shortly after war was declared, the Royal Navy took control of all commercial traffic into and out of the UK and, together with the Royal Canadian Navy, instituted the convoy system. It became compulsory for all vessels crossing the Atlantic to deliver cargo to the UK to travel in convoy from east coast Canadian ports. Vessels that could steam above 15 knots were exempt from sailing in convoys, but tankers slower than 15 knots (the great majority at that time) were forced to sail in "HX" or "fast" convoys from Halifax or "SC" or "slow" convoys that departed from Sydney, Nova Scotia.

A tanker headed for the UK from Trinidad, for example, would have to make its way to Halifax. When it arrived, it had to wait until a convoy was formed. Once the convoy sailed, the tanker was forced to stay with the convoy at the convoy's best speed, which was determined by the slowest ship. When the tanker arrived in UK waters, it had to proceed in a local convoy to a port as far from German bomber bases as possible. The oil would then be offloaded into local storage facilities or railway cars, or the tanker would join a coastal waters convoy that could take as long as three weeks to travel from, say, Northern Ireland to the farther destinations on the UK coast. All this additional waiting and convoying added literally weeks to the normal journey.

Thus, turn-around times for tankers increased dramatically. In the spring of 1940 a tanker might be expected to make an average of six round trips a year between the UK and the Caribbean; that dropped to 4.5 trips by winter. This 25 per cent reduction in carrying capacity could only be made up by adding at least one extra tanker for every four already in service. By the end of May 1941, oil stocks in the UK had fallen "below the level that had been declared to be the absolute minimum for safety."[10]

More tankers were needed; the United States stepped in decisively. At the end of June 1941, the Americans made tankers available to cover the "Canadian trade" (carrying oil from Venezuela to the major refinery complexes at Montreal), thus relieving eight Canadian and eight Norwegian vessels chartered to Canadian companies. They also made tankers available to cover the trade of five long-charter Norwegian tankers working in South American waters. Later in July, 19 more long-charter Norwegian tankers were relieved. These moves effectively freed up 30 tankers for the North Atlantic. Then 26 US-owned, Panamanian-flagged tankers were pressed into service bringing oil from the Caribbean to New York or Halifax, where the cargo was transferred to British-chartered vessels. The United States paid the entire cost for these charters; by the fall of 1941, British oil stocks were recovering nicely.[11]

* * *

Even before Britain effectively lost access to Middle East oil, it had become highly dependent on oil from the Caribbean; by 1940, some 40 per cent of its petroleum requirements came from Trinidad and Venezuela.[12] In less than four decades, the Caribbean basin had emerged as one of the fastest growing oil-producing regions in the world. By the late 1920s, Colombia's oil fields were pumping from 40,000 to 69,000 barrels a day, most of it for Standard Oil of New Jersey or one of its affiliates. By 1940, Trinidad, a British colonial possession, was lifting some 58,000 barrels a day. Its refineries, including the Empire's largest at Pointe-à-Pierre, were churning out more than 28 million barrels a year, much of it from Venezuela.

In fact, by the outbreak of World War II, Venezuela had become the third-largest oil producing country in the world. Its daily output was over half a million barrels, 80 per cent of it produced by Standard Oil of New Jersey and Royal Dutch Shell in fields under and near the eastern shore of Lake Maracaibo. Due to a lack of deep-water ports on the Venezuelan coast, the shallowness of Lake Maracaibo, and the turbulence of Venezuelan politics, American and British producers had long ago decided not to refine Venezuelan crude locally. Instead, it was transported to refineries in the Dutch West Indies (Aruba and Curaçao) and Trinidad via slow, shallow-draft, tankers. Purpose-built for the Lake Maracaibo-Dutch West Indies trade, these ships were terribly vulnerable, as was the entire sea-borne line of supply. Cut that line and the refineries on Aruba and Curaçao would have closed in short order.

When taken together, three Caribbean islands – Aruba, Curaçao, and Trinidad – were home to the largest refining complex in the world. The Lago Oil & Transport Co. at San Nicolas, Aruba, a subsidiary of Standard Oil of New Jersey, produced about 300,000 barrels of refined product per day. It employed between 8,000 and 10,000 people, most of the island's adult population. The company ran the island almost like a private preserve, building its own grocery stores, restaurants, movie theaters, tennis courts, and golf courses. The American executives, engineers, and other professionals who lived on Aruba had their own American schools with American teachers and an American-style hospital staffed with American doctors. There were a few Britons there as well. They and Netherlanders ran the small Royal Dutch Shell Arend (or Eagle) refinery at Oranjestad with a through-put of some 8,000 barrels daily.

On Curaçao, the Royal Dutch Shell Santa Anna refinery produced 200,000 barrels daily, the crude arriving via pipeline from the deep-water terminal at the Bay of Caracas. Together, Aruba and Curaçao had a refining capacity of slightly more than half a million barrels daily. Added to this was the 80,000 barrel-a-day refining capacity of Trinidad.[13] This surpassed some of the world's other major refining complexes at the time: 280,000 barrels a day at the Anglo-Iranian refinery at Abadan, Iran; 230,000 barrels a day at the Soviet plant at Baku; and 100,000 barrels a day at the American refineries along the Gulf of Mexico coast.

The Caribbean was highly important to Britain because the Standard Oil of New Jersey refinery on Aruba was one of the key global sources for the newly developed 100-octane aviation gasoline, a product obtained by a complex process known as catalytic cracking, first developed experimentally in the 1920s. Through a variety of production processes pioneered by Standard, Shell, and other companies, a sort of hybrid gasoline was developed from ordinary gasoline that could be used in high-compression engines. The gasoline was given a rating of "100 octane," which was a measurement of its anti-knock capability or its ability to fire high compression engines without roughness or engine knocking, which occurs when the fuel-air mixture fed in to the cylinder is too imperfect to burn cleanly, quickly and with the maximum push. The 100-octane gasoline enabled a great increase in engine power without increases in the size or weight of an aircraft engine. Higher compression ratios "enabled a plane to achieve greater speed, climb at a higher rate, and fly at higher altitudes … the extra power [also] increased a plane's carrying capacity."[14]

The US Army Air Corps adopted 100-octane gasoline as the standard for all its combat aircraft – fighters and bombers – in 1937. The Royal Air Force did the same shortly after. The Germans chose not to. Instead, their fighter aircraft manufacturers, such as Messerschmitt, concentrated on producing fuel-injection systems versus the carburetors used in the early models of British Hurricanes and Spitfires. Prior to the outbreak of war, refineries in the United States began to produce substantial quantities of aviation gas – in June 1935, the US Army Air Corps purchased its first million gallons of 100-octane gasoline. Two years later the RAF contracted with Standard to produce aviation gas at its refinery on Aruba because the British were worried that, in the event of war, Washington might adopt a policy of strict neutrality and not allow Britain access to US aviation gas.

That did not happen. When war broke out in September 1939, the United States continued to sell aviation gas to Britain. When this was combined with the aviation gas supplied by refineries in the Caribbean, at Abadan and the Dutch East Indies, and in the UK itself, the RAF was able to maintain sufficient stocks to defend the British Isles and fight the Luftwaffe in the Battle of France. After the Mediterranean was effectively closed to the UK, Britain was forced to rely entirely on the United States and the Caribbean for aviation gas. But as the United States began to build up its own air force after the surrender of France, even US supplies were not completely guaranteed.

Britain relied on the Caribbean for more than oil and aviation gas. The southern shore of the Caribbean and Central and South America were a treasure trove of strategic materials such as tungsten, manganese, chromite, copper, tin, industrial diamonds, mica, platinum, nickel, and quartz.[15] All of these minerals were necessary for the production of modern weapons and the machine tools and modes of transportation that would produce and carry them. Many or most were in short supply even in the United States, and even though Canada had large deposits of nickel, both countries needed to look south for the rest. The virtual total lack of any means of land transportation from either the east or the west coast of Central and South America to North America, let alone the Caribbean, meant that these minerals had to be transported by sea lanes that were vulnerable to German submarine attack.

One of the most important of these minerals – the raw material from which aluminum is made – was bauxite. It was absolutely vital for American and British aircraft industries. British Guiana and Dutch Guiana (Suriname) together produced 1.5 million tons of this strategic mineral each year, most of it through a virtual monopoly exercised by both the Aluminum Company of America (ALCOA) and the Aluminum Company of Canada (ALCAN). The two colonies accounted for close to 40 per cent of total global production. The rapid increase in the manufacture of warplanes in the UK after September 1939 and another increase in the spring of 1940, and US plans for a multifold increase in the US Army Air Corps, drove up the value of bauxite from this region. Britain's entire requirement of 302,000 tons came from British Guiana. US President Franklin D. Roosevelt's audacious plans, announced in December 1941, to increase US aircraft production a hundred-fold (the United States in fact accomplished it in only three years) could only be fulfilled if both British Guiana and Suriname increased shipments to at least 2 million tons each per annum. Suriname was the source of some 60 per cent of the US aluminum industry's supply of bauxite.[16]

By the summer of 1940, Britain had come to rely heavily on the Caribbean for crude oil, aviation gas, and bauxite, among other crucial commodities. Put simply, Caribbean crude kept Britain in the war. The fortunes of war had made this beautiful and tranquil sea a key theater of war. And yet, the Caribbean trade stayed virtually untouched for the first 18 months of the war. This was partially because the United States and 20 other American states on September 23, 1939, issued the Panama Declaration, which not only proclaimed their neutrality but also announced the formation of a Maritime Security Zone to extend 480 kilometers into the Atlantic from the coasts of the United States and Central and South America. The area, which included much of the western Caribbean and all the American-owned islands such as the US Virgin Islands and Puerto

Rico, was then patrolled by US planes and ships. To avoid conflict with the United States and to ensure that the neutral American states were not dragged into war, the Germans observed the zone and kept their warships clear. Until Pearl Harbor.

* * *

The United States produced sufficient oil for virtually all its needs before Pearl Harbor. Imports amounted to just 2.9 per cent of requirements, or roughly 140,000 barrels a day (daily domestic production pumped out 3.8 million barrels).[17] But the United States still relied heavily on ocean tankers to supply those parts of the nation that were far away from the producing regions; 95 per cent of the Atlantic coast's requirements came by sea. There were no large-capacity pipelines from California, Texas, Oklahoma, or the Gulf of Mexico to the east coast. Most of the oil that heated the homes and fed the factories and refineries of New York, New Jersey, and Pennsylvania, among other industrial states, for example, came by sea. In the fall and winter of 1940–41, and as winter approached again at the end of 1941 – always peak demand seasons for oil products – the U-boat offensive in the Atlantic caused serious disruptions in supplies as Allied, neutral, and even some American tankers were sunk in increasing numbers. Neither the railroads nor the highways were capable of handling enough petroleum supplies to make up the difference. In the words of one historian of the period, "The Administration's response to such problems was slow and haphazard."[18]

The major fault lay with Congress. No doubt swayed by the powerful petroleum lobby, it was reluctant to interfere in the allocation and distribution of oil. As a result, President Roosevelt delegated his executive authority to deal with the nation's war-related supply problems to Secretary of the Interior Harold Ickes, a lawyer and a political reformer who was no friend of business. In May 1941, Roosevelt appointed Ickes Petroleum Coordinator for National Defense (later known as the Petroleum Administrator for War or PAW). In usual Roosevelt fashion, the title was more impressive than Ickes' powers, which remained vague and ill-defined. But Ickes was determined to define his own powers by creating a virtual government-business alliance to coordinate production, allocation, and distribution of petroleum products. His first move in making peace with the industry was to select Ralph K. Davies, vice president of Standard Oil of California, as his deputy and vest him with power equal to his own. That went a long way to win the oil industry over. On June 19, Ickes and Davies met with some 1,500 oil men and told them the government was determined to build a partnership between the oil business and the government and that no measures would be imposed on the industry that had not been agreed upon beforehand.[19] Ickes also managed to convince the US Attorney General to exempt the oil industry from antitrust charges as they pooled resources and equipment or cooperated in coordinating supply.

Under Ickes' stewardship, the companies voluntarily reduced deliveries to gasoline and fuel oil retailers by 10 per cent; the diverted product was used to produce aviation gasoline and other defense-related products. Ickes sought to rationalize the overland delivery of oil and petroleum products by pooling and coordinating the movement of railway tank cars. The result was a dramatic increase in rail shipments of oil to the east from 40,000 barrels per day in June to 140,000 per day in October 1941.[20] But even this was not enough; the east coast remained about 100,000 barrels a day short of its requirements. Ickes thus also proposed the construction of a 22- to 24-inch crude pipeline, capable of delivering at least 60,000 barrels per day from the Southwest to the Atlantic coast. Due to Roosevelt's plan to rapidly expand war production, however, there was not enough steel available to begin production before Pearl Harbor.

On the eve of its entry into World War II, the United States was in far better shape than Britain to ride out the blows that were about to fall on its oil supply. But due to the lack of a national pipeline network, it was still far from being able to meet its own requirements for both civilian and defense needs, let alone offer the increased aid to Britain that would be so necessary for victory. And like Britain, the United States was almost totally dependent on South America for bauxite and other important strategic materials. Any widening of the war at sea – in the North or South Atlantic or in the Caribbean – would add significantly to the strains the United States was already under in the late fall of 1941.

* * *

The Panama Canal was a vital strategic interest of the United States, whose fate it had been deeply involved in determining since the turn of the century. Completed in 1914, the canal was greatly beneficial to much of the world's shipping for obvious reasons, but particularly to the United States, which used it heavily for sea transport from coast to coast. It also allowed the United States to quickly move its fleet (larger US warships were specifically built to allow them to pass through the canal's locks) from one ocean to another. Not surprisingly, as Germany and Japan began to build powerful fleets in the interwar period and grew more aggressive on the world stage, the United States took steps to shore up its defenses in the Canal Zone and on US-held islands and bases to the east of it.

Initially, the Americans worried more about sabotage of the canal by potential enemy agents. By the late 1930s, however, that concern had been surpassed by the possibility that enemy aircraft carriers might attack the canal from one ocean or the other, or that islands or airbases held by unfriendly countries might allow their territories to be used for air attacks against it. As soon as war broke out in Europe, additional US troops and aircraft were deployed to both Panama and Puerto Rico. The bulk of the aircraft were obsolete open-cockpit P-36 monoplane fighters and Douglas B-18 bombers that were limited in bomb load and had an operational radius of less than 600 miles. The task of defending the Canal Zone and Puerto Rico fell to the US Army, while the US Navy was responsible for defending the Caribbean. Not only did the two services clash often about both strategy and priorities, but the facilities available for defense in the fall of 1939 were very limited. The navy had a base at Guantánamo on the eastern tip of Cuba, a radio station at San Juan, Puerto Rico, and a small Marine Corps airfield at St. Thomas in the Virgin Islands.

During the first six months of 1940, the US garrisons and air contingents in Panama and Puerto Rico expanded rapidly; new airstrips were built, new barracks constructed, new radio and radar facilities created. But the surrender of France and the conquest of the Low Countries by June 1940 changed the entire Caribbean defense picture. The French Caribbean islands of Guadeloupe and Martinique fell under the control of the new pro-Axis Vichy government. Martinique possessed an excellent harbor and naval base; at the very moment of the French surrender, the aircraft carrier *Bearn*, with 106 US-built planes, was anchored there.[21]

The British took responsibility for the defense of Aruba and Curaçao from the Dutch government-in-exile. On May 24, British ambassador to the United States Lord Lothian sent a cable to London suggesting that the UK make a formal offer to the United States to ask it to lease lands for air bases in Trinidad, Newfoundland, and Bermuda. At first the British Cabinet balked at this suggestion. After all, the United States had done virtually nothing at that point to help the Allies. But with the French surrender on June 22 and the first arrival of American rifles, ammunition, and some artillery to the UK, the Cabinet's thinking shifted. On June 29, it agreed to offer leases to the United States for base sites in Newfoundland, Bermuda, and the British West Indies.

On September 2, 1940, US Secretary of State Cordell Hull and Lord Lothian signed an arrangement[22] whereby the United States would transfer to the UK 50 World War I flush-deck destroyers in return for leases for bases in Bermuda, the Bahamas (Great Exuma Island), Antigua, St. Lucia, Trinidad, British Guiana, and Newfoundland. The United States had chosen well. Newfoundland, already under Canadian protection, flanked the first 1,000 miles of sea route from the east coast of the United States and Canada to the UK. Bermuda sits near the main oil tanker routes from the Caribbean to the UK. The other bases extended on an arc from the Bahamas, close to the coast of Florida, to northern South America – an outer ring of defenses for the Panama Canal.

But the Caribbean bases-to-be faced significant challenges. Local governors had to be won over to the American choice of base sites. Local populations consisted of multiple cultures – Blacks, Asians, South Asians, who were both Muslim and various Christian denominations, and small but socially and economically dominant White minorities. Each of these peoples had long-developed cultures and most – even the non-American Whites – clashed significantly with the American way of doing things. The social issues were not helped by a climate in British Guiana and Trinidad that was hot, rainy, and humid and thus tended to exacerbate and magnify cultural differences.[23]

The US military began preliminary site visits and initial engineering work shortly after the destroyers-for-bases deal was signed, but British agreement to negotiate leases was just that – agreement to negotiate. The negotiations turned out to be long and protracted with arguments and disagreements over a wide range of issues from base sites to postal author-ity to jurisdiction over criminal matters (extra-territoriality). Roosevelt's announcement of his "lend-lease" offer to the UK on December 17, 1940, did not, at first, smooth the discussions. In March 1941, however, the new US ambassador to the UK, John G. Winant, told the British that Con-gress might not approve Roosevelt's Lend-Lease Bill without a successful conclusion to the lease negotiations. Churchill quickly broke the deadlock by essentially ordering the British negotiators to concede on almost all of the contentious points. The leases were signed on March 17, the same day Roosevelt signed the Lend-Lease Act.

In anticipation of the successful conclusion of the lease negotiations, the US Army began to consider a command structure for the new bases as early as the fall of 1940. Eventually, a Caribbean Defense Command was established under General Daniel Van Voorhis; on May 3, 1941, the new command was officially approved. It had three departments, one each for the Panama, Puerto Rico, and Trinidad sectors. The Panama Department was responsible only for Panama; the Puerto Rico Department included commands for Puerto Rico and the US bases in the Bahamas, Jamaica, and Antigua. The Trinidad Department would command the US bases in Trinidad, St. Lucia, and British Guiana. The Caribbean Defense Com-mand was officially inaugurated on May 29.

Although the US Army with its Army Air Corps (soon to become the Army Air Forces) was responsible for the defense of the Canal Zone, Puerto Rico, and the US bases themselves – and the air defense of the sea lanes between them – the US Navy had the responsibility to patrol the waters of the Gulf of Mexico, the Caribbean, and the 400 miles of ocean to the east of the Windward and Leeward Islands – the neutrality zone proclaimed at the Panama Conference in September 1939. It must be remembered that at this point – mid-1941 – the United States still thought the main threat to the Panama Canal and other US territory in the region (including the new bases) would come from German aircraft carriers – even though Germany had none. No one was thinking very seriously about a U-boat threat. It would be the Navy's responsibility to deal with any threat from an enemy fleet. Thus, in June 1941, the Navy set up the Caribbean Sea Frontier, which extended from the Yucatán Peninsula to an area west of the island of Grand Cayman, northward to Cuba, out through the Bahamas, eastward into the Atlantic north of Puerto Rico, then southeast to the coast of Brazil. Eventually, the Caribbean Sea Frontier was divided into the Panama, Trinidad, Puerto Rico, and Guantánamo sectors.[24] The islands, and especially Trinidad, awaited a tsunami of Americans and their culture.

"RUM AND COCA-COLA":
THE YANKEES ARE COMING!

Allied antisubmarine warfare defenses in the eastern Caribbean were anchored on two islands – Puerto Rico and Trinidad – and the American air and naval base at Guantánamo Bay on the eastern tip of Cuba. By 1940, the Americans had become familiar enough with Guantánamo – which they referred to as "Gitmo" – and Puerto Rico, having taken possession of both with the Treaty of Paris (1898), ending the Spanish-American War. But Trinidad was another matter. Though a small contingent of British troops and seamen were based there, they and much of the island's population would soon encounter thousands of American troops – some 16,000 by October 1941, many more thousands after Pearl Harbor – who knew little or nothing of Trinidad's unique history, culture, government, or social institutions.

Although most middle- and upper-class Trinidadians were loyal to Britain and the anti-Nazi war effort, they soon found that the irritation arising from life with their new ally was almost more than they could bear. That, combined with dashed expectations that Britain would reward Trinidadians for their war service, gave way to cynicism, if not bitterness. In 1943 the much-respected and highly educated physician-mayor of Port of Spain, Tito P. Achong, wrote in his annual report:

> We, West Indians, are passive onlookers of the great game of power politics. We are not supposed either to think or to express any opinion on what is going on. The role assigned to us, in the British Colonial Empire, is to shout hosannas at the amoral exploits of the mighty Aryans [*sic*] into whose hands

Jehovah has delivered us for safe-keeping, and then to get back to our natural task of hewers of wood and drawers of water.[1]

Before the war, Trinidad, in the words of the journalist and labor activist Albert Gomes, "was a remote and forgotten back-water of the world. It lay deep and still in its sweaty sleep."[2] Sugar, cocoa, and oil were Trinidad's major exports; ownership was mainly in British, French, American, and South African hands. Wages were low, poverty a way of life. Malaria, hookworm, tuberculosis, and venereal disease were rampant in parts of the island. Trinidad's racial make-up was a colored patchwork of white and black, East Indian and Chinese, East Indian Creole, and Chinese Creole, Syrian and Jewish, and people of "mixed" heritage, referred to as "colored."

There had always been high demand for cheap labor in Trinidad, first by the Spanish, then by the French, and finally by the British plantation owners. Black slaves had been brought in from Africa, and on the eve of World War II their descendants constituted about 46 per cent of Trinidad's population of half a million.[3] The British abolition of the slave trade in 1807 and the formal emancipation of the slaves on Trinidad in 1834 prompted plantation owners to turn to India for cheap labor. From 1845 to 1917, about 143,000 indentured laborers, both Hindu and Muslim, were imported to the island. In 1939 the "East Indian" population on Trinidad stood at 158,000.

Trinidad's few cities were served first by a small Portuguese commercial class, and then by the Chinese, who soon controlled its many general stores, leaving the Portuguese to run the lucrative rum business. In the 1930s the Chinese community, the so-called "Coolies," numbered about 5,000; not quite 1,000 Syrians came after the Chinese. There was also a small Jewish contingent, mainly engaged in banking and business. Finally, there was a substantial South American migrant population. Distinct and apart from these social groups, and almost autonomous in every way, stood the White, powerful, and largely foreign oilfield communities. The overnight demands for cheap labor occasioned by the arrival of the first American military and civilian authorities in the spring of 1941 resulted in the further influx of thousands of West Indian migrant workers, raising the black contingent on Trinidad to just over half the total population.

As was to be expected, given this patchwork of races and cultures, socio-economic and racial relations were both ambiguous and complex. The dominant European cultural – as opposed to business – community on Trinidad contained Spanish, French, and British elements.[4] While the operative language was English, the French Creole dialect, known as *patois*, was widespread, especially among the poor, mostly black population. Spanish and French forms of Catholicism remained the near universal religion, although large segments of the black and the black Creole populations had turned to religious syncretisms such as African Shango or Orisha and Shakerism (Shouting Baptists). Roughly, the racial composition of Trinidad during World War II broke down as follows: white 2.7, black 46.8, "colored" 14.1, and East Indian 35.1 per cent.

The locals had a permissive attitude toward assimilation. As a result, Trinidad's social elites were an "association" of white, black, and "colored" communities. "Miscegenation, acculturation, and assimilation," in the words of one scholar of Trinidad's "plural society," established "a single continuum in racial, cultural, and social terms." That "continuum" was simply labeled "Creole." Consisting of people born within the West Indies, but excluding East Indians, the term generally refers to white, black, and mixed white-black ancestry. These groupings, along with mainly British businessmen and a small "mulatto bourgeoisie," coexisted – sometimes uneasily, and largely free from external pressures.[5]

The most obvious symbol of authority – and for many, of colonialism – was the British administration, from the august figure of the governor down to the most junior civil servant. In the early years of the war Sir Hubert Young ran Trinidad aristocratically, almost as a feudal fiefdom handed him by his liege in London. He was there to exploit the island for the Crown, not to develop and much less to enrich it. He had little sympathy for its chronic fiscal, labor, political, and social problems as well as injustices and was mainly interested in upholding the dominant financial, political, and social position of the planter aristocracy. Sir Hubert and the ruling white elite patronized the swank Union Club on Marine Square, which offered billiards and cards as well as lager beer and crab-backs. They took tea at St. Benedict Monastery. They shot birds in the Caroni marshes.

The British upper crust viewed the indigenous population with what can only be called disdain and arrogance. The governor's wife, Lady Margaret, in 1940 expressed the feelings of many of the British ruling elite in a private letter to Secretary of State for the Colonies Malcolm McDonald:

> Local white creoles have no conception of manners, loyalty or any other civilized virtue. They simply do not live in the same box as ordinary human beings … they are as strange and remote morally as the African and low-caste Indians.[6]

The war merely validated Governor Young's aristocratic inclinations – and gave him a splendid opportunity to clamp down on the few freedoms that the native Trinidadians enjoyed. And he was not about to share power, graciously or otherwise, with the Americans who arrived in early 1941 and to whom the government of Prime Minister Winston S. Churchill had given 99-year leases on parts of his fiefdom. The Americans would ultimately conclude that "Trinidad proved to be the most difficult of all the British colonies to deal with."[7]

Poverty defined Trinidad. It was the glue that held the poorer segments of its society together. Gomes put it thus:

> Poverty in Trinidad is not an extremity of coldness. On the contrary, it is suffocatingly hot and humid, bug ridden and flea-infested. Its olfactory characteristics consist in the main of emanations from the ubiquitous cesspits, stale piss and the aromatic goat flavour of sweating, unwashed bodies.[8]

Wages, whether in the oil patch, on the sugar and cocoa plantations, or in the small service sector, had been kept at bare subsistence levels. Wild fluctuations in the global sugar market in the 1930s and a precipitous plunge in 1940 in the price of raw sugar to below one cent per pound had brought more economic misery and uncertainty. Overpopulation made an already bad situation even worse. Last but not least, most food staples such as rice, wheat flour, salt fish, and lard had to be imported by ship. With the arrival of the U-boats of Operation New Land, Trinidad's huddled poor literally lived from ship to ship bringing food, mainly from the United States.

Labor unrest had shaken the island just before World War II in what Gomes called "a crude surgery of murder, riot and arson."[9] On June 19, 1937, police officers had attempted to arrest the labor activist T. Uriah Butler while he was addressing a large crowd of workers of Apex (Trinidad) Oilfields Ltd. at Fyzabad. Butler was born in Granada and had seen service in World War I with the West India Regiment. He had come to Trinidad after the war to seek employment in the oil fields; he was badly injured on the job and left with a permanent limp, but received no compensation. It is not known whether it was this experience that set him off on a second career as the "Chief Servant" of Trinidad and prompted him to establish the British Empire Workers and Citizens Home Rule Party, but he quickly became a central figure on the island, combining religious fervor, showmanship, and anger at social injustice at his public meetings. As he drew larger crowds, his oratory grew angrier and more violent, possibly even seditious. On this particular afternoon, warrants were issued for his arrest and the bungling police decided to serve these at a large public meeting where Butler was speaking.

The mob rushed the police and gunfire broke out; the police chief was shot dead and another constable was severely beaten and burned to death.[10] Within 48 hours, the wildcat strike had spread to the other oilfields and refineries on Trinidad, most notably the United British refinery at Point Fortin and the Trinidad Leaseholds plant at Pointe-à-Pierre. From there, the labor unrest had moved to the sugar mills and asphalt works. Stevedores and lighter-men in the ports had refused to report for work, thereby disrupting shipping of vitally needed food. Over the next weeks, the violence and looting had escalated and spread to the sugar and cocoa plantations. By early July, about 15 people had been killed and 45 seriously injured in the riots.

British authorities reacted swiftly and forcefully, for Trinidad's three refineries provided 63 per cent of the Empire's fuel oil. HMS *York*, a heavy cruiser with 8-inch guns and flagship of the Royal Navy's America and West Indies Squadron, had been dispatched from Bermuda to Trinidad at once, as had troops of the Sherwood Foresters. They, along with about 200 hastily armed civilian guards, had helped to put an end to the violence and to provide security for the oilfields and refineries. As war clouds gathered over Europe, London had been developing plans to build massive plants on Trinidad to produce high-octane aviation fuel for the Royal Air Force and hence had been in no mood to tolerate organized unrest in the oilfields. To ameliorate its heavy-handed military response, the Government had agreed to raise the pay for non-skilled, non-agricultural laborers in the cities, and it had promised to regulate the price of food for staples such as rice, coconut oil, salt fish, and flour.

A Royal Commission established to examine the causes of the 1937 unrest placed most of the blame on the island's appalling living conditions:

> Fyzabad, a village which has grown up on the edge of the oilfields without any apparent regulation or control or observance of elementary rules as to structure, space or sanitation ... forms a suitable rendezvous for all the undesirable elements which congregate in the neighbourhood ... similar examples of the worst housing conditions adjacent to the oilfield exist at Frisco Junction, Point Fortin and Cochran Village Guapo.[11]

In addition, lack of machinery to promote collective bargaining and a general belief that it was time for greater representation for the islanders in their government all combined to create conditions ripe for social explosion. Although improvements in living conditions and labor relations slowly followed, little was done about the political situation until well after the war.

Wages on Trinidad in 1939 remained abysmal[12] by any standard.[13] What made the lot of Trinidadian workers utterly unbearable was the fact that the British colonial administration had set up a taxation system designed, in the words of the journalist Arthur Calder-Marshall, who visited the island in 1938–39, "to spare the rich and to soak the poor."[14] Additionally, colonial authorities had established a network of nefarious customs tariffs – at an *ad valorem* rate of 10 to 20 per cent – on imported building materials, clothing, coffee and tea, condiments, flour, household utensils, meat, medicines, shoes, and oils of every kind. A vast array of exemptions – for the Colonial Government, the Church, the diplomatic corps, and even the Constabulary Sports Club – made perfectly clear the thrust of the legislation.[15]

In September 1939, Trinidad was automatically sucked into the vortex of war against this backdrop of fiscal and labor inequity. Not surprisingly, for many Trinidadians – and especially the well-to-do who had been educated in New York, Massachusetts, and Pennsylvania – the United States was a beacon of liberty and enlightenment. They found American society, especially in the North, to be more open, less class-ridden, and not as patronizing as that of Britain. They listened to American short-wave radio, kept up with American sports, read American magazines, saw Hollywood movies, and dreamed of someday owning a second-hand Ford, Chevrolet, or Pontiac. Many spoke openly of "secession" (from Britain) and of "union" (with the United States).

Trinidadians showed their pro-American and anti-British sentiments in numerous ways before the war. They cheered at the cheap movie theaters when newsreels showed photos of Adolf Hitler and Benito Mussolini. They refused to stand for "God Save the King," their small way of "getting the better" of their British overlords. And they openly welcomed American visitors to Trinidad. The "Yanks" spent their money freely, more of them (than British) came to the Island, and when there, more of them hired taxis and tipped handsomely.[16] They swayed to the music of Trinidad's famous calypso singers, such as "Attila the Hun," "The Lion," "Lord Invader," and "Radio." They enjoyed their sojourns in the island paradise.

President Franklin D. Roosevelt's stirring announcements of the Four Freedoms (of speech and worship, from want and fear) and of the Atlantic Charter (the rights of all peoples freely to choose their form of government) resonated in Trinidad. The historic "destroyers-for-bases" deal of September 1940 stirred many hopes for better days ahead in Port of Spain – for the inevitable clashes between the brash Yanks and the crusty Governor Young could only play into the hands of Trinidad's political activists. Broad sections of the population anticipated that there would be the proverbial pot of gold at the end of the rainbow. Washington initially estimated the costs for military bases in the Caribbean basin at roughly $200 million, with almost half earmarked for Trinidad.[17] There would finally be jobs, real jobs at good American wages. New highways would crisscross the island. New docks would bring the world's commercial traffic to Trinidad. New airfields would connect it to the rest of the Americas.

Thus, hopes ran high in the fall of 1940 when the first United States mission arrived in Trinidad on the light cruiser USS *St. Louis* and immediately made it clear that its members were coming, not to defend the British Empire or enrich the locals, but to advance American security needs. Not surprisingly, a bureaucratic "cold war" broke out at once among Governor Young, the US War Department, General Frank M. Andrews, chief of the Caribbean Defense Command, and Rear Admiral J. W. Greenslade, who headed the mission.[18]

Governor Young officially welcomed his "guests for such a long time" – 99 years, to be exact – and then launched into a plethora of concerns about what he believed the Americans wanted, including a significant expansion of the fleet anchorage and bases "dotted about in different localities all over the Colony." He claimed that the Americans seemed to leave no role at all for Britain concerning ASW measures to be taken. They viewed Trinidad alone as being worth "forty out of the fifty destroyers that had been handed over by the United States Government" as part of the September 1940 deal. They intended Trinidad, rather than being the center for the fight against German U-boats, to be the primary "jumping-off ground for operations by the United States Army in South America." Given that the U-boat attacks in the Caribbean were still a year and a half away, there was no doubt much to this, but then Young had no better foreknowledge of the U-boat campaign than did the Americans.

The US mission demanded vast tracts of Trinidad as sites for air bases. The testy governor at first offered what the Americans deemed to be a "large, miasmic swamp" between Port of Spain and San Fernando.[19] Greenslade rejected this outright and insisted on the greater part of Trinidad's northwest peninsula, and especially an area known as the Cumuto Reserve west of the town of Sangre Grande, for the US Navy. In effect, Young had offered the Americans some of the worst land on the island and the Americans insisted on some of the best.[20] Young would have none of it. On December 4, he flew to Washington for discussions with US Secretary of State Cordell Hull and the British Ambassador, Lord Lothian. President Roosevelt then appointed a special commission to resolve the base sites problems and sent them to Trinidad. The commission sided with Greenslade, but Young dug in his heels. On December 20, the president announced that the Trinidad bases question was to be a matter of direct negotiations with Prime Minister Churchill.[21] The Americans carried their case in March 1941 – in part because Roosevelt let it be known to British authorities that if he "leaked" the details of these desultory negotiations, the result might be defeat of the Lend-lease Bill, then before Congress.[22]

Thus, the Americans acquired the Cumuto Reserve, where they built Waller Air Field and the Fort Reid army base on 18 square miles. They also received the entire northwest peninsula east of Arima, the five "quarantine islands" off Port of Spain for their fleet anchorage, a small recreation strip on Trinidad's eastern coast, "supply and gun wharf facilities" in the capital itself, and an auxiliary airfield east of Longdenville. As well, Rear Admiral Greenslade insisted that the US Navy occupy and arm the islands of the Dragon's Mouth and the Serpent's Mouth to protect the entrances to the Gulf of Paria.

The Americans selected Port of Spain to be their primary naval base and materials shipment center. Although its harbor had to be dredged regularly, it provided good docking facilities at King's Wharf, which, in turn, had a decent rail connection to the city. But the vast amount of men and materials scheduled to garrison Trinidad against the U-boat threat soon overtaxed Port of Spain's facilities, and hence the Americans established a second base, Docksite, adjacent to King's Wharf. Extending along the Gulf of Paria for some 1,000 meters, Docksite in 1941 was an undeveloped, tidal mudflat of about 28 acres. It would eventually be expanded to include 183 acres and to reach as far west as Chaguaramas.

Preparations for the site of the US Navy base at Chaguaramas began March 1, 1941; on the 31st the Americans formally took possession; on June 1, they commissioned the base. Under the existing Defense Regulations and the Trinidad Base Agreement, they expelled local residents to construct the naval base. By mid-March, the last 25 families had been given notices to leave their homes at Nicholas. By mid-December, residents at Staubles Bay, Saline Bay, and Tetron Bay had received similar notices. Their homes were demolished and, to add insult to injury, they were denied use of their former beach clubs and holiday homes.

The first American contingent of six officers, 995 enlisted men, and ten civilians arrived on May 5, 1941.[23] Within months, contractors at Port of Spain and Chaguaramas threw up a plethora of buildings: general depots, warehouses, repair shops, seaplane hangars, administration buildings, a theater, a hospital, and even a new, large army wharf. Concurrently, work gangs labored around the clock to dredge deep channels through the mudflats for use by ocean-going tankers, bauxite carriers, and merchant steamers. Local black labor had long worked the mudflats. British journalist Calder-Marshall left a vivid description: "Fivepence an hour, ten hours a day.... Nightshift, dayshift, nightshift. Ten hours on, fourteen off.... The noon sun blazing, rain like gravel on the back, the sudden cold, the steam of drying."[24]

According to island legend, the Americans at Chaguaramas gave birth to the steel band. Base personnel threw out garbage in empty steel (mostly oil) drums and burned the contents at noxious dump sites. Trinidadians working on the base observed that as the drums heated up in the fire, they gave off peculiar sounds, "and so began the long and laboured experimentation that resulted in the unique music from empty steel containers."[25]

The dramatic expansion of port facilities between Port of Spain and Chaguaramas did not sit well with Trinidad's educated elite. Eric Williams, the future first prime minister of an independent Trinidad and Tobago, lamented both the length of the leases (99 years) and the United States' selection of Chaguaramas, the natural site for any future expansion of Port of Spain, for its major naval base.[26] He argued that no formal deed of lease had been registered at Port of Spain. He remonstrated that many residents of Chaguaramas had received inadequate compensation when the US Army expropriated their homes for base construction. He complained that islanders had suffered from the spiraling inflation brought about by this massive infusion of "Yankee dollars" – the cost of living had escalated from a base of 100 in 1939 to 170 by 1942.[27] Yet, in the end, "wartime necessity" overruled such considerations.

Having been rebuffed by Washington on the matter of strategic bases, Governor Young turned the discussions toward the environment. The Aripo River at best supplied 4 million gallons of fresh water a day. He offered the Americans 20 to 25 gallons per man per day; they demanded 100 gallons because Americans "were accustomed to take shower-baths." With as many as 40,000 soldiers, sailors, contractors, and construction crews expected to arrive soon, Young calculated that the Aripo River reserves would be totally exhausted.[28] His argument fell on deaf ears.

Young then returned to his earlier offer of the Caroni Swamp. His "naval, military and air advisers," he allowed, had calculated that this vast site south of the capital would permit the Americans to place all their air and naval assets in one central area; would obviate the need to punch new roads through the mountainous terrain of the Northwest; would permit use of existing shore batteries to protect the oil refineries; would facilitate the building of two airfields; and would "eliminate the possibility of constructing any form of British naval base in Trinidad for the next hundred years." The last argument, especially, was hardly attractive to Greenslade. The admiral's consulting engineers countered that reclamation of the Caroni Swamp would "be quite impractical for military purposes" as it would "take 15 years to complete," given that the swamp would have to be built "up to a height of 10 feet on the shore line and 15 to 20 feet inland." More, the mud of the Caroni River would not support "the heavy weights necessitated by military requirement." In short, any military development of the swamp would "be fighting against nature."[29] Greenslade insisted that his naval base be sited at Chaguaramas. It was.

The indefatigable Governor Young then shifted his diplomatic offensive to the fiscal and customs privileges extended to the American forces as well as to military and civil jurisdiction on the bases. He demanded that British laws and taxation prevail. He lost the battle on all fronts: the Americans simply were unwilling to place base security in British hands or to recognize British civil courts. They insisted on (and received) complete extra-territorial rights at all base sites. The final settlement between Washington and London was clear on the matter. "His Majesty's Government agree that the United States may exercise ... all such rights, powers and authority as may be necessary for conducting any military operations deemed desirable by the United States."[30] The document left no room for Anglo-American "joint" efforts, or even for mutual consultation.

Governor Young's final gambit was to demand that British contractors be allowed to bid for construction of the American bases. In this, too, he lost. The Eastern Division, US Corps of Engineers, made certain that contracts went to American firms on a negotiated cost-plus-fee basis. It also decreed that most of the construction materials had to come from the United States. Its commander, Colonel Joseph D. Arthur, Jr., dispatched the first construction crews to the Docksite area at Port of Spain in March 1941. As for Governor Young, continued labor unrest combined with deteriorating Anglo-American relations on Trinidad prompted London to recall him (on grounds of "ill health") in June 1942 and to replace him with the more diplomatic Sir Bede Clifford. Eventually the US Corps of Engineers spent roughly $82 million ($993 million in 2010 dollars) on construction in Trinidad, second only to what was spent defending the Panama Canal.

By mid-May 1941, construction was well underway at all the base sites. Barracks and mess halls, hangars and runways, taxiways and control towers sprang up as local labor was hired and construction workers poured in from the United States. By the end of June, temporary runways were in use on St. Lucia, Antigua, and British Guiana. Heavy rain delayed construction on Trinidad, but a 5,000-foot runway was completed there by October. Almost immediately, labor problems erupted on Jamaica and Trinidad when local trade unions protested wages and working conditions and the lack of housing for domestic workers. The American commander of the Trinidad sector attributed the labor troubles to "Nazi sympathizers and Fifth Columnists in the Guianas." No evidence was ever found to substantiate these charges. Most of the friction that arose from time to time between the islands' peoples and the American military stemmed from the completely different cultural backgrounds of the two groups, the boisterousness of young men far from home seeking drink, women, and a howling good time when off duty, and the military's failure to foresee the immense social strains that would arise. As the official history of the US Army in World War II concluded:

> Too little cognizance was taken of the incapacity of Americans generally to adapt their ways to those of strangers or to take comfort or serious interest in unfamiliar surroundings. Too little attention was given to preparing the men for the antipathy of a local populace, however friendly, toward any foreign garrison, however well-intentioned.[31]

Despite the difficulties of climate, distance, differences of culture, and differences of nationality, the base construction went remarkably quickly. On Sunday, April 20, 1941, US infantry and coast artillery units arrived in Bermuda; four days later the men of the 1st Bomber Squadron arrived in Trinidad from Panama. Their aircraft were flown in eight days after that. Then, on May 5, infantry and artillery units arrived from New York. On all the islands and in British Guiana, base airstrips were completed, coastal defense installations were manned, radar stations were put into operation, and planes – Navy Catalinas and Army bombers and fighters – arrived by the score. Great Exuma and Antigua hosted air strips while a small army base was established on Jamaica. St. Lucia had an extensive base capable of housing an entire army division along with a large air base. British Guiana hosted a small US Army unit that guarded the Georgetown airport where both American and British military aircraft were based. Trinidad was home to Fort Reid, a major army base and a large airfield. In all, 189 bombers and 202 fighters[32] were based across the islands, backed by a handful of Royal Navy ships and United States Navy destroyers. In August 1941, discussions began with the Dutch government-in-exile and the British to station US troops on Aruba and Curaçao, to replace the two British infantry battalions that had been stationed there the year before and to establish a US garrison in Suriname. Close to 1,000 US soldiers began to arrive in Suriname in late November 1941 with artillery, bombers, and fighters. By December 1941, several squadrons of Army reconnaissance and bomber squadrons, equipped primarily with twin engine B-18 "Bolos" and A-20 "Havocs," were deployed around the Caribbean as well as some Navy patrol squadrons equipped with twin engine Catalinas.

The sporadic air patrols carried out by these aircraft were better than nothing, but the planes were not equipped for antisubmarine warfare, the crews were untrained in spotting or attacking subs and, quite simply, there were not enough of them. A number of old World War 1-era US Navy and Royal Navy destroyers plied the Caribbean and the odd Dutch naval vessel. But as one major study by the US Army on the antisubmarine war in the Caribbean later recorded:

Douglas B-18B (S/N 37-530, originally a B-18A) with the Magnetic Anomaly Detection (MAD) tail boom. B-18s were frequently used for anti-submarine warfare in the Caribbean theatre (U.S. Air Force Photo). Source: National Museum of the US Air Force, http://www.nationalmuseum.af.mil.

> That the U-boat menace would grow to gigantic proportions in this area was not predicted in the War Department or in the Caribbean Defense Command, and extensive development of procedures and materials to offset any underwater campaign was not included in the early preparations. The problem of antisubmarine measures at that time concerned the Caribbean Defense Command primarily as it should affect the Canal, although enemy U-boat operation in the whole Atlantic area had already begun to cause concern.[33]

Beginning shortly after the United States entered the war, air and ground forces were dispatched to key islands in the Caribbean (Trinidad, Panama, Puerto Rico). Trinidad had the largest contingent of US forces: 12,000 ground troops and 4,000 aircrew and command and maintenance personnel. One bomb group of 55 aircraft, one pursuit (fighter) group of 130, and

Douglas A-20A of the 58th Bomb Squadron over Oahu, Hawaii, on May 29, 1941. The United States Army Airforce deployed several A-20 "Havocs" for anti-submarine patrols in the Caribbean (U.S. Air Force photo). Source: National Museum of the US Air Force http://www.nationalmuseum.af.mil.

one reconnaissance squadron of 13 were also based there. In the Panama Canal Zone, the US Navy operated Patrol Wing Three composed of 26 PBY Catalina flying boats while the Army Air Forces' 59th Bombardment Squadron consisted of 12 A-20 "Havocs."

The aircraft operated by the US Navy and the Army Air Forces in the Caribbean theater were not well suited to the job of hunting and killing U-boats. The PBY was first flown in 1935 as an amphibious naval reconnaissance aircraft. It was originally designed as a torpedo bomber but was almost never used in that role. The main armament on early models was two .50-caliber machine guns in large waist blisters, one .50- or .30-caliber machine gun in the nose, and one .50-caliber in the bottom rear aft of the hull step. Early PBYs had no searchlights or purpose-built depth bombs. They were slow, with maximum speed less than 200 miles per hour, but did have a maximum range of 3,100 miles. The B-18 was a military version of the Douglas DC-2 transport with a thicker forward

Consolidated PBY-5A Catalina in white camouflage for hunting submarines. iStock photo.

fuselage and a glassed-in nose. It carried only three light machine guns but could accommodate up to 4,000 pounds of bombs or depth charges in its bomb bay. It was about 20 miles per hour faster than the PBY but had half the range. The A-20 was by far the newest and best of the aircraft, far more heavily armed with a crew of three and a maximum bomb (or depth charge) load of 4,000 pounds and at least 100 miles per hour faster than the PBY or the B-18. Unfortunately, the A-20 light bombers were in great demand in all theaters of the war, and in this early period only very limited numbers were available to Caribbean Defense Command. Moreover, the bulk of the new four-engine B-17s and B-24s, with much longer ranges and carrying capacity than the navy planes or the B-18s and A-20s, were used to patrol the Pacific side of the Panama Canal.

The available ASW aircraft were dispersed throughout the Caribbean, with a squadron each at Jamaica, Puerto Rico, Saint Croix, Antigua, St. Lucia, Trinidad, Georgetown, and Paramaribo – when Brazilian permission was obtained. Each base was to mount regular air patrols in seaward sweeps while naval vessels in Guantánamo Bay, Cuba, and San Juan, Puerto Rico, supplemented the coverage. As usual, competition between the army and the navy hindered a joint defense; less than a week after Pearl Harbor, the commanding general of the Caribbean Air Force refused to assume responsibility for long-range reconnaissance since "that was the mission assigned the Navy forces."[34]

Prolonged and somewhat difficult negotiations between the Dutch government-in-exile and the United States – at one point the Americans began to prepare for an invasion of the Dutch islands from Trinidad – delayed the arrival of American aircraft on Aruba and Curaçao until mid-January 1942, when units of 59th Bombardment Squadron arrived in the Netherlands West Indies from the Canal Zone. One flight was stationed at Dakota Field, Aruba; the other at Hato Field, Curaçao. These were the first US combatants on the Dutch islands and they found the facilities less than perfect. Dakota Field (today Queen Beatrix International Airport) was on the west side of the island, about 15 kilometers from San Nicolas harbor and the Lago refineries. Its 2,500-foot gravel runway, like that at Hato Field, was too short for the A-20s and had to be doubled in length and paved. In the early days, officers and men lived in temporary wooden barracks; flight crews slept in tents near the aircraft; and almost all had to do with saltwater showers.[35] The land defenses of the two Dutch islands in January 1942 consisted of Royal Netherlands Marines and Dutch military police, shored up by a British infantry battalion on each island. Little joint training was undertaken and almost no preparations were made for any possible invasion. As one US report put it, "British troops were anxious to leave and Dutch troops were equally anxious to have them do so."[36] American troops began to arrive on the islands on February 11.[37] Their first task was to move into the facilities that the British had left.

In the Caribbean as elsewhere, the German victory over France forced a rapid and thorough reconsideration of local defense measures and heightened American concerns about the security of the Panama Canal. This turn of events, along with Churchill's desire to entice the United States to take more of the burden for the defense of shipping in the Battle of the Atlantic, led to a rapid buildup of US forces from Suriname to the Bahamas to Bermuda. When the U-boats arrived in Caribbean waters, there was at least some defense in place to meet them. It was ironic, however, that the U-boat threat against the UK that had largely prompted the British to offer up the leased bases in the first place was so little considered in US defense planning for the Caribbean. When the United States went to war in early December 1941, its troops and aircraft were ready to take on the nonexistent German aircraft carriers but were completely unready for German submarines.

* * *

Back in Lorient, the U-boat crews wasted not a moment in preparing for their departure. As soon as the work details at the Kéroman bunkers had completed repairs to *U-161* on January 20, Albrecht Achilles took his boat northward into the Scorff River. The technical gang stowed 25 torpedoes, 15 below and ten in pressurized tubes under the upper deck's wooden planks. Next, they hoisted on board 110 rounds of 10.5-cm shells for the deck gun as well as 2,625 rounds for the 3.7-cm and 4,250 for the 2-cm anti-aircraft guns (FLAK). It was hard work in a wet, cold January. Then *U-161* bunkered 214 tons of fuel oil, six tons of lubricating oil, and five tons of drinking water. Finally, it took on food supplies for the long journey by a crew of 49 officers and men. It was an awesome sight.[38] Below decks disappeared literally mountains of crates with canned goods: beef, pork, lamb, ham, sausages, sardines, herring, lentils, cauliflower, spinach, sauerkraut, asparagus, kale, mixed fruit, apple sauce, as well as an abundant supply of salt and sugar, coffee and milk. The arrival of 50 tropical pith helmets aroused a good bit of conjecture among the seamen. Last but not least, *U-161* took on board 15 bottles of "medicinal" cognac, to be rationed out by Achilles for persistent "colds" among the crew.

Finally, the moment of departure was at hand. *U-67*, *U-156*, and *U-502* were first to put to sea on January 19, 1942, followed five days later by *U-161* and *U-129*. On *U-161* Werner Bender reported: "All hands present and accounted for! Engine room crew ready! Upper and lower decks cleared for departure!" Achilles snapped a brisk, "Thanks. Heil *I WO*."[39] Then to the crew:" "Eyes front! At ease!" Bender ordered, "Let go all lines!" The fenders were hauled in, hawsers cast off from the old hulk *Isère*. A military band struck up the navy's unofficial anthem, *Wir fahren gegen Engeland.* Officers waved good-bye and wished the Kaleus "Good Hunting!"

"Engines ahead one-third!" The electric motors began to hum. Slowly, *U-161* glided into Lorient's main harbor channel. The water was a turgid dark brown, a nauseous mix of oil, seaweed, tar, and sewage. An armada of dilapidated tugs and fishing boats, prams and ferries, lighters and oil barges, flitted about. A blast from the submarine's "Typhon" signal horn summoned one of the patrol boats from the so-called "bedbug flotilla." "Starboard engine slow ahead! Rudder hard to port!" The nine-cylinder MAN diesels roared to life, spewing out their gray-blue exhaust fumes.

The hull and deck plates began to vibrate. Cold sea spray greeted the watch on the bridge. It was just past 1 p.m., January 24, 1942.

Off the starboard side, the deck crews could make out a small number of Vice Admiral Karl Dönitz's staff officers waving their caps from the bunker roof of the Villa Kerillon.[40] As *U-161* passed the narrows between Kernével and Port-Louis, it picked up its escort boat, a minesweeper. Then it headed for Rendezvous Point *Luci-2*. The day was gray and overcast. Force 4 winds. Moderate sea swells. Achilles gave the order to shape a course: "Destination San Miguel (Azores)." It was the first rendezvous point. Ahead lay what the Kaleus joyously referred to as the "Golden West."

2

ATTACK ON ARUBA

As January gave way to February 1942, Group Neuland shaped course for
the Azores. The weather was a mix of clouds and light rain, with moderate
seas. Day after day, the boats, running on just one engine, averaged about
170 nautical miles. The men referred to these long runs out to and back
from the operations areas as "garbage tours." It was boring, monotonous
work. One by one, the cases of fresh provisions disappeared: meat after
eight days, bread after 12, eggs after 21, and even potatoes as well as
smoked hams after 35. Soon, crates of canned meats and vegetables were
retrieved from the boat's second toilet, much to the relief of the 49-member
crew. Every crate eaten meant more living space. As the weather slowly
warmed, the men took turns to sunbathe on the deck, to try their luck
at fishing, to play cards and chess, and to listen to the shortwave radio
– including the forbidden BBC, the Voice of America, and the "Black"
radio programs such as Siegfried Eins and Radio Atlantic, emanating
from London. There were also 200 records on board.

The leisurely crossing gave the mostly green crew time to take the
measure of their new surroundings.[1] At 1,541 tons fully loaded, the Type
IXC boats had almost twice the displacement of the standard Type VIIC
boats. They were sturdy craft of double hull construction, with the diving
tanks and main fuel oil bunkers in the outer hull. Five watertight compartments protected the boat against cracking of the pressure hull – by depth
charges or aerial bombs. Two nine-cylinder MAN supercharged "Jumbo"
diesels, each capable of producing 2,200 horsepower, gave the boats a best
speed of 18 knots on the surface. Sixty-two AFA batteries housed in boxes
set on rubber shock absorbers and stowed underneath the interior deck
plates powered two Siemens-Schuckert double electric motors at a best

The aft facing twin 2cm antiaircraft guns of a Type IXC Uboat. Source: Ken Macpherson Photographic Archives, Library and Archives at The Military Museums, Libraries and Cultural Resources, University of Calgary.

speed of seven knots submerged. Surface range was 14,035 nautical miles at ten knots; submerged range was 63 nautical miles at four knots. The boats had a safe diving depth of 100 meters (328 feet), but skippers often doubled that during particularly severe depth-charge attacks.

Armament consisted of six 53.3-cm (21-inch) torpedo tubes, four in the bow and two in the stern. For the long journey to the Golden West, the boats carried more than their standard allotment of "eels": six in the firing tubes, nine strapped under bunks, and ten stored in watertight containers under the wooden planks of the top deck. Given that the new electric G7e T2 "Eto" torpedoes had often failed due to their faulty proximity and contact fuses, the boats had also been outfitted with the older but more reliable G7a T1 "Ato" torpedoes. Finally, the boats carried one

10.5-cm^2 deck gun mounted on the forward deck; one 3.7-cm anti-aircraft gun on the after deck; and small 2-cm anti-aircraft cannons on the bridge.

Only some 30 feet longer than the Type VIIC boats, the IXC class offered little more in the way of comfort for their crew of roughly 50 officers and ratings.[3] The boats were divided into four main sections. Moving from the bow to the stern, the first section was the forward torpedo room with its four firing tubes and reloads stashed under the bunks which were shared, or "hot-sheeted," by the ratings depending on which were off-duty. Stepping back through a heavy watertight bulkhead, one entered the petty officers' quarters; these lucky few had their own bunks. Just behind the petty officer's quarters was the small galley, the domain of the cook, or *Smutje*. It consisted of a small refrigerator, two small ovens, and three hot plates. Moving back from the galley through another watertight bulkhead, one encountered the brain and nerve center of the boat: the officer's wardroom and the captain's cabin, adjoined by the radio room and the underwater sounding station. Most importantly, the radio room housed the Enigma cipher machine. This was where messages were encoded as well as decoded ("officers only"), and where courses and positions were plotted on the secret grid charts (*Quadratkarten*). Hydrophone readings were taken in the underwater sounding station, where the war diary (*Kriegstagebuch*, or KTB) was also maintained. Clocks were always set on Berlin (GMT+1) time.

Amidships, directly under the conning tower, was the mechanical heart of the boat, the control room – a bewildering array of gauges, switches, meters, valves, hand-wheels, pumps, magnetic and gyro compasses, rudder and hydroplane controls, as well as chart closet and mess table. Appropriately named *die Zentrale*, this section was dominated by the two periscopes – the large sky scope and the smaller attack scope. Between the periscopes were a ladder and hatch that led to a small conning tower above the control room. From there, the executive officer worked the attack calculator, compass repeater, and attack periscope to aim and arm the torpedoes during an attack. Directly above him was a watertight hatch leading to the bridge.

Moving still further back inside the boat through yet another watertight bulkhead, one entered the engine room, the noisy, grimy nether world of the chief engineer and his "black gang." It housed the two massive MAN diesels, mounted side by side and with a narrow pathway between them to allow the engine crews to service the monsters. Just behind the diesels were the two Siemens-Schuckert electric motors, aligned on the twin shafts that ran from the diesels to the two three-bladed propellers. Finally, through yet another watertight bulkhead one reached the aft torpedo room. Aside from the two stern torpedo tubes and reloads, it had eight bunks for a crew of sixteen and an auxiliary steering wheel.

The boats turned into a veritable sewer within the first two weeks of a war patrol. Men wore what they called "whore's drawers," black underwear to hide the sweat stains that daily grew once the boat reached its operations area. Humidity often approached 100 per cent inside the steel hull; dripping condensation turned paperback novels into paste. Especially in the tropics, the temperature inside the steel hull reached 40 degrees Celsius. The bilge became a sluggish rivulet of oil, urine, and spilled battery acid. Mold was commonplace. Cheeses stored in the torpedo rooms and sausages and slabs of smoked ham hanging off the bulkheads further fouled an already odiferous air. Human waste was launched through an empty torpedo tube as the toilets could not be flushed at depths greater than 100 meters. The men's only hope to breathe fresh air was that their skipper would call them up on deck in warm climes and during non-combat hours.

The Neuland boats had received two of the German Navy's latest technical innovations before leaving Lorient. The standard Enigma (or *Schlüssel M*) machine had been given a fourth "*alpha*" rotor to make it virtually impossible to crack. Already with just three rotors, each with 26 contacts, 16,900 live posts were possible. The fourth rotor expanded those possibilities to 44,000 live posts. U-Boat Headquarters confidently projected for this new M4 machine with its "Triton" cipher circuit a theoretical total of 160 trillion settings for a complete radio transmission.[4] Surely, no human brain could possibly unravel such staggering combinations! The second innovation was reserved solely for *U-156*: a fixed array FuMO 29 radar detector mounted on the front of the upper conning tower. The device had a range of just under five miles and a field of view of but 60 degrees forward.[5] Neuland was to be its first operational test.

Albrecht Achilles and *U-161* had left Lorient the evening of January 24, 1942, on a course of west-by-southwest. Destination: Trinidad, 3,600 nautical miles from Lorient. Precisely one week later, they spotted their first two enemy destroyers. Luckily, they remained undetected.[6] Early in the morning of February 1, *U-161*'s lookouts spied the Arnel lighthouse on San Miguel Island in the Azores. Thereafter, Achilles altered course due west for Trinidad.

For the crew of *U-161*, their new skipper remained a mystery. They knew only the barest details of his career. Born on January 25, 1914, he had just passed his 28th birthday. He was regular navy, having entered the service in April 1934. He had been assigned first to the old battleship *Schleswig-Holstein* and then the new 11-inch battleship *Gneisenau*. In April 1940 Achilles had transferred to the U-Boat Service and after three war patrols as executive officer on *U-66* had received his own boat.

Second Watch Officer Götz Roth recalled his skipper in 1942.[7] "He was still young, only a few years older than the crew, which on average were between 20 and 22 years of age.... Achilles was not tall – about 1.74 meters. He quickly won our confidence." The Old Man was "friendly," a man who set the right tone for both the command bridge and the mess. He maintained strict, but not unbending, discipline on board and was a professional through and through. "Political topics were taboo on U 161."

As soon as the Neuland boats reached the mid-Atlantic, Vice Admiral Karl Dönitz reminded his skippers not to give their position away by attacking single freighters, however tempting they might be. As *U-161* crossed the line 40 degrees west longitude, it radioed in its oil situation, as ordered by U-Boat Headquarters: "Still have 190 cbm fuel." On February 10, Kernével sent out news that "a Spanish naval officer" had informed U-Boat Headquarters that the harbors at Curaçao as well as Trinidad were "open, not mined, no blackout." Shortly before midnight on February 15, Achilles sighted the lighthouses of South Point and Ragged Point on Barbados. Twenty-four hours later, Trinidad hove into sight. *U-161* had reached its operations area. Grid Square ED 9596.

* * *

U-156 also had an uneventful crossing. After leaving Lorient on January 19, 1942, Kapitänleutnant Werner Hartenstein, the Neuland group leader, radioed the top-secret Operations Order 51 to his fellow Kaleus:

U-156 returning home from a patrol. Source: Ken Macpherson Photographic Archives, Library and Archives at The Military Museums, Libraries and Cultural Resources, University of Calgary.

"Surprise, concentrated attack on traffic in the immediate vicinity of the West Indies islands of Aruba a[nd] Curaçao. Codename: Neuland. Attack 5 hours before sunrise [on February 16]."[8] Hartenstein's destination: Aruba, nearly 4,000 nautical miles from Lorient. The weather also held for *U-156*: moderate seas, light winds, overcast skies with occasional rain squalls.

As the days went by, the crew got to know their skipper. Born on February 27, 1908, at Plauen, Hartenstein spoke with a heavy Saxon accent. He was a stern taskmaster. A regular navy man, he had entered the service in 1928 and had been assigned mainly to torpedo-boats.[9] In March 1941, he had transferred to the U-boat arm. He proudly wore the Iron Cross, 1st and 2nd Class. On February 2, he received word that he had been awarded the German Cross in Gold for "extraordinary bravery" while serving with the destroyers.[10] The loudspeakers on *U-156* blared out martial music, the Old Man donned his dress blues, and the *Smutje* outdid himself with oxtail soup, stuffed rolled flank steak, and chocolate pudding. Hartenstein reciprocated with three shots of the "medicinal" cognac for the mess. He was now just a few weeks away from his 34th birthday. He had a wiry but solid frame of medium build. His high cheekbones and eagle-like nose reminded some of the men of Dr. Joseph Goebbels, the Nazi propaganda minister. A rapier thrust from a duel during his two years at university had left a permanent scar on his left cheek.

The men knew that he was a "loner." The story had made the rounds that, as a newcomer to the U-Boat Service, he had run into Erich Topp at a bar in France. Topp already was an "ace," a commander who would end his career ranked third on the list of all-time greats, with 35 ships of just under 200,000 tons destroyed. He was one of only five skippers to receive the Knight's Cross with Swords and Diamonds. When Topp had taken a seat next to him at the bar, Hartenstein had merely growled: "My name is Hartenstein. I don't give a damn what yours is."[11]

At 2 p.m. on February 4, the Enigma board lit up: "Neuland 186." *U-156* would be free to begin operations against Aruba five hours before sunrise on Monday, February 16, 1942. The next day and again five days later Hartenstein spied single freighters, unmarked and unescorted in the mid-Atlantic. Easy targets, but off limits. "Did not attack according to operations orders," he dryly noted in the war diary. By now, *U-156* was

Werner Hartenstein. One of the most senior U-boat commanders, Korvettenkapitaen
Hartenstein began his career on the light cruiser *Karlsruhe*, and in March 1941 transferred
to the U-Boat forces. He undertook four patrols with *U-161*, sinking 19 Allied ships of
97,489 tons. A Knight's Cross holder, he became most famous for his actions to rescue
the survivors of the 19,695-ton liner *Laconia*; his "bag" in the Caribbean included the
destroyer USS *Blakeley* off Martinique. Source: Deutsches U-Boot-Museum, Cuxhaven-
Altenbruch, Germany.

U-156. Commissioned at Bremen in September 1941 and commanded by Korvettenkapitaen Werner Hartenstein, *U-156*, a Type IXC boat, played a major role in the German attack on Aruba; it was sunk in March 1943 east of Barbados by depth charges from a US Catalina flying boat. Source: Deutsches U-Boot-Museum, Cuxhaven-Altenbruch, Germany.

skirting the southern edge of one of the strangest and most notorious bodies of water on the planet – the Sargasso Sea, just above the Tropic of Cancer. Vast islands of green seaweed lazily floated on the ocean. As his boat sliced through the mats, Hartenstein noted that *U-156* was now in the heart of the Bermuda Triangle.

Under cover of darkness on February 10, 24 days out of Lorient, *U-156* passed the Guadeloupe Channel south of Antigua in the Leeward Islands, being sure to hug the friendly Vichy-French coast of Guadeloupe. It was now time to power up the second of the supercharged MAN diesels. These were halcyon days for the crew. The gentle Caribbean breeze and warm sea were welcome relief from the cold and rainy mid-Atlantic. Swims in the deep blue waters were much appreciated, as were showers from the fire hoses rigged up on deck. Few bothered with bathing suits. Hartenstein broke out the tropical gear: soft canvas shoes, tan tropical shorts, sleeve-less light tan jerseys, and pith helmets. The men detested the latter. Many had brought along their own khaki field caps, similar to those worn by the Afrika Korps.[12] In the 40-degree-Celsius heat inside the boat, the jerseys and caps were quickly stowed and replaced by a sweat rag worn around the neck. To the joy of the crew, Hartenstein introduced them to the exotic family *exocoetidae*, or flying fish. For some time, the men had seen blue and silver fish with long pectoral fins accompany the boat, oftentimes gliding through the air on these "wings." Fishing was easy and breakfast now consisted almost exclusively of fried flying fish. Watches competed with one another in this piscatorial sport; the record was 60 fish in a single 24-hour period.

As the island of Curaçao came into view, the Old Man decided that it was time to inform the men of their mission. "Soon it will be time for action. It is our job to upset as far as possible the course of tankers between Venezuela and the islands of Aruba and Curaçao, *and* to attack the oil re-fineries – with our guns." He informed them that *U-67* and *U-502* would also join the hunt and that all action against non-tankers (battleships and aircraft carriers excepted) was "forbidden" by headquarters. "Now – I wish you a good night. Dismiss."[13] Later that night, the officer on watch sighted a brightly lit passenger liner through his periscope and could make out couples dancing on the upper decks. It seemed from another place and another time.

At 7:16 p.m. on February 13, Hartenstein brought *U-156* to the surface.

Grid square EC 9347. The northeasterly trade winds were blowing a fresh breeze, Force 4. Through the Zeiss binoculars, he could make out the Colorado light and the cliffs of Colorado Point on the southeastern tip of Aruba. *U-156* had arrived at its operations area ahead of schedule. The surprise attacks on Allied oil in the Caribbean were about to be unleashed.

* * *

Off Aruba on February 13, it was time for Hartenstein to reconnoiter.[14] Slowly, *U-156* pointed for the Colorado light and then made its way up the west coast of the island. An "oily haze" hovered above the surface of the sea. Shortly before 11 p.m., Hartenstein, having carefully noted Aruba's treacherous coral reefs on his chart, approached San Nicolas Harbor to within 900 meters. An awesome sight spread out before him: "4 large tankers inside the harbor; 3 small tankers in the offshore roads." What he called "factory installations" were "brightly lit up." In fact, this was the giant Lago "Esso" Refinery, over whose 2,716 acres Standard Oil of New Jersey in 1928 had received a 99-year lease. It was a beehive of activity, working around the clock to refine Venezuelan crude and reship the desperately needed high-octane fuel to the United States and Britain. The bright lights were a wonderful change from blacked-out Europe. The skipper let the crew come up top in small groups to take in the sight.

Hartenstein made quick mental notes. He gauged the configuration of San Nicolas harbor, the location of the tanker piers, the tank farm and refinery, and the lake tankers anchored just outside the lagoon. He especially reconfirmed the deadly coral reef that protected San Nicolas, creating a bright turquoise lagoon between it and the shore. The lagoon was too small to maneuver the 240-foot-long boat; too shallow to submerge and hide; and without sufficient range (180 meters) for the torpedoes to arm once fired.[15] Finally, he noted the high phosphorescence of the warm sea, a possible give-away to alert hostile aircraft.

Hartenstein continued his course northwest. He arrived off the capital, Oranjestad, just after midnight, and passed through the gap in the reef. Entering the mouth of the harbor, he saw little traffic. In the darkness he easily evaded several light aircraft. He also took note of another valuable reference point – the 541-foot-high, cactus-studded Hooiberg ("The Haystack") that towered over Oranjestad.

Satisfied, Hartenstein slipped away from the capital to scout the Caribbean Sea approaches to Aruba. Early on February 15, he returned to Oranjestad. A Dutch patrol craft was flitting about the harbor. Stealth was his best friend; he submerged *U-156* until dusk. At 5 p.m., he spied a large ocean-going tanker leaving the refinery. Two-engine aircraft were buzzing overhead. Three hours later, Hartenstein continued on to Colorado Point and shaped a course for the sea lanes between San Nicolas and Maracaibo, where he undertook a practice attack on an unsuspecting tanker.

Sunday, February 15, 1942, dawned bright and clear on Aruba. It was the middle of the dry season: blue skies, brilliant sunshine, perfect weather for picnics and beach parties. The ever-present trade winds blew at a steady 15 knots out of the northeast and soon the temperature climbed above 30 degrees Celsius. By late morning, the residents of Seroe Colorado at San Nicolas emerged from their $24-per-month company rental bungalows and started off to the Lago Community Church. Some headed straight for the pleasures of Rodgers Beach and Baby Beach, both safely inside the reef. Others searched out the fish markets at Savaneta, the island's oldest town and former capital, for the Sabbath meal of fresh barracuda, crabs, sailfish, wahoo, blue and white marlin, or black and yellow fin tuna. Small wooden boats had come across the Caribbean Sea from Venezuela to sell their tropical bounty: avocados, bananas, guava, kiwi, mangoes, melons, and Jalapeño peppers. Many Seroe Colorado residents lunched on fresh crab and fish, followed by lazy strolls on the beaches or along the high ground on the dry, bare island.

Few even took notice of the newly arrived American coast defense gunners who lounged lazily in the barracks around San Nicolas harbor, or watched as crews began the laborious job of unloading their artillery, the American 155-mm "Long Toms." Almost a thousand American troops under the command of General Frank Andrews had descended on the island just four days earlier to relieve the Queen's Own Cameron Highlanders, British veterans of Dunkirk who had departed on February 13. What little news penetrated Aruba from the outside world came by way of American shortwave radio stations or the island's weekly newspaper, the *Pan Aruban*. That past Sunday, the paper had reported the most important American college football scores: Alabama 29 – Texas A&M 21, Oregon State 20 – Duke 17, Georgia 40 – TCU 26. And it had informed anxious pugilists that in the prize fight of the year, Joe Louis, "The Brown Bomber," had knocked out Buddy Bear at 2:56 of the first round at Madison Square Garden in New York. Hollywood had entertained the island throughout January with *In the Navy* starring the Andrews Sisters, *The Captain Is a Lady* with Charles Coburn and Beulah Bondi, and *New Moon* with Nelson Eddy and Jeanette MacDonald. Given that the United States had curtailed all exports of automobiles after the Japanese attack on Pearl Harbor, Arubans eagerly scanned the *Pan Aruban* for used cars: most popular in the advertisements were 1934 Fords and 1936 Chevrolets for 150 florins each (three-months salary for a lake-tanker sailor) and upscale Buicks for 250 florins.[16]

* * *

Hartenstein for a third time reconnoitered San Nicolas harbor after dark on February 15. Again, it was perfect. Houses and warehouses were brightly lit up, as were the refinery and the ships in the inner harbor. He noted laconically in the war diary: "Brisk traffic. Harbor well populated." Suddenly, his Enigma machine sprang to life. Final orders from Vice Admiral Dönitz:

> Main task is the attack on shipping. Once such attacks have taken place, artillery may be deployed against land targets already on the morning of the [first] day of Neuland, should the opportunity for this prove favorable. In case no ships can be targeted, authorize use of artillery against land targets beginning evening [first day of] Neuland.

The "Great Lion" had decided to ignore Grand Admiral Raeder's orders to concentrate on the refineries.

Satisfied with his reconnaissance of San Nicolas, Hartenstein took *U-156* west to the Los Monjes Islands. "The Monks," which were in Venezuelan waters, consisted of gray, barren rocks far removed from the major shipping lanes. Time for some last minute ersatz Jan Maats smokes, card games, and swims. Chief Engineer Wilhelm Polchau studiously measured the water temperature: 29 degrees Celsius, or 84 degrees Fahrenheit. It was a welcome relief from recent temperatures in the boat that had reached 37 and even 42 degrees Celsius.

U-156 retraced its course to Oranjestad, using the Hooiberg as reference point. Hartenstein evaded several sailboats and a large passenger freighter. Surprise was critical. Moderate seas and a fresh breeze continued unabated. It was a bright, starry night. He ignored a single tanker taking on product at the piers and continued down the coast. Trimmed to where just the conning tower was above water, Hartenstein proceeded outside the reefs until he was off San Nicolas harbor. It was 1:35 a.m., February 16. The time for reconnaissance was over. His orders were clear: "Attack 5 hours before sunrise."[17]

Silhouetted against the bright lights of the Lago Refinery were two lake tankers riding at anchor.[18] "Battle stations!" Remembering that he was regular navy, Hartenstein put on his dress-blue blazer. The bow tubes were loaded with a mix of G7a and G7e torpedoes. Executive Officer Paul Just counted down to zero hour: "Just three more minutes." At 1:59 a.m., Hartenstein barked out, "Tubes I and II ready to fire!" He then counted off the range: "800 – meters – 700 meters – 600 meters." The tension in the boat was electric. It was, after all, their first hostile engagement. Every eye strained for a glimpse of the Old Man. How would he hold up under pressure? How much longer the wait? Some stared at their watches. It was 2:01 a.m., precisely one minute into the official start of Operation Neuland.

"Tube I ... Fire!" Lieutenant Just slammed down the firing knob. The G7a "Ato" torpedo shot out of the tube at 44 kilometers per hour. There was a slight jolt to the boat. Hartenstein could watch the torpedo's tell-tale wake of bubbles on the surface of the sea. Good thing that it was dark. He then ordered the boat to come about slightly to starboard. Range to second target: 700 meters. Time: 2:03 a.m. "Tube II ... Fire!" Again, Just slammed down the firing knob. This time there were no surface bubbles, for the skipper had opted to fire the G7e electric torpedo. Anxiously, the boatswain's mate counted off the seconds. The men on *U-156* were painfully aware of the miserable track record of the navy's G7e "eel," which all too often failed to explode on impact.

After what seemed an eternity came the welcome cry: "Detonation after 48.5 sec. Tanker burns immediately. 3080 t[ons]." *U-156*'s first torpedo had penetrated the sides of the 4,317-ton British tanker *Pedernales*[19] amidships. There followed a thundering explosion that seemed to lift the vessel up in the water. Two minutes later, a similar fate befell the second tanker: "Detonation after 53.2 sec. Tanker burns immediately. 2740 t[ons]." The 2,396-ton British lake tanker *Oranjestad* also had been torpedoed amidships. Burning oil poured out through the gaping holes of the two tankers and spread over the water. Screams could be heard across the burning sea as terrified sailors scrambled into lifeboats or jumped off the slanting decks.

Hartenstein watched the destruction with satisfaction. He had thirsted for action, and now the moment had finally arrived. He called his men up to the conning tower in small groups to watch the *Oranjestad* roll over and settle in the waters just outside the lagoon; it was burning furiously. The *Pedernales*, also seemingly settling into the shallow sea just off the reef, likewise was a raging fountain of fire. Thick, black smoke drifted over the reef and harbor. Herbert White, an Associated Press photographer assigned to cover the arrival of the first contingent of American troops on Aruba, caught the moment: "The harbor scene was like a raging forest fire right in our own front yard.... The blaze was shooting up high over the waterfront.... I could see the decks of [one] ship as a mass of flames."[20]

Hartenstein ordered *U-156* to come about to 300 degrees. "Course: harbor!" He could clearly make out the refinery silhouetted against the well-lit furnaces and the yard lights. Having carried out his primary mission against shipping, he was now free to attack land targets. "Clear the decks! Prepare to fire artillery!" The gun crews clambered out of the boat under the command of their artillery officer, Lieutenant Dietrich von dem Borne. They slammed the first shell into the breech of the 10.5-cm deck gun. Borne trained it on the refinery. Hartenstein laid *U-156* parallel to the coast, 500 meters away. It would be a turkey shoot. It was now 2:11 a.m.

"Both engines stop! Fire at will!" Hartenstein yelled down to the gun crews.

Seconds later he saw a bright flash from the big gun and heard a deep rumble. The 3.7-cm gun was also beginning to fire on the tanks filled with precious aviation fuel. Hartenstein spied two bright tracer shells speeding toward shore. But there was no fire from the 10.5-cm gun. "Fire 10.5-cm!" he screamed down to Borne. Nothing. Only the small afterdeck cannon continued to bark out. "Cease Fire!" Furious with his artillery officer and gun crew, Hartenstein leaped down from the conning tower and raced along the forward deck. As he reached the gun platform, he heard a low murmur. "10.5 out of action – Explosion in the bore!"

It was Lieutenant von dem Borne. He was propped up against the conning tower, his right lower leg shattered. Next to him lay the motionless body of Seaman Heinrich Büssinger; his stomach was lacerated and exposed, his thighs scored. Both men had been struck down by red hot pieces of metal. Blood covered the deck. In their excitement to destroy the refinery, Borne and his crew had forgotten to unscrew the tampion – the muzzle cap that kept salt water out of the barrel when the boat was submerged. The muzzle had splayed open like an "overcooked cauliflower"; more than a foot of the barrel was missing. An artillery expert, Hartenstein at once realized that the gun was beyond immediate repair. He had the technical staff put a clamp on the barrel at the point of the blast.

The 3.7-cm cannon fired 16 rounds at the refinery. Several landed in its compound. Hartenstein detected what he called a "tongue of fire" in the tank farm: one shell had struck Tank 111, making a four-inch by six-inch dent, without rupturing it. Twice more, the small gun spat out fire – without effect. No use to continue. "Cease fire!" Hartenstein roared. With no night sights and with black smoke enveloping San Nicolas Bay, there was no sense risking the boat.

By sheer coincidence, Frank Andrews, Commanding General, Caribbean Defense Command, was spending the night on Aruba when Hartenstein attacked. The explosions in the harbor woke him up.[21] His aide, Captain Robert Bruskin, later remembered:

> [A]n explosion knocked me out of bed.... I looked out the windows. Flames were shooting straight up and seemed mountainous. The ship [*Pedernales*] just seemed to break apart. Flaming oil spread over a wide area under a steady wind. I could hear cries out in the water, which I learned were badly infested with barracudas.[22]

No shore installations returned fire. The crews of the 37-mm guns recently set up at San Nicolas Wharf could not see through the smoke and fire, while those at Camp Savaneta, about five kilometers up the coast, could not be brought to bear. The larger 155-mm American guns with ranges of at least 20 kilometers were still lying on the docks. A complete blackout of the island followed in short order.[23]

At sea, the British tanker *Hooiberg* had just arrived off San Nicolas when suddenly, in the distance, the crew saw the *Pedernales* and the *Oranjestad* blow up. Its captain realized immediately that his ship was in danger and ordered the helmsman to turn the ship about to steam back in the direction of the Gulf of Venezuela. After running southwest for about three hours, *Hooiberg* turned around and returned to Aruba, making port at 8 a.m. The crew was aghast to see the immense wreckage of the two ships and the smoke and damage at some of the installations.

Up the coast at Dakota Field, word of the attack was slow to arrive. A Dutch guard reported that guard posts at the army base near the refinery at San Nicolas were under fire and that the refinery was also under fire and apparently burning. A single A-20 Havoc from 59th Bombardment Squadron took off at 2:30 a.m. The pilot soon radioed back that he could see ships burning "in the harbor and oil spread over the sea was also burning," but the crew did not see any tracer or other signs of gunfire from the sea.[24] Acting for General Andrews, Captain Bruskin ordered the A-20 to keep flying over the harbor and the refinery until daylight. At about 3:10 a.m., a second A-20 took off on a submarine search to the west of the island. Reports came in to Dakota Field that a tanker was on fire about 45 miles southwest of Aruba. The second A-20 was sent over to investigate; the crew saw "a ship on fire [the British lake tanker *Monagas*] and being abandoned."[25] The tanker had been torpedoed by Jürgen von Rosenstiel's *U-502* about an hour after Hartenstein's attack on San Nicolas. The long night of the tankers had already begun.

Hartenstein's attack was a shocking surprise, though as early as January 26, 1942, a US Navy naval intelligence report had reached headquarters, Caribbean Defense Command, in Panama that a large number of German submarines were entering the Caribbean. Although there had been no solid intelligence as to where they were headed, it warned that "attacks on tankers from Venezuela, Curaçao and [the] vicinity of Trinidad [are] possible."[26] The warning went unheeded; when war came to the southern latitudes in February 1942, German submarines, not Japanese aircraft carriers, had proved the real menace.

3

LONG NIGHT OF THE TANKERS

Kapitänleutnant Jürgen von Rosenstiel had taken *U-502* well out into the Caribbean Sea, close to the Venezuelan shore. It proved to be a target-rich environment. At 2:05 a.m. on February 16, some ten miles off Point Macolla, he spied a bright flame in the direction of Aruba: Hartenstein's attack on the two tankers at San Nicolas. "Free to attack," Rosenstiel noted in the war diary.[1] Two hours later, the lookouts reported a tanker as well as what they took to be a Venezuelan gunboat at a distance of 1,000 meters. At 3:44 a.m., Rosenstiel fired.

The "eel" slammed into the tanker. "High column of fire, the entire ship is aflame under a developing cloud of strong smoke and sinks." He watched intensely as its boilers exploded. The ship heeled over on its side and then sank. It was the 2,650-ton British lake tanker *Monagas* en route from Maracaibo to Aruba with a full load of crude. Twenty-one survivors were later picked up by the Dutch tanker *Felipa*.

The *Monagas* was not alone in the waters off the Paraguana Peninsula in the early hours of February 16. The previous evening four lake tankers had left Maracaibo at regular intervals: *Monagas* was followed by the 2,395-ton British *Tia Juana*; close behind came the 2,391-ton *San Nicolas*, also British-flagged; and the last ship in line, leaving about four hours after the others, was the British *Yamonota*, about 2,300 tons.

Less than an hour after attacking *Monagas*, Rosenstiel spied the *Tia Juana* at a range of 1,400 meters – as well as a dark shadow just behind the brightly lit tanker. That shadow was *San Nicolas*. As Rosenstiel closed in, his radio operator picked up a distress call from Aruba: "To all ships. U-boat in vicinity of Aruba." This was no time to be timid. Rosenstiel was determined to sink both tankers as quickly as possible. At 4:28 a.m., he attacked. The torpedo struck *Tia Juana* amidships, in the engine room. But it refused to go under. Fourteen minutes after the first torpedo, Rosenstiel decided to deliver the *coup de grâce* for fear that the approaching dawn would deny him a clear shot at the third tanker. The torpedo ran true for 300 meters, and then suddenly veered off to the left. It struck *Tia Juana* in the stern. Looking through the periscope, the Kaleu thought the tanker was staying afloat, but in fact, it was not.

Rosenstiel immediately went after his third kill in less than four hours: the *San Nicolas*. His KTB recorded the action at 5:34 a.m. "Tube V ... Fire! Inexplicable miss, assume it went under the target." He loosed a shot from Tube IV. This time the torpedo ran true. "Hit admidships. Tanker flies completely into the air." The torpedo sliced through the ship's sides and exploded in the engine room. Seven of its crew of 26 died instantly. Witnesses on the *San Nicolas* later reported:

> About 3:00 A.M.[2] February 16th, the S.S. *Tia Juana* (British) which was immediately ahead, received two torpedoes in her engine room, exploding the after end of the ship. The ship sank immediately. We were picking up survivors of the SS *Tia Juana* when at 4:00 A.M. February 16th, we were struck by two torpedoes.

The *Yamonota* was still somewhere behind in the dark. The ship's captain recorded: "[A]t about 4:00 A.M. February 16th ... I observed a red flare immediately ahead and on coming closer I saw that a ship was on fire. I realized probable cause of the fire and immediately veered off to the west." Three hours later, *Yamonota* returned to the scene and found the bow of the *San Nicolas* protruding from the water. They picked up four of the crew while a "Venezuelan Government craft was searching the area for survivors." *Yamonota* steamed for Aruba, arriving at 4:00 p.m.[3]

As the first rays of sunlight were breaking over the horizon, Rosenstiel decided to break off the kill. About an hour later, *U-502* was running on the surface when one of the patrolling American A-20 light bombers spotted it and attacked. The Havoc dropped four 300-lb. demolition bombs – completely unsuited for antisubmarine operations – which fell into the sea and exploded about 100 meters from the submarine. No

U-502. Commanded by Kapitaenleutnant Juergen von Rosenstiel, the Type IX C *U-502* was part of the initial German assault on Caribbean oil off Aruba; it was lost in the Bay of Biscay due to depth charges from a UK Wellington aircraft in July 1942. Source: Deutsches U-Boot-Museum, Cuxhaven-Altenbruch, Germany.

damage was done. Returning to the surface just before 10 a.m., Rosenstiel spotted another tanker, this one empty and on its way to Maracaibo. He worked his way to within 800 meters of the target and then fired. Another "Inexplicable miss. Assume [torpedo] ran under the target." A Venezuelan gunboat suddenly appeared and fired five shells, all well off target. It was time to dive. And to learn first lessons. "Decide not to fire any more eels at empty tankers. Artillery of little use during daytime due to aerial reconnaissance."[4]

* * *

While Rosenstiel was picking off lake tankers, Hartenstein's *U-156* was the subject of intense air activity – most of it in the wrong direction. When the American A-20, which had been dispatched to the vicinity of the

Juergen von Rosenstiel. Kapitaenleutnant von Rosenstiel fought in the light cruiser
Karlsruhe during the Spanish Civil War, and in March 1940 transferred to the U-Boat
service. He commanded *U-502* on three war patrols, sinking 14 Allied ships of 78,843
tons, mostly off Aruba in the Caribbean Sea. On 6 July 1942, returning from his third
patrol, Rosenstiel and *U-502* were sunk with all hands by an aircraft in the Bay of Biscay.
Source: Deutsches U-Boot-Museum, Cuxhaven-Altenbruch, Germany.

sinking *Monagas* at around 3:00 a.m., began to return to Aruba, a radio message from Dakota Field informed it that a long-range SCR-268 radar, hastily set up on the south coast of the island, had detected a submarine about 20 miles due south of Aruba. The pilot turned in the direction of the contact report and, after a few minutes, dropped a flare. He and his copilot thought they saw a submarine, but the bombardier saw nothing; with nothing to aim at, the bomb run was aborted. In fact, Hartenstein was nowhere near the spot. With the shore defenses alerted and aircraft buzzing about in the night sky, he decided to quickly leave San Nicolas harbor and head up the coast. *U-156*'s crew was amazed by their skipper's tenacity and coolness under fire. From now on, their nickname for him was "Crazy Dog."

The dark of night, the dense smoke from the burning tankers, and plain bad luck denied Hartenstein two further "kills." For inside San Nicolas harbor was the *Henry Gibbons*, a US Army ship loaded with 3,000 tons of TNT. Its crew had insisted on taking a coffee break before sailing, and, by the time the ship finally eased away from the pier shortly after 2:00 a.m., the sky was lit up by the explosion on the *Pedernales*. While the *Henry Gibbons'* skipper wanted to continue full ahead, the Aruban pilot refused the command and instead returned the ship to its berth.[5]

U-156 now steamed northwest to Oranjestad. Perhaps new targets had arrived from Maracaibo. The submarine slid past the tanker *Hooiberg* in the last remaining dark, unaware of its proximity. Hartenstein disappeared below deck to check on the two injured men. There was little hope for Seaman Heinrich Büssinger; the 19-year-old died 45 minutes later without ever regaining consciousness. Lieutenant Dietrich von dem Borne presented a grisly sight.[6] His right knee and leg had a large open wound with multiple splinters. Blood spurted from the arteries. There was no choice but to amputate. Borne was given some of the Old Man's special cognac, and, while four sailors held him down, one of the wireless operators, on temporary sick-bay duty, began to saw just above the knee. The leg, with its shoe and sock still attached, fell into a bloody pail.

At 3:16 a.m., Hartenstein maneuvered *U-156* into the opening in the reef just outside Oranjestad harbor. He was in luck. A tanker lay at the Eagle Refinery pier. Range: 600 meters. It would be a simple surface shot from Tube III. Executive Officer Paul Just slammed down the firing knob.

Once more, the boatswain's mate counted off the seconds. Nothing. After sixty seconds, Hartenstein noted the obvious in the war diary: "No detonation. Inexplicable miss." Had the contact pistol failed again? Or had the depth-keeping mechanism malfunctioned?

Cool as at San Nicolas, Hartenstein left the harbor and prepared for a second attack approach. At 3:30 a.m., he fired his last bow torpedo at the seemingly hapless tanker. Yet again, the boatswain's mate called out the seconds. Yet again: Silence. "No detonation. Inexplicable." Furious and with all bow "eels" expended, Hartenstein ordered *U-156* to come about for a stern shot. At 3:43 a.m., he fired a third torpedo at the tanker. The crew waited anxiously for the sound of a detonation. Again: Silence. Ruefully, Hartenstein entered into the war diary: "Miss. After 1 min. 29.5 [seconds] detonation on the beach." At 70,000 Reichsmark per torpedo, this was a costly misadventure. Morale in the boat, Executive Office Just recorded, had "plummeted to zero." Unbeknown to Hartenstein, not all three "eels" had missed their target. One had struck the side of the empty and degaussed 6,452-ton American tanker *Arkansas*, formerly the *Aryan* and hastily renamed in 1940, which suffered only the force of the explosion of the torpedo's 600 pound warhead.[7] Four Dutch demolition experts died on Eagle Beach two days later when they tried to disarm the beached "eel."

The bridge watch suddenly heard the sound of an aircraft approaching from nearby Dakota Field. It was time to leave Oranjestad. Staying on the surface, Hartenstein shaped a course for the northern end of Aruba and then headed out to sea to raid the traffic bound for Mona Island, off Puerto Rico. At 6:28 a.m., he informed Vice Admiral Karl Dönitz of his actions:

> Sank 2 tankers 5800 tons, 2 misses at tankers at the pier. Explosion in the bore, 2 men seriously injured, including Second Watch Officer. May I head for Martinique to hand over [Borne]? 159 cbm [fuel oil]. v. Hartenstein.

Once well out to sea, Hartenstein ordered Büssinger's corpse sewn into a canvas sheet, brought up on deck, and covered with the navy's battle flag. The skipper intoned the *Lord's Prayer*; the crew sang the traditional lament, *I Had a Comrade*; and then the corpse was quietly delivered into the waters of the Caribbean. At Kernével, Dönitz's staff investigated the legality of landing Borne on the Vichy-French island of Martinique. *U-156* received permission to do so around midnight on February 17.

* * *

As *U-156* and *U-502* carried out their attacks off Aruba, Oberleutnant Günter Müller-Stöckheim's *U-67*, another of the large Type IXC boats, had been sizing up the port facilities on Curaçao. There, Maracaibo tankers offloaded their crude at Caracasbaai, a deep-water terminal on the southeast coast of the island, from where it was carried by pipelines to the Santa Anna refinery at Willemstad. The latter was a daunting target. A menacing coral reef guarded the waters off the Curaçao capital and a long, narrow channel, Santa Anna's Bay, connected the Caribbean Sea to the inner harbor, the Schottegat, where the Royal Dutch Shell refinery and a fleet of tankers were located. It would be suicide to attempt to enter the inner harbor, guarded by three forts. Thus, Müller-Stöckheim made the only decision possible: to attack the fully laden tankers anchored about a mile off Willemstad.[8]

As he approached the targets at 2 a.m. on February 16, Müller-Stöckheim saw a flickering tongue of flame 290 degrees on the horizon. "Apparently Hartenstein is active there," he noted in the war diary. At 3:52 a.m., he fired a double spread at the nearest tanker, a mere 500 meters away. "Inexplicable miss." Caracas radio was broadcasting warnings to all shipping concerning U-boats. "Not clear to me why, because they cannot possibly see me from there," he wrote in the KTB. Obviously, the radio reports pertained to Rosenstiel's *U-502*. Müller-Stöckheim pursued the tanker, which remained oblivious to the danger lurking nearby. At 4:11 a.m., he fired a third torpedo. The crew on the bridge counted off the seconds. Silence. Then they heard several slight rings, like those of bells. Duds. Again, the "eel" had struck its target but had not exploded. The chronic problem with the contact pistols on the G7e electric torpedoes continued to dog the U-boats.

Another loaded tanker hove into sight. Angrily, the skipper turned *U-67* until its stern pointed at the new target. This time he fired one of

U-67. One of 54 commissioned 1200-ton ocean-going boats, *U-67* was commanded by Korvettenkapitaen Guenther Mueller-Stoeckheim off Curacao. It was sunk by aircraft in the Sargasso Sea in July 1943. Source: Deutsches U-Boot-Museum, Cuxhaven-Altenbruch, Germany.

the older "Ato" torpedoes. Time: 4:30 a.m. The torpedo ran true. "After 22 seconds hit just abaft midships. Column of fire and smoke. Tanker ... breaks in half, up 15–20 degrees by the bow and stern. Slowly tanker begins to sink. Fires on deck are extinguished after 15 minutes." Müller-Stöck-heim decided to deliver the *coup de grâce*. "Nothing." The tanker had run aground near one of the forts guarding the entrance to the harbor. He was mystified why the "eel," even if it missed the target, did not explode "against the coast or the bottom of the channel." Then the crew on the bridge heard a tremendous explosion: the 3,177-ton Dutch tanker *Rafaela* burst into flames, lighting up other tankers as well as *U-67*.

Shells from the shore batteries began to splash all around *U-67*. Its hydrophone operators reported the sound of rapidly approaching "high-pitched propellers." Warships. His six tubes empty, Müller-Stöckheim decided to leave the scene of destruction. But then, the Enigma machine lit up. The "Great Lion" from France ordered him to shell the Royal Dutch

Shell refinery with the deck gun. Müller-Stöckheim could not maneuver *U-67* close enough to attack because of the warships buzzing all around the harbor. A second order next morning to attempt again to shell the refinery likewise foundered on enemy warship activity. This would later earn Müller-Stöckheim a stinging rebuke from Dönitz. "The commander should have pursued more energetically the attempt to attack the oil refineries on Curaçao."[9] Still, the tally for the first morning of Operation Neuland now stood at six tankers. Refinery authorities on Aruba and Curaçao immediately ordered a temporary halt to all further shipments of crude from Venezuela. The Associated Press reported that 14 lake tankers had been recalled to Maracaibo.

* * *

Back off San Nicolas, the grisly rescue attempts proceeded in fits and starts in the darkness. Within an hour of being torpedoed, the *Oranjestad* started its final descent to the bottom of the sea. Its skipper, Herbert Morgan, and a small group of three officers and sailors had huddled until the last moment on a section of the bow not yet on fire. They waved pieces of clothing for nearly an hour in hopes of attracting attention on shore. No one saw them. They all had lifejackets, except the second officer. As the bow slipped beneath the waves, they were washed off their safe haven. For nearly an hour, Morgan and his group drifted through the burning oil. Finally, at 3:30 a.m., they and the remaining six others of the 25-man crew were fished out of the oily waters by a Dutch patrol boat and a local fishing boat. The careless second officer had drowned.

The *Pedernales* continued to burn, but remained on the surface. The torpedo's explosion had broken its back and both stern and bow stuck out of the water at 20-degree angles. Herbert McCall, the ship's master, gathered up a small group of five sailors and guided them to the *Pedernales'* port lifeboat. They lowered it, but at an uneven keel, with the result that the boat's oars were lost. Once in the burning waters, they ripped up some floor boards to use as paddles. To no avail. The boat drifted out to sea, northward off Oranjestad, where they were sighted by a fishing boat and towed to shore. Eight of the crew of 26 had died that night.

Still, the *Pedernales* refused to go down. "Charred, twisted and crumpled," it lay in the waters off San Nicolas.[10] By late morning, the fires had burned themselves out. Tugs towed the hulk to Oranjestad and grounded it on the beach. Thereafter, shipyard crews dynamited it in two and tugs towed the bow and stern sections back to San Nicolas. Lago shipyard crews fitted the two sections together and the plucky *Pedernales*, now 124 feet shorter, steamed to Baltimore. There it again was cut in half, a new extended mid-ship section was floated in, and the three parts were reconnected. *Pedernales* returned to service, just in time to take part in the Allied assault on North Africa in late 1942.

For the residents of the 600 Lago Colony bungalows at San Nicolas, it had been a night of sheer terror. Who was the intruder? Was it just a submarine or a warship? Had the Germans mounted a full-scale invasion of the island? Or had the refinery staff simply been careless in handling the highly volatile aviation fuel? Only one thing was certain: these could not have been the acts of fifth-columnists, for Dutch authorities had removed all Germans from Aruba back in May 1940 and interned them on nearby Bonaire Island.

At first, few could believe the fiery hell that spread before their eyes just off the reef. The war was thousands of miles away, in Europe and in Asia. Surely, they were not in harm's way! As the noise of the explosions rattled windows and shook some of the bungalows near the refinery, and as the bright light of the burning tankers flickered through their windows, most residents of Seroe Colorado reacted with both curiosity and indifference. A case in point was the Fred C. Eaton family in waterfront Bungalow 12.[11] The glow of the burning *Pedernales* awakened Mrs. Eaton who, in turn, roused her husband from his sleep. Someone must have been careless, he reassured her, it must have been an accident. When the *Oranjestad* also exploded, Fred Eaton reassured his spouse that the brisk Trade Winds must have blown a spark across the water and ignited the second lake tanker. As streaks of white light flashed across the refinery compound, he was certain that the flares in one of the rocket boxes on the tankers had ignited. Then reality hit home: those were not flares but tracer shells fired by some enemy lying offshore. Eaton gathered the family in his car to take them to the shelter of the Lago Community Church. The only injury that the family suffered came when Fred's blacked-out car collided with another vehicle en route to the sanctuary.

John B. Teagle watched the inferno from Bungalow 77. "I had a front row seat to watch the burning of the three [*sic*] Lake Tankers."[12] His son Lenny also recalled: "We had a full view of the burning ships. At 10 years of age this left an everlasting impression." A neighbor, Nancy MacEachern, was witness to one of *U-156*'s errant 3.7-cm shells slamming into the radiator of a car parked at Bachelor Quarters. Another neighbor, Jane Andringa, remembered what she called a "conspiracy of silence" immediately after the attack. "I do not remember any discussion at school, from teachers or students. [Parents] discouraged questions."

Most residents of Seroe Colorado were intensely curious. Something big was happening on their little island and they wanted to be part of it. They turned on the lights in their bungalows and then piled into their Fords and Chevrolets and drove down to the lagoon area, lights blazing. Only slowly did they realize that what they saw was no accidental fire at the refinery, but an act of war against two ships off the reef. Most extinguished their car's headlights and went home to pack up their most precious belongings in order to head out for relatives in other, more tranquil, parts of Aruba. They were spurred on by the sound of aircraft engines droning high above. Were they friendly or hostile?

Countless others, panicked with fear, had but one thought: out into the *cunucu*, the Papiamento term for "countryside." Quickly, a mass exodus ensued. It was a poor decision. The *cunucu* was a rock-strewn desert punctuated by dry, nasty forests of kadushi, yatu and prickly pear cacti, aloe, and small, wind-bent Divi-divi trees. It was a dangerous place to be. Ankles turned and cacti needles punctured arms and legs. In the morning, more than two dozen residents sheepishly returned to San Nicolas to be treated for minor wounds in the Lago hospital. There had even been talk of seeking refuge in the many caves on the north side of the island, but thankfully no one opted for this course of action.

One of the few who kept their head was the refinery's general manager, Lloyd G. Smith.[13] In horror, he discovered that there was no way to switch off the row of lights that illuminated the boardwalk that ran from the Main Dock to the Lake Tanker Dock. It provided perfect lighting for whoever lurked out there beyond the reef. With cool resolve, Smith walked the length of the boardwalk and doused the lights one by one by throwing rocks at them. Today, the main street of Oranjestad bears his name.

At around 8:00 a.m., Caribbean Defense Command finally reacted. It dispatched two flights of B-18 bombers to Dakota Field, one from Puerto Rico and the other from Trinidad. They arrived at about 1:30 p.m. After refueling, the planes took off again to patrol the approaches to the island. The B-18s spotted and attacked one submarine at 10:15 a.m. and another an hour later, but in neither case did they do any damage. Two A-20s of 59th Squadron already had attacked a submarine at about 11:33 a.m. some 100 kilometers southwest of Dakota Field. They had dropped eight bombs and seen "an oil slick and air bubbles," leading Squadron Headquarters to conclude that a U-boat had been destroyed. Neither Hartenstein nor Rosenstiel recorded any such attack in their KTBs. At the end of a very busy day, 59th Squadron's war log recorded ruefully:

> This unit was not notified of enemy action for sometime after it started, slowing down the attack of this unit. The submarine shelling the refinery would have been an easy target with proper notification. The lack of depth bombs caused this unit to be very uncertain of the damage done to submarines attacked with three hundred (300) lb. demolition bombs.[14]

* * *

Some semblance of normality returned to San Nicolas late on February 16. Lago Hospital took in 27 sailors fished from the burning seas, while San Pedro Hospital at Oranjestad cared for others. The fires at Lago Refinery were kept low until sheet-iron blackout shields could be rigged up for the furnaces. The lake tankers resumed their Maracaibo runs several days after Hartenstein's assault. US Navy warships and Army Air Force planes escorted the Maracaibo-to-Aruba runs, which were temporarily diverted through Amuay Bay on the Paraguana Peninsula. About 140 women and children (from 58 families) elected to return to the United States, leaving their husbands to work at the Lago Refinery. Just to be on the safe side, for almost a year following the attacks the US government refused to issue passports for family members to travel to Aruba. Full censorship was introduced. In its February 21, 1942, issue, the *Pan Aruban* informed its readers with "regrets" that it could not report on the week's big news.[15] Nor did the island's other paper, the *Aruba Post*, report on the San Nicolas attack.

* * *

Hartenstein and *U-156* resumed their war patrol. The skipper was still furious over the "Ato" torpedoes that had run errant at Oranjestad. En route to Fort-de-France, Martinique, the bloody "eels" continued to bedevil him. At 10:35 a.m. on February 19, two-thirds of the way on his northeasterly course from Aruba to Martinique, he sighted what he estimated to be a 3,500-ton freighter.[16] Given that a Force 5 fresh breeze and three-foot waves were buffeting his slender craft, he opted for a submerged shot. "Missed! Checked targeting data! Probably fired under him!" He at once launched a second torpedo at the freighter. Same result. "Missed! Not clear why!" Angrily, he noted in the war diary: "That is now the 3rd inexplicable miss, after 2 torpedoes had already found their targets." Both of the "misses" were "Eto" electric torpedoes. The first failed to detonate under the freighter; the second simply missed the target. He cursed the Torpedo Directorate.

In the first dawn light of February 20, west of Martinique, *U-156* came across another lone freighter.[17] Course: 300 degrees. Range: 5,000 meters. Hartenstein ordered battle stations and approached the target submerged. At 6 a.m., he fired a first "eel." The "Eto" struck the freighter amidships, but apparently did little damage. It stopped and opened fire with its stern gun. As it was then almost full daylight, Hartenstein had but one choice: to attack submerged. He fired another "Eto" at the enemy. "Miss! Inexplicable!" Livid, he ordered a third shot, this time with the older, more reliable "Ato" torpedo. "Miss! Freighter turns off to the left!" Hartenstein then switched back to the electric "eels." The fourth torpedo ran for 35 seconds and seemingly struck the target, but the Kaleu was not able to see the explosion because the hostile kept firing at his periscope.

Hartenstein had emptied all four bow tubes. By now in a towering rage, he repositioned *U-156* and fired a fifth "Eto" torpedo, this one from the stern tubes. "Miss! Inexplicable!" The freighter continued to fire at the U-boat's periscope and it put out SOS calls. In short order, an unidentified PBY flying boat dropped four depth charges near *U-156*, but well off target. The Catalina dogged *U-156* for the next seven hours. "Can't seem to shake this character!" With Borne slipping in and out of consciousness and vomiting up all food offered him, Hartenstein decided to shape course for Martinique. An Enigma message, "A son born, mother and child healthy!," brought little joy to *U-156*'s executive officer, Lieutenant Just.

* * *

The German assault on Aruba ended on a note of mystery and suspense. According to panicky local reports, a submarine surfaced in the still waters of Oranjestad's side harbor, just inside the reef, the day after Hartenstein's attack at San Nicolas.[18] Soldiers guarding the piers at first took it to be an American craft. But residents near the shore could clearly make out the intruder: it was German and there were men on its bridge looking at the city and its protective fort through binoculars. At once, the cry of "Nazi submarine" went around the capital. Much of the population, including the students returning to Juliana School after lunch, flocked down to the docks to get a closer look at the invader; others headed out of town for the safety of the *cunucu*. It was *U-502*, coming to Aruba to have a look. But the schoolchildren were not the only ones to spot *U-502* – which had run aground not more than 400 meters from the end of Dakota Field.[19]

Rosenstiel was certainly brazen. Using the Hooiberg as his reference point, he maneuvered *U-502* half submerged into Oranjestad roads. It was 10:30 a.m. Well aware of the reefs that guarded the port, he rang up "Dead Slow." But the strong current was driving him off course. "Hit bottom. Boat rises quickly. Full speed reverse!" For two hours, *U-502* refused to budge. Rosenstiel peered through his sky periscope. "Precisely at the entrance of Oranjestad. Impossible drift due to the current." He ordered the hatch to the bridge opened and the diving tanks blown. From the bridge, he barked out, "Hard a-starboard, Full speed head!" Both diesels roared up to full power. Slowly, *U-502* began to move off the sandbank. The current drove the boat into a harbor buoy. Rosenstiel ordered the machine-gun crews up on deck in case of hostile air attack. He took a look around the harbor. "No escorts in sight. Harbor empty. My first somewhat involuntary view of Aruba," he cheekily noted in the war diary.

Suddenly, at 1:16 p.m., "Alarm!" A "warship" (in reality, a Dutch motor launch) had come out of nowhere and was driving for *U-502*. Rosenstiel ordered an emergency dive – just in time. At Dakota Field every available plane was scrambled. As the first A-20 Havoc took off, the onlookers at the airfield saw the submarine make a crash dive apparently with his diesel engines running judging from the blue smoke that followed the craft. The A-20 pulled into a tight turn and made its bombing run, followed by at least one other craft. Bombs suddenly began to explode near the U-boat: "2 depth charges in the vicinity. 2 closer. 1 relatively close. 1 really close. 1 far away," the KTB recorded. The surface attacker crossed *U-502*'s wake, but then disappeared. Rosenstiel escaped without damage and shaped course for Los Monjes Islands, off Venezuela. His action proved to be the last German "attack" on Aruba for the duration of the war.

Yet, the island's rumor mills could not be silenced. Dan Jensen recalls his father telling him how after the raid on the Lago Refinery, "neutral Spanish tankers" would put into San Nicolas and purchase fuel – to resell at great profit to German U-boats. And how French boat operators would buy diesel fuel on Aruba, fill 55-gallon drums at St. Barts Island, stow the drums on "fishing boats," and take them out to sea – to sell to the German raiders in exchange for gold.[20] Other stories circulated about submarines pulling alongside inter-island schooners to take on supplies of fresh oranges, lemons, bananas, avocados, and bread fruit. And about daring U-boat officers coming on shore and attending local movie theaters.

The island defenders were so jittery that just before dark on the morning of February 19 "shell fire was heard and flares observed over the Lago Refinery." The base defenders reported two star shells (illumination rounds) and explosive shells fired at the refinery. Three A-20s were dispatched and one reported spotting and attacking a submarine. In fact, the star shells had been mistakenly fired by the destroyer USS *Winslow*, newly arrived at San Nicolas after the attacks. Two of the "shells" did slight damage to some houses in the Lago Colony.[21]

* * *

The attacks that began on February 16, 1942, were but the opening rounds in the fight for Allied oil in the Caribbean. In the last twelve days of February, Hartenstein and his comrades sank *Monagas, West Ira, Tia Juana, San Nicolas, Oranjestad, Nordvangen, Delplata, Scottish Star, Kongsgaard, Circle Shell, J.N. Pew, Kennox, Thalia, West Zeda, Lihue, George L. Torrain, La Carriere, Esso Copenhagen, Cabadello, Macgregor, Everasma, Oregon,* and *Bayou* and damaged *Pedernales* as well. Seventeen Allied tankers or cargo vessels of 115,856 tons went to the bottom.[22] Disaster loomed. Shares traded on the New York Stock Exchange dropped sharply as a result of the attacks, while President Franklin D. Roosevelt warned Americans that the Axis could hit the United States and that the shelling of New York or even Detroit was possible.[23] The all-important tanker traffic that was sustaining Britain and supplementing US stocks was in danger of complete paralysis. But the attacks had also galvanized the defenders. Venezuela abruptly granted the United States permission to use its airfields for ASW patrols and the Dutch government-in-exile placed its armed forces under American control. Ships, men, and aircraft headed to the Caribbean to bolster the local defenses. But for now, it was too little, too late. The U-boats were the masters of the Caribbean.

MARTINIQUE

Shortly before 8 p.m.[1] on February 20, 1942, Kapitänleutnant Werner Hartenstein brought *U-156* up to the surface. The sea was relatively calm.[2] A welcome breeze was blowing and large wavelets rippled off the boat's gray steel hull. He could smell tropical blossoms and wood fires. Off the right bow, he spied the lights of Martinique. He had drafted his own chart of the west coast of the island from the commercial map that the Kriegsmarine had given him at Lorient. A cautious man, he ordered battle stations. He was not "fully certain," he noted in the war diary, whether he could "trust the loyalty" of the French colonial authorities on the island, "far away from Vichy." For "security reasons" he ordered the boat's emblem to be covered and the new experimental FuMO 29 radar detector to be stowed below decks.

U-156 approached the coast at half speed. Bright moonlight illuminated the broad sweep of the bay of Fort-de-France. The lights in the harbor had been extinguished and all shipping buoys removed. Hartenstein ordered running lights and set lanterns to light up the battle flag raised on the extended periscopes. It was 9 p.m. He signaled French shore authorities in his best high-school French. "German vessel. Please dispatch a boat for a wounded man." After a while, a small lighter appeared. Its captain spoke no German. The Kaleu tried his high-school English. No response from the lighter. The Frenchman disappeared into the darkness.

At 10:35 p.m., a patrol boat approached *U-156*. On its deck, Hartenstein could make out three officers and six black sailors, all smartly turned out in crisp white uniforms and caps. The officers, the skipper happily noted, were "very cordial." One of them, an Alsatian, spoke some German. The boat also had a medical doctor on board. The French officers asked Hartenstein to douse his lanterns. Under the cover of darkness, Lieutenant Dietrich von dem Borne, his bleeding staunched but his fever out of control, was brought up on deck and handed over to French sailors. Fortunately, Vichy had alerted Martinique to Hartenstein's arrival. A few last farewells and best wishes for a speedy recovery for Borne,[3] and *U-156* headed westward back out to sea at flank speed.[4] The entire undertaking had taken less than three hours.

* * *

The lush tropical island of Martinique lies between St. Lucia and Dominica, both British possessions in 1942. It prides itself on being "the queen of the Caribbean islands." Its Indian name, "Madinia," translates into "the island of flowers." Distinguished manor houses and old rum distilleries evoke the splendors of the past. Nestled in the turquoise and blue sea, fantastic white beaches graced by king palms dot its shores. One of Martinique's major claims to fame is that it was the home of Napoleon Bonaparte's first empress, Joséphine. In February 1942, it was the most feared western bastion of Vichy France.

The island's capital, Fort-de-France, in the words of a US Navy officer, was "a potential Gibraltar of the Caribbean."[5] A protective semicircle of high mountains rings the capital and its port. The bay that fronts Fort-de-France is a massive 13 square miles in area; the entire 1942 US Navy could have anchored in its calm, yet deep waters. The entrance to the bay was well guarded: Fort Tartenson to the west had four 16-cm guns and two 8-cm mortars as well as anti-aircraft guns; a second installation farther east and right near the coast was Fort Desaix. It was "sturdy, easily defended, well armed and well supplied." Its 17th-century walls were "solid rock hewn out of the mountain surrounded by a dry moat 150 feet deep and 50 feet wide." One easily defended road accessed the fort, which counted two 9.5-cm guns and five 8-cm mortars as well as a battery of eight anti-aircraft weapons. It was well camouflaged by vegetation and trees and was difficult to spot from the sea. Below the fort were a military hospital and the artillery headquarters and barracks for the French West Indies and French Guiana.[6] Roughly 5,500 Vichy French troops with artillery, ten warships, and 106 fighter and bomber aircraft were based on Martinique.

The US War Department's BUNGALOW Plan for a possible war with Vichy France estimated that it would take a reinforced infantry or Marine division of 21,100 men – augmented by 75 fighter aircraft, 30 medium and ten light bombers, four cruisers, and 16 destroyers – to take Martinique. It wildly anticipated casualties at anywhere between 250 and 18,000 men.[7]

Presiding over this large garrison was Admiral Georges Achille Marie Joseph Robert, High Commissioner for the French Antilles, serving at the pleasure of the Vichy government of Field Marshal Henri Philippe Pétain. Robert, a crotchety bachelor, had retired in the grade of vice admiral in 1937 but had been recalled to service in September 1939 and promoted admiral as the Republic's commander in the western hemisphere. He had made the transition to the Nazi puppet state based at Vichy seamlessly. He saw Pétain as the very "incarnation of Eternal France." He viewed both the "Anglo-Saxons" and "Gaullism"[8] as France's primary enemies. He seemed to share the visceral dislike that his boss, Admiral Jean-François Darlan, Commander in Chief French Fleet, had for British "dishonesty" and "untrustworthiness and treachery" in general, and for the "drunkard," Prime Minister Winston S. Churchill, in particular.[9]

At various times, Admiral Robert commanded a formidable fleet of warships totaling 70,000 tons, based on Martinique and Guadeloupe.[10] Shortly after the collapse of France in June 1940, its only aircraft carrier, the rebuilt 28,000-ton *Béarn*, had put into Fort-de-France. On board were 106 American Brewster Buffalo and Curtiss fighter aircraft. They had been sold by the US government to the Anglo-French Purchasing Commission, then moved to Halifax, Canada, and in June 1940 loaded onto the *Béarn*. Robert also had command of two 6-inch light cruisers, *Jeanne d'Arc* and *Émile Bertin*, both displacing about 6,000 tons. As well, Robert could call on the service of the 5-inch destroyer *Le Terrible* and the armed merchant cruisers *Esterel*, *Barfleur*, and *Quercy*. Six tankers and nine freighters of 80,000 tons rounded off his flotilla.

Of special interest to the Allies was the *Émile Bertin*, then the pride of the French navy. It had arrived at Martinique on June 24, 1940 – with a precious cargo estimated as high as $300 million in gold. On June 16, the French government had ordered all gold reserves held by the Bank of France, including those of occupied Belgium and Poland, to be evacuated. When Premier Paul Reynaud's request for an American cruiser to be sent to Bordeaux to pick up the gold fell through, it was decided to put the bullion on the French cruiser and dispatch it to Canada. But the French armistice was signed, and the allegedly neutral – but in fact collaborationist – Vichy government under Marshall Pétain was installed after the *Émile Bertin* reached Halifax on June 18.

The Canadian government of Prime Minister William Lyon Mackenzie King was challenged by the cruiser's presence. Canada was at war with Germany, but not with France, or what remained of it, rump or not. Mackenzie King tried to persuade the French Ambassador to Canada, René Ristelhueber, to let the *Émile Bertin* stay in Halifax and transfer the gold to the Bank of Canada,[11] but the Vichy government insisted that the ship be allowed to leave. The Canadian Minister of National Defence, Colonel J. L. Ralston, and the deputy head of the Bank of Canada, Donald Gordon, urged Mackenzie King to restrain the ship by force. But the prime minister was afraid that armed action against the ship ("folly and injury") would cause "no end of trouble throughout Canada" because "the British Admiralty had expressed the view we should not let the ship go."[12] King and his government cogitated for three days, then decided to allow the ship to depart. The captain of the *Émile Bertin* seized the initiative and left under his own steam on June 21.[13] He then sailed straight for Martinique, trailed all the way by the heavy cruiser HMS *Devonshire*, which did not interfere with its passage.[14]

Admiral Robert was delighted. As soon as the cruiser arrived at Fort-de-France, he ordered the gold unloaded. Fearing what he slyly called "the eventuality of ulterior transport … (if the need arose)," he ordered the colony, "under duress" as he maliciously noted in his memoirs, to build 8,000 crates to store the "precious cargo" in uniform weights of 35 kilograms (77 pounds) each. Thereupon, it disappeared into the rock caverns of Fort Desaix.[15] Robert's gold reserve is generally estimated at 12 billion francs – which translates into $3.85 billion in 2012 US dollars[16] Robert, with his fleet of ten warships, 106 airplanes, and 12 billion francs in gold, was a man to be reckoned with in the western hemisphere. Solidly in the hands of Vichy, Martinique was perceived as a major threat to Allied interests in the entire Caribbean. In the words of the official history of the Trinidad sector, "it was in the position of being able to inflict considerable damage on American bases as well as to supply a headquarters for enemy intelligence, operations or supply in the Western Hemisphere."[17]

Well before Hartenstein showed up, stories circulated around the Caribbean basin about German activity on the island. In late April 1941, US Army Intelligence in the Panama Canal Department received a report that German pilots flying for commercial airlines in South America were stationed in Martinique and preparing to carry out bombing missions against the Panama Canal. Apparently, this story followed the unloading of "airplane equipment" at Martinique.[18] Other tales indicated that all the customs inspectors were German and that the roads on the island were being mined.

After the United States entered the war, surveillance of Martinique was increased. Intelligence was gathered from Pan American clipper pilots flying the Martinique-Trinidad route. People who escaped from the island were closely questioned, but their information was often contradictory. Not long after the Pearl Harbor attack, many officers and men began leaving Martinique to join Free French forces. On December 21, two men from Martinique landed at St. Lucia in a canoe. One was a French artillery officer, the other a student. Their hands were raw after paddling for 24 hours.

On February 21, the US consul in Martinique reported Hartenstein's arrival in Fort-de-France and the transfer of "a wounded man" ashore to US Secretary of State Cordell Hull. Two hours later, the State Department instructed Ambassador William D. Leahy, a recently retired US Navy admiral and personal friend of President Franklin D. Roosevelt, to inform the Vichy regime that the United States "cannot permit any of the French possessions in the Western Hemisphere to be used as a base for Axis operations." Apparently, the landing of a wounded sailor had already escalated into a German assault on the Caribbean. Leahy poured fuel on the fire on February 26 by informing Hull that the wounded sailor in question was "the son of an officer high in the German Admiralty."[19] And Anthony Eden, British Secretary of State for Foreign Affairs, lectured Hull "that in his opinion the United States Government should immediately have proceeded to occupy Martinique as a result of this incident."[20]

Given the state of developments on Martinique, it is no wonder that Hartenstein's landing of his wounded officer on February 20 set off a flurry of diplomatic activity. Secretary of State Hull was still fuming over what he termed the "so-called Free French" seizure of the Vichy-controlled islands of St. Pierre and Michelon off Newfoundland on Christmas Eve 1941, and especially the storm of protest his comments had aroused in the American press.[21] He was in no mood for further French transgressions in the western hemisphere, be they Vichy French or Free French. To make matters worse, Admiral Ernest J. King, Commander in Chief US Fleet, warned the Secretary that the aircraft on the *Béarn* were still "useable and might be used very effectively against us."[22]

Hartenstein's arrival in Fort-de-France harbor further set off a new and more frightening cycle of rumors and stories about German submarines and Vichy collaboration. A Pan American crew reported to the US Army on Trinidad that people on Martinique were very irritated because the Vichy authorities were permitting U-boats to refuel there. Two stowaways who landed at St. Lucia in mid-March asserted that there was a large number of Germans on the island. They also reported flying over a submarine lying in harbor at Portsmouth on the eastern side of Dominica, 100 kilometers northwest of Martinique. One of the more fantastic stories alleged that bauxite carriers coming to Martinique, which were owned by the American aluminum company ALCOA, were actually bringing food and provisions to the island in violation of a British embargo on the French West Indian islands proclaimed after the establishment of the Vichy state. It was alleged that, although ALCOA owned one of the largest fleets in the Caribbean, it had suffered relatively small losses. Was ALCOA somehow in cahoots with the Germans? One captain of a torpedoed ship claimed that a U-boat had slipped alongside his life raft and had provided provisions to the survivors, including two cans of milk "bearing the name of a Martinique distributor."[23]

Something had to be done to defuse a tense situation. Hull turned to Admiral John W. Greenslade, who had been on Martinique since August 1940, to deal with Robert about the US-built aircraft. The American admiral had found Robert extremely arrogant, even by French standards. Robert acted as virtual head of state and sovereign in Martinique. But Greenslade understood power. He ordered two US Navy destroyers operating out of the American base still under construction at St. Lucia to patrol off Fort-de-France, at all times within sight of its inhabitants. Robert had relented, to a degree, and agreed that his warships would not leave the West Indies and that he would give Washington 48 hours notice before he moved the larger units. But out of a sense of "honor," he refused to return the 106 aircraft on the *Béarn* to the United States.

Greenslade then stepped up the pressure. The US destroyers now patrolled routinely off Fort-de-France by night, and PBY Catalina flying boats, also based on St. Lucia, by day. Robert relented some more. He agreed to remove the Brewster Buffalo and Curtiss airplanes from the carrier and to put them ashore, where, presumably, the "elements" would "take their toll" on the craft. When the State Department sent a special delegate, Sam Reber, as well as Admiral John H. Hoover, Commander Caribbean Sea Frontier, to Martinique to press the American case, Robert collapsed completely. He agreed to immobilize his warships, and American crews removed engine parts as well as reduction gear pinions from the *Béarn* and shipped them to San Juan, Puerto Rico. And when it seemed that American forces might actually land on Martinique in the face of the grave U-boat threat, French flyers, under orders to "destroy the planes" in such an eventuality, chopped off the tail assemblies of numerous aircraft and set countless others on fire.[24] It was an inglorious end to Admiral Robert's once-proud fleet and air arm.

It was not until early March 1943 that the Americans finally received some reliable information about the state of affairs on the French island. On March 6, three young men from prominent island families made their way to Trinidad to join Free French forces. They reported that they were certain that the army forces on Martinique would fight if the island was invaded and the French naval ships scuttled. There was an acute food shortage on the island, which was having a profound effect on its politics by undermining remaining support for the Vichy government. In fact, they claimed, nearly all inhabitants had rejoiced at the news that the British and Americans had taken North Africa. When asked if there had been any contact between German submarines and Martinique with the exception of Hartenstein's entry into Fort-de-France harbor, all three were "very positive" that none had occurred. In light of the serious shortages of both food and fuel on the island, the notion that the Germans were using it as a supply base was "extremely foolish."[25]

The news should have come as no surprise. The embargo on food and fuel shipments to the French islands, proclaimed in November 1942, had caused massive unrest. French Guiana deserted the Vichy cause under American pressure in March 1943, and there were anti-Vichy riots on Guadeloupe in late April. That month, the Americans finally cut diplomatic relations with the island. Robert's position was becoming increasingly unstable. The US Army and Navy had already initiated plans for an invasion of the island using American troops from Guantánamo, Trinidad, and San Juan. The Trinidad Mobile Force, as it was called, would consist of the 33rd Infantry Regiment, accompanied by a regiment of engineers. In May 1943, these troops were concentrated in Trinidad and began to practice amphibious landings.[26] But the landing never happened. What the unpublished official history of US Army operations in the Caribbean in World War II termed "probably the most inflammable situation of the war in the West Indies" was resolved peacefully and the invasion "at the last moment became unnecessary."[27] Local resistance forced Robert to resign at the end of June; he was replaced by a representative of the Free French. The gold was recovered and the *Émile Bertin*, ironically, was refitted in New York and later put to sea as part of the revived Free French fleet.[28]

* * *

After leaving Martinique, Hartenstein resumed the hunt for ships. He headed west to search the waters between Trinidad and the Anegada Passage just east of the British Virgin Islands; and thereafter the Trinidad–Mona routes. He was champing at the bit. By now, Kernével would have had time to assess the disaster with the deck gun's tampion, and thus his failure to destroy the Lago refinery. As well, they would have heard from enemy reports that he had not spotted the *Henry Gibbons* inside San Nicolas harbor and that he had missed the *Hooiberg* just arriving from Maracaibo with a full load of crude. A reprimand was sure to result.

Shortly after midnight on February 21, Hartenstein fired off a terse after-action report to Vice Admiral Karl Dönitz at Kernével:

Transfer [Borne] without incident. Day before yesterday in CM 55 empty freighter 2 misses Eto. Secure targeting data. Depth 3 and 2 meters. Yesterday JK 65 2 hits. 3 misses against 4,000-ton stopped freighter. Could not observe sinking due to flying boat. 140 cbm.[29]

He could only hope that better times lay ahead.

In the evening of February 21, Hartenstein, on a course for the Trinidad–Mona sea lanes, stopped to reload the torpedo tubes. It was damned hard work. Not even the cool of the evening eased the torpedo gang's sweaty labors. Each of the seven-meter-long and 1.5-ton-heavy "eels" with their 300-kilogram warheads had to be hoisted out of its watertight compartment under the Prussian Scotch pine planks that covered the upper deck and ever so gently hoisted by block and tackle down into the bow of the boat, where it had to be placed on tracks and moved forward into the tubes. The torpedo gang then worked hours to load and arm the "eels." In the always terse words of the war diary: "Transferred 4 Atos from the upper deck into the boat. Duration: 2 hours, 40 minutes from the start of the operation until 4 Atos under deck."[30]

But this was just the start of heavy-duty operations that night. As the last of the torpedoes disappeared into the forward torpedo compartment, Hartenstein called the remaining technical staff up top to deal with the deck gun, whose muzzle had exploded at San Nicolas. Yet again, the boat's sparse war diary gives no sense of the ardor of the task. "Sawed off destroyed 10.5-cm gun muzzle's 53cm length with hacksaw. Duration: 7 hours 20 minutes." In fact, what Hartenstein and his crew accomplished quickly became legendary even in the daredevil "Volunteer Corps Dönitz."

The Old Man was an artillery expert. Ever since his cadet days at the Navy School in 1931, he had rotated through a series of peacetime artillery assignments and at the start of the war had been assigned artillery officer to destroyers. It was now time to put that training to work. Hartenstein had a good sense of humor. The deck gun was cumbersome and unwieldy, so he dubbed it the "lazy Susan."[31] Moreover, he liked to poke fun at convoluted German naval terminology, referring to the gun as "number-so-and-so-many-centimeter rapid fire cannon in submarine-mounting C 36."

More seriously, how to remove the splayed muzzle? "By acetylene burner or with the hacksaw?" he mused aloud. Puzzled faces. Was the Old Man suffering from heat stroke, Executive Officer Paul Just wondered? It had to be a joke. Surely, no one could seriously consider cutting through Krupp-hardened steel in the middle of the Caribbean – and expect the mutilated gun to fire with any degree of accuracy! Hartenstein merely shrugged his shoulders. "But if we get close enough to a steamer, we can hit it even with the shortened barrel." Given that the burner would give off an intense bright light, he opted for the hacksaw. "How many on board?" the Old Man asked. The reply was 14. "Well then, we will use the better half of those." With tape and chalk, Hartenstein measured off the length of barrel to saw off. Two men began the job. After half an hour the blade was dull. Hartenstein measured: 4-mm into the barrel. Hour after hour, the men worked in shifts. Blade after blade gave out. Shortly after 5 a.m., the deformed muzzle fell onto the deck. Mission accomplished.

It was time to get out of sight of possible prowling aircraft. First Watch Officer Just, a former navy flyer, agreed. The next day brought home to Hartenstein the fact that enemy air assets were being daily enhanced. Twice he had to crash dive to avoid wheeled aircraft. When he finally spotted a lone tanker at 5:30 p.m. on February 23, it had air cover. Twice more, *U-156* had to seek safety by emergency dives. It was maddening work. The Kaleu turned the boat into the trade winds on "Slow" in order to preserve precious fuel. *U-156* was now 34 days out of Lorient.

At 7:30 p.m., Hartenstein resurfaced just west of the Dominica Channel.[32]

"Steamer in sight!" The moon sparkled on the waves. The target was making 12 knots and steering a zigzag course. After four hours, the moon dipped below the horizon. It was pitch black. Every time the Old Man made a run at the vessel, it veered off. "I suspect detecting [*U-156*] with sound locator."[33] Finally, Hartenstein maneuvered *U-156* to a position 900 meters behind the target. It presented a sliver of a shadow. He fired a surface "Ato" torpedo, hoping against hope that it would hit the propellers. "Miss! Attack broken off." Was it another defective firing pin, or had the depth mechanism failed again? It was like shooting with blanks. The radio room took it to have been the 5,127-ton British freighter "*Del Plata*." It had cost *U-156* 14 hours.

Just after noon on February 24, the bridge watch spied two tankers in line-ahead on a southerly course. One was a small Aruban lake tanker, the other an ocean-going tanker. They were moving along smartly at 13 knots. "Both engines full ahead!" The chase was on. Hartenstein pursued the targets for the rest of the day and well into the night.[34] He was proud of the Jumbos. "Performance of the diesels most delightful: both engines at three-quarter speed 17 knots against Force 3 [seas] and, despite tropical cooling water, can maintain this for hours."

Since the full moon was out early, around 7 p.m., he positioned *U-156* four kilometers ahead of the large tanker and submerged. At 8:18 p.m., an "Ato" torpedo leaped out of Bow Tube III. While the boatswain's mate counted down the seconds, Hartenstein struck again, this time with another "Ato" from Tube IV. He ordered Chief Engineer Wilhelm Polchau to maintain *U-156* at periscope depth. No guessing what else might be up there. Then the first welcome news: "Heard detonation! 1 min. 43 sec." More anxious seconds. Then, more welcome news: "Hit stern of ship between mast and engine! Thus first hit must have been amidships. Running time 1 min. 43 sec." Through the periscope, he saw the tanker stop, flood the decks with lights, blow off steam, and settle into the sea.

Just then the Enigma machine lit up. A press report out of Lisbon confirmed the sinking of the 5,127-ton American freighter *Delplata*, en route from Buenos Aires to New Orleans with a general cargo, last Friday in the eastern Caribbean. An American airplane had spotted the wreck, called in a US Navy warship, and as a result 52 survivors had been safely landed in an unnamed port in the Antilles. Hartenstein was delighted. "That one goes to our credit from 20 [February]." Apparently, two of the five torpedoes he had fired at the pesky freighter that day had, in fact, found their target.

No time to gloat. After a ten-minute chase, he spied the small lake tanker, course 350 degrees, speed nine knots. It was in ballast. Hartenstein decided to cross its wake to the starboard side, forge ahead four kilometers, submerge, and repeat the attack. Just before midnight he was ready to press the kill. There were still two "Ato" torpedoes in the stern tubes. Again, the agonizing countdown. Again, silence. "Both Misses!" Then, after 40 seconds, a bright flame on the tanker's stern. Perhaps one of the "Atos" had hit? "Joy premature," the Old Man ruefully noted in the war diary. The tanker was firing at the head of the bubble trail of the G7a "eel." All the while, it was turning away hard. One torpedo was left in the bow tubes; no need to save it for another day.

By now, the moon had dipped over the horizon. Hartenstein raced ahead of the tanker for a third time and submerged. He fired at the incredible distance of 1,800 meters. Time: 0:21 a.m. He immediately ordered a hard turn to starboard to avoid ramming the tanker, which was coming straight on. Again, the seconds ticked away. Silence. Then:

> Hit in the fore-ship! Running time: 1 min. 33 sec. Tanker sinks quickly. Stern out of the water. Lifeboats swung out. Fore-ship slips under. Tanker rises into the air and then falls away. 2 min. after hit! All torpedoes fired off!

God bless the last "eel," Hartenstein thought. Its victim was the 5,127-ton British steam tanker *La Carriere*, in ballast from New York to Trinidad. Fifteen of its crew of 41 died that night.

It was time to shape a course for Lorient, still a very long 3,000 miles away. In the distance, Hartenstein could see the lights of Puerto Rico. Mona Passage and the open Atlantic lay just beyond. Still, there was one piece of unfinished business: the sawed-off deck gun.[35] What if some easy target came into sight? The Old Man called the repair crew back up on deck at 1 a.m., February 25. Years of ballistics training now paid off. Hartenstein quickly calculated that the barrel had lost 40 kilograms (88 pounds) weight, and hence the breech tilted down toward the deck. He ordered two cast iron ballast weights each of 25 kilograms (55 pounds) to be removed from under the floor plates and brought up on deck. After arduously drilling two holes through each, the crew mounted them onto the barrel with long connecting rods and nuts to counterbalance the heavy breech. Still, it was jerry-rigged. Hartenstein had the ballasts and connecting rods welded onto the barrel.

Sailors held blankets to form a protective screen around the electric welder. Then it was time to submerge. The first rays of light were just breaking over the eastern horizon.

In the morning light of February 26, the Old Man decided to test the gun.

What would the new muzzle velocity and windage be? Would the gun's recoil mechanism still function? Would the two ballasts attached to the barrel hold under the jolt of firing? Hartenstein ordered the repair crew below deck. He then hooked up a line to the gun's trigger and from the bridge gave it a sharp pull. The 10.5-cm shell flew out of the barrel. A water column rose some 500 to 600 meters downrange. "Fired gun. Gun seems to be combat worthy. Windage unaltered. Recoil mechanism sufficient." He would later recall, "One just has to want it badly!"

The killer instinct in "Crazy Dog" refused to let go. He decided to scout the roadways of Aquadilla on the west coast of Puerto Rico, and then to slide past Arecibo on the northern shore as far as San Juan in the hope of encountering strays. After all, he still had the deck gun and more than 200 shells. And German naval intelligence had intercepted an Allied radio signal instructing merchantmen to hug the coasts of the islands as U-boats were operating in the central Caribbean. En route, the Enigma again lit up: the neutral press had reported the sinking of SS *La Carriere* off Puerto Rico the previous Wednesday. Hartenstein registered his "disappointment" over its tonnage (reported as 5,685) in the war diary.

At 4:35 a.m. on February 27, *U-156* stood off Silver Bank, north of Haiti. Just barely visible in the falling moon, Hartenstein spied the dark shadow that he had been pursuing for the better part of ten hours. He had the 10.5-cm ammunition brought up on deck by a human chain. Then he ordered "Hard-a-port" and brought the boat on a parallel course with the shadow. Range: 1,000 – 800 – 700 – 600 meters. Calm Sea. Moderate wind.

"Clear the decks. Prepare to fire artillery! Open fire with all guns!" Now the acting artillery officer, Hartenstein was in his element. The incendiary shells landed amidships, ripping apart side hull plates, while the smaller guns sprayed the hostile's bridge. It returned fire with its stern deck gun and then tried to send out an SSS signal, which *U-156*'s radio crew managed to jam. Soon, the bridge watch heard a loud explosion and saw bright flames shoot up into the darkness. The burning vessel began to list to port. All the while, Hartenstein pumped more shells into the wreck: 92 10.5-cm and 111 3.7-cm shells, of which 25 to 30 per cent found their target.

Suddenly, a brightly lit ship off on the horizon began to open fire on *U-156*. "Crazy Dog" was no fool. It would be suicide to engage what obviously was a hostile warship in an artillery duel with a sawed-off, jerry-rigged deck gun in the approaching daylight. Since the burning ship was swinging its lifeboats out, Hartenstein ordered "Cease Fire!" He gave the crew, gathered in four lifeboats, sailing instructions to the nearest land. As the ship slipped beneath the waves with a loud hissing noise, he ordered Lieutenant Just to shape a northeasterly course at full speed.

Hartenstein would later learn that the "Lazy Susan" had dispatched the 2,498-ton British steam freighter *Macgregor*, loaded with 2,621 tons of coal and bound from Tyne to Tampa. It was a nice birthday present for the 34-year-old Hartenstein. True to form, the Old Man allowed no reference to his birthday in the war diary on February 27.

The sinking of the *Macgregor* only whetted Hartenstein's appetite. Instead of continuing on to Lorient, he ordered the boat to make one more pass at the traffic coming out of the Mona Passage – or possibly even to scout the north coast of Puerto Rico. It was a wise choice. Just before 6 p.m., the bridge watch reported: "Tanker in sight!" It was fully loaded and running on a zigzag course. No deck guns visible. No escorts in sight. Hartenstein could not make out its grimy flag. Again, the arduous chase – nine hours this time. Again, a human chain brought the heavy shells up on deck. As far as the men were concerned, "Crazy Dog" was fully justifying their nickname for him.[36]

At 5:17 a.m., Hartenstein opened fire on the tanker. The first two shells slammed into the bridge. Range fell to 400 meters. The target's decks soon were enveloped with flames. But its captain threw the wheel hard-a-port. He was heading straight for *U-156*! In the excitement of the action, Hartenstein had not foreseen this clever move. Several thousand tons of flame and smoke were bearing down on the slender U-boat. The men on deck could feel the intense heat and smell the putrid black smoke of the tanker.

"Starboard engine full ahead! Port engine full reverse! Rudder hard-a-port!" But before hunter and prey could separate, the two collided in a screech of iron and a shower of smoke and fire. *U-156* heeled over to port, then righted itself again. The forward diving plane on the starboard side scraped against the burning hulk, sustaining considerable damage. It could not be repaired. Thankfully, there were no casualties among the gun crews up on deck.

As soon as the two antagonists ground past one another, Hartenstein ordered the guns to blaze away. The jerry-rigged ballast that he had attached to the deck gun flew off the barrel under the strain of rapid firing and collision. This notwithstanding, Hartenstein fired at the tanker at almost point-blank range. Later, he recorded the night's action in the war diary:

> Continued to fire. Expenditure: 58 rounds 10.5-cm, 304 rounds 3.7-cm, and 101 rounds 2-cm shells. Observed about 25–30 hits by the 10.5-cm and 200 hits by the 3.7-cm. Tanker brightly flaming amidships. 10.5-cm and 3.7-cm ammunition expended. Expect later sinking [of tanker].

The victim turned out to be a rich prize: the 7,017-ton American steam tanker *Oregon*, en route from Aruba to New York with a load of high-octane gasoline.

After setting a decoy course of northeast, Hartenstein at last shaped a course for Lorient. *U-156* was down to 101 cbm[37] fuel. Still, it had been a daredevil act: the B-18 bombers of No. 45 Squadron on Puerto Rico were within easy range of *U-156*, and he had risked all for another kill. But he still had on board 1,300 rounds of 2-cm anti-aircraft shells. To the disbelief of the crew, "Crazy Dog" was not yet content. Like a Weimaraner that has chased down and killed its first wounded deer, the scent of blood was in his nostrils.[38] He plotted anew:

> Plan: in case a freighter is encountered on the march home during the day, steam directly ahead of it, dive, observe if it is armed, surface at a distance of 500 meters at an angle of 100 degrees, suppress its stern gun with the 2-cm cannon, signal it with semaphore or radio: 'Stop and Surrender.' Sink with blasting charges.

The adrenaline slowly ebbed. Hartenstein returned to his professional, rational self. Late on March 2, he refused to attack a freighter with his 2-cm anti-aircraft guns. The next morning he chased another freighter but then saw that it had a heavy aft deck gun manned by four men in uniform. He broke off the chase. "No, this is not right. They will open fire before my antiaircraft crew is ready to fire. Abandon attack. Surface. Proceed on course 51 degrees."

Day after day, the "garbage tour" proceeded on its tedious course. The Old Man had a shower rigged in the engine room and ordered each man to wash off the sweat and grime of two months at sea – and to shave off his beard. As well, he had dress blues pressed. Shortly after 9 a.m. on March 17, *U-156*, flying six pennants from the extended periscope tubes, tied up alongside the hulk *Isére* in Lorient harbor. Precise as ever, Hartenstein made his final entry in the KTB: "58 days at sea, 10050 nautical miles. 3 tankers, 1 freighter sunk; 1 tanker, 1 freighter probable. 23632 tons."

It had been a highly successful first war patrol. Overall, the Neuland boats had destroyed 22 ships in ten days in the Caribbean; 17 had been tankers. In terms of surprise and impact, Operation New Land had outstripped the earlier Operation Paukenschlag assault on the US eastern seaboard.

5

"THE FERRET OF PORT OF SPAIN"

Werner Hartenstein's *U-156* was the first boat to return from the Caribbean on March 17, 1942. He and his crew received a tumultuous welcome, with the band of 317th Infantry Regiment playing rousing marches and the naval female service, the *Blitzmädchen*, assembled near the *Isére*. Second Flotilla Commander Viktor Schütze was waiting as the sub tied up. "*U-156* back reporting from war patrol!" Hartenstein declared, throwing Schütze a snappy salute. "*Heil* to the crew of *U-156*!" The men roared out in unison: "*Heil* Herr Korvettenkapitän!"

At that moment, a black Mercedes rolled up to the pier. Out stepped Karl Dönitz – the "Great Lion" to the German submarine service. "Attention! Eyes Right!" The men, smartly turned out in their blue dress uniforms, snapped to attention. "At ease!" Dönitz and Schütze walked down the gangplank onto the deck of *U-156*. His eyes narrow slits and his thin lips pursed, Dönitz reviewed the assembled crew. He had a few words for each man. He stopped in front of Hartenstein and pinned the German Cross in Gold – mockingly dubbed "Hitler's Fried Egg" because of its overly large round shape – onto the Kaleu's coat. Then he had Hartenstein show him the deck gun with its temporary clamp still on. Well done! Only then did Hartenstein recognize the third narrow, gold "piston ring" above the broad gold bands on Dönitz's jacket sleeves: he had been promoted to the rank of admiral three days earlier.

With two months' wages tucked in their pockets, the men of *U-156* headed for some much-needed and well-deserved shore leave. A third would be assigned to new boats; another third would go on furlough; and the remainder would stay behind to supervise repairs on *U-156* for the next war patrol. Midshipman Max Fischer, who was about to turn 20, had

A crowd welcomes the return of *U-67* from a patrol. Source: Ken Macpherson Photographic Archives, Library and Archives at The Military Museums, Libraries and Cultural Resources, University of Calgary.

already been selected as a replacement for Lieutenant Dietrich von dem Borne. For rest and relaxation, the U-Boat Service had requisitioned two houses each in Carnac and Lamor-Plage for officers, and three large hotels in Carnac for the ratings.

Then it was off for the post-patrol banquet at Staff Headquarters – and mail call. Next came lunch, glorious lunch! Hot sausages, white bread, fresh vegetables, French desserts – and, of course, real Beck's and Falstaff beer. The dining room soon became a blue-gray haze as the men lit up pipes and their favorite *Atikah*, *Memphis*, and *Gold Dollar* cigarettes. Later that night, it was off to the "Street of Movement," where the action was. "Decadent" American jazz, French *chansons*, scents of perfume, and glorious *mesdemoiselles*. "Come in, sailor! Good music, *beaucoup dance*, good drink, *amour*."[1] The Kriegsmarine had established special "houses" with "ladies of pleasure" in Lorient's Rue de Sully for its U-boat crews; civilians were not allowed. The "houses" were routinely inspected by medical personnel. After sex, each "lady of pleasure" had to hand her customer a special card that noted her name, date, and time of tryst for later "inspection" of the "client" by the medical authorities.[2]

That "inspection" was undertaken before the next war patrol during a mandatory one-week health cure at the Institute of U-Boat Diseases at Carnac, where medics meticulously searched the men for the "Luftwaffe antelopes" that they might have picked up from the *mesdemoiselles*.[3] Syphilis and gonorrhea – which the Germans called "the French disease" – initially were treated with Albucid tablets. If these brought no relief, then the doctors reverted to the older method of injecting permanganate of potash through a urethral syringe. And when all else failed, they applied the "Kollmann treatment," using a metallic expandable instrument to dilate urethral strictures.[4]

Hartenstein, meanwhile, settled into the officers' billet at the Hotel Majestik – and then checked in at the officers' mess at the Hotel Beau Sejour. Wide French beds with clean white sheets were a welcome relief after two months in a 52-inch-long bunk with an artificial leather mattress. There was steaming hot food and rivers of vintage red wine – at a paltry 20 francs (or one Reichsmark) per bottle. He walked the beach at Carnac and reveled in the thunder of the surf. And there was a small château nearby for special encounters: "Officers Only!"

One final item of business: the after-action report. It was well known in the U-Boat Service that the "Great Lion" read every war diary after a patrol, so the Old Man had to be precise, yet careful, for much had gone wrong during Operation Neuland. Hartenstein began his report with the first-ever observation of conditions in the Caribbean. These waters were so translucent, he wrote, that even at full speed with a strong bow wave, one could clearly see the bow hydroplanes; and at periscope depth, the boat's fore- and afterdeck appeared as a "gray shadow" through the periscope. Suggestions: deck and bridge had to be painted black; the light pine decking had to be replaced with dark, impregnated wood; and the conning tower had to be painted to resemble teak. And space had to be made available on deck to store more torpedoes, for all Neuland boats had returned due to lack of "eels," not shortage of fuel.

Finally, Hartenstein turned to "lessons learned." First, the next wave of boats desperately needed precise sailing schedules for individual ports and roads in the Caribbean. Second, they needed civilian airlines' schedules to avoid accidental sightings in congested areas such as the waters around Aruba, Curaçao, and Trinidad. Third, they needed a manual on international maritime law, as every encounter raised new issues of legality. The boats also "desperately needed" some form of air conditioning. Aircraft had sometimes forced them to stay underwater for as long as 13 hours off steaming-hot Caribbean ports with temperatures in the boats rising to 40° Celsius and with unbearable humidity. Thirteen hours under water in the tropics resulted in "severe diminution of the crews' efficiency."[5]

As always, Dönitz carefully read *U-156*'s war diary and Hartenstein's after-action report. After all, here was a senior group commander just returned from a first, potentially war-winning, operation. Initially, Dönitz had censured Hartenstein's action off San Nicolas as too timid and evanescent; the skipper should have kept his eye on the prize, the tankers and not the refinery. But how could he argue with sheer tenacity and eventual success? He penned his evaluation:

> Very well executed first operation of a commander with a new boat. Dash, aggressive spirit and calculated actions on the part of the commander brought the boat a very nice initial success, unfortunately impaired by inexplicable [torpedo] misses. It is regrettable that the artillery deployment against the oil refinery at Curaçao [*sic*] was thwarted by a misfire (firing with tampion in place). Later measures for renewed deployment were recognized.[6]

He summoned Hartenstein to the Villa Kerillon, and congratulated the Kaleu on the tankers and freighters torpedoed, on the repair of the deck gun, and on his attempt to shell the "Esso" refinery. Later that night, at his favorite restaurant, Le Moulin de Rosmadic à Pont-Aven, Dönitz mused about "Crazy Dog" Hartenstein. He liked the wiry, eagle-beaked commander and thought him "an excellent officer, tough in battle and humane when it was over."[7] He would need Hartenstein to head a second wave of assault boats in the Caribbean.

* * *

Hartenstein was not the only one of the five Neuland captains to chalk up an impressive list of kills. While Hartenstein in *U-156*, Rosenstiel in *U-502*, and Müller-Stöckheim in *U-67* had focused mainly on the waters around the Netherlands Antilles, Kapitänleutnant Albrecht Achilles in *U-161* had tackled the difficult job of attacking shipping in the Gulf of Paria. Achilles arrived on station off Trinidad at 6 p.m.[8] on February 16, 1942. "Trinidad in sight."[9] For much of the next day he circumnavigated the island. Both Achilles and Executive Officer Werner Bender took careful notes of its many bays and peninsulas, its sand banks and mud flats, and its air and surface defenses. It brought back memories of when both men had sailed these waters for the Hamburg-Amerika Line.

* * *

Trinidad's northern and southern peninsulas reach out like "arms" due west toward Venezuela, almost in the shape of a "reverse C." Between the two "arms" and the coast of Venezuela lies the Gulf of Paria, a shallow body of water no deeper than 275 meters (900 feet) with an overall area of 7,800 square kilometers. The Gulf has two entrances. The southern is the Boca de la Serpiente (the serpent's mouth) separating the southern tip of Trinidad from the northern coast of Venezuela. The strait is less than 20 kilometers wide at its narrowest point. The northern is the Bocas del Dragon (the dragon's mouth), about 25 kilometers wide. Vessels entering the Bocas del Dragon had to transit one of four channels interspersed with small islands; most used the largest, the Boca Grande – the main shipping channel to the northwest coast of Trinidad and its major harbor and capital, Port of Spain. The Boca Grande and the other channels were rock strewn and noted for their strong currents. Most ships transiting the Dragon's Mouth did so slowly and carefully. By February 1942, most of the islands in the Bocas had been fortified with British or American coastal guns, some of them newly arrived US-built 155-mm "Long Toms." About halfway down the west coast of Trinidad was the largest oil refinery in the British Empire – Trinidad Leaseholds Ltd. at Pointe-à-Pierre, outputting 21 million barrels in 1941.[10] It was protected by a mere brace of 6-inch coastal guns. Port of Spain was the home of a British naval station, HMS Benbow, with a single unit of the Royal Naval Volunteer Reserve commanding a few yachts and minesweepers.

The defenses of Chaguaramas, near the western tip of the northern arm of Trinidad, were more formidable. The port had been included in the September 1940 Anglo-American "destroyers-for-bases" deal, and by May 1941 was home to US 11th Infantry Regiment as well as several units of 252nd Coastal Artillery Regiment. Six PBY5 Catalina flying boats were scheduled to be based at Chaguaramas, once its silt-laden harbor had been dredged. For the time being, the US Navy berthed two 1,000-ton World War I "flush-deck" destroyers, the USS *Barnley* and the USS *Blakeley*, at Chaguaramas. Built in 1918, they mounted four 4-inch guns but had terrible "sea legs" owing to their slender beam, great length, and high superstructure.[11]

Further inland, the US Army Air Forces had stationed squadrons of twin-engine Douglas B-18 "Bolo" bombers of 1st Bombardment Group at Waller Field in the Fort Reid complex near San Rafael on the Caroni plains. It had also selected Trinidad's civilian Piarco Field, 20 kilometers south-southeast of Port of Spain, as the training site for these units. While the effort was impressive on paper, American "disorganization was so complete" that it would take seven months before the patrol planes and bombers were fully operational.[12] Moreover, there was a confused command structure: President Franklin D. Roosevelt had designated the Navy as the service with overall responsibility for the defense of the central and eastern Caribbean but had also chosen the Army to defend the Panama Canal Zone and its approaches. But most of the fighting forces in Puerto Rico, the Lesser Antilles, the Leeward Islands, Trinidad and Tobago, the Netherlands Antilles, and Jamaica were army personnel (with Dutch, British and Canadian troops also present) while the great majority of the aircraft were Army Air Forces. General Henry C. Pratt, US Army, commanded the overall Trinidad Sector; Captain S. P. Oineder the US Naval Air Station; and Lieutenant Colonel Waddington of the Army Air Forces Waller Field. Admiral Sir Michael Hodges, Royal Navy, outranked all three Americans; and above him towered the difficult British Governor, Sir Hubert Young, who despised anything "American."

*　*　*

Achilles and Bender were eager to exploit the situation. On February 16, Kernével infomed *U-161*: "Achilles Free to Attack!" Hartenstein had already assaulted Aruba, and so there was no reason to hold off. Achilles monitored the sea and air traffic while he planned his approach through the Dragon's Mouth. Running on the surface, he guided *U-161* safely through the Boca Grande at 3 a.m. on February 18, in good part thanks to well-lit navigation buoys and shore lights. A British-laid "antisubmarine loop" under the sea detected the intruder, and soon the bridge watch heard the drone of aircraft overhead – B-18 bombers out of Waller Field sent to investigate the contact – along with a few patrol boats. Incredibly, few of Trinidad's defenders had heard anything at all about the attacks around Aruba and north of Lake Maracaibo. They were far from alert.

At 10:34 a.m., Achilles set *U-161* down on the bottom at 60 meters off Chaguaramas under the very noses of the patrol craft and bombers. He did not know it at the time, but the two American destroyers *Blakeley* and *Barney* had left their moorings to charge westward toward Curaçao. The antisubmarine alert was called off in mid-afternoon, but Achilles chose to stay on the bottom until dark, and then surface to charge the batteries. Then, and only then, would he attack the freighters riding at anchor in Port of Spain. He made a mental note not to approach the shore at depths less than 15 meters for fear the torpedoes would not run true, since they first dove upon leaving the tubes, then rose to their preset depth. "Intention: Before operating in the entire area, will drive into the harbor [Port of Spain] since unnoticed until now and defenses hardly anticipating this."[13]

Achilles brought *U-161* to the surface at 7:32 p.m. The hatch was cracked and cool evening air rushed into the conning tower. Achilles sprang up through the narrow opening, followed by the rest of the bridge watch. Bender stayed below. The sea was calm. Search lights from Chacachacare Island played on the waters. A few clouds danced overhead. An aircraft flew by. The U-boat went completely unnoticed. There were targets in the roadstead off Port of Spain. "Freighters are well defined before the well-lit town," Achilles noted in the KTB.[14] All were brightly lit, oblivious to the lurking danger. "Ajax" could clearly make out vintage

Albrecht Achilles. "Ajax" Achilles first served on the German battleship *Gneisenau* and joined the submarine forces in April 1940. On six patrols with *U-161* he destroyed 14 Allied ships of 104,664 tons, and damaged the British light cruiser *Phoebe*. His actions off St. Lucia earned him the nickname "ferret" for his ability to destroy enemy vessels in their ports. He died on 27 September 1943 when *U-161* was sunk with all hands off Bahia, Brazil. Source: Deutsches U-Boot-Museum, Cuxhaven-Altenbruch, Germany.

American cars moving slowly along the shore as well as fishing boats bobbing up and down in port. The piers were littered with tarred fishing crates and nets. The watch took in the bewildering tropical scents: the sweet smells of damp earth and tree bark, of warehoused sugar and nutmeg; the fragrant scent of orchid and cocoa trees as well as oleander; but also the sulfurous stench of mangrove swamps, tar pits, and cesspits.

Achilles was ready. To guard against an unexpected attack, he positioned the boat with its bow facing out into the Gulf of Paria for a quick escape. Range: 3,700 meters. Sea: calm. Time: 11:32 p.m., February 18. Bender fired both stern tubes. After four minutes, "Hit freighter amidships. 20 m[eter]-high water column, thereafter thick, dark cloud of smoke." The 7,460-ton *Mokihana*, en route from Baltimore to Suez with general cargo, heeled over and began to settle almost immediately. Twenty-eight seconds

U-161. Another large Type IXC boat, *U-161* under Kapitaenleutnant Albrecht Achilles operated most dramatically off St. Lucia, Caribbean; it was destroyed by depth charges from a US Mariner aircraft near Bahia, Brazil, in September 1943. Source: Deutsches U-Boot-Museum, Cuxhaven-Altenbruch, Germany.

later, a second detonation. "Hit freighter amidships, very high fire column, parts of the superstructure fly into the air." The bow rose above the sea, the target "broken apart amidships."[15] It was the 6,900-ton tanker *British Consul*, bound for the United Kingdom with a load of fuel oil. The burning vessels lit up Port of Spain – as well as *U-161*.

"Both engines full ahead!" *U-161* sliced into the Gulf of Paria. Achilles ordered a test dive. The bow almost immediately sank into the soft mud of the harbor floor. With engines in full reverse, the sub was freed and raced toward the Bocas on the surface. "High-speed escape in the direction of Chacachacare Island!" Achilles noted for the KTB. Flare bombs lit up the Boca de Navios. "Emergency Dive!" But the strong current prevented progress with the electric motors. "Prepare to surface!" Achilles opted to drive *U-161* west toward neutral Venezuela "to seek protection from searchlights." Then, in the midst of the confusion caused by his attack, he would attempt a daring escape northward, through the Bocas, on the surface.

U-161 had scored a complete surprise and plunged Port of Spain into chaos.

Coastal defense gunners fired their weapons off without ever spotting a target. Patrol boats raced out of Chaguaramas, without any idea of what they were looking for. The island's overburdened telephone system broke down. Air raid sirens wailed. Someone cut off the city's power supply, cutting radio communications and throwing military installations into utter darkness. Antisubmarine personnel at HMS Benbow had to work by flashlight.[16]

How to make good the escape through the treacherous Dragon's Mouth? Achilles decided that audacity was his best chance.[17] He kept *U-161* on the surface, reduced his speed, trimmed the boat to where the decks were just awash, set only a few running lights, and mixed in with a small flotilla of fishing boats and US Navy and Royal Navy launches. No one would suspect such cheek. He passed the guns of the Royal Artillery sited 300 feet up on Caspar Grande Island, so close to shore that they could not be depressed sufficiently to fire at him – even if its gunners had spotted the U-boat. There remained a last hurdle: eight US 155-mm coastal defense guns on Chacachacare Island. Beyond lay open sea. Achilles stuck to his daring plan and coolly ran for three miles under the gunners' noses. Not a shot was fired at *U-161*. Its crew took full notice of their skipper. At home, the German propaganda machine celebrated Achilles as "the ferret of Port of Spain."

Achilles' bold penetration of the Gulf of Paria has been likened to the earlier feat of Günther Prien, who on October 14, 1939, had slipped his *U-47* into the closely guarded British fleet anchorage Scapa Flow and had sunk the 29,150-ton battleship HMS *Royal Oak*. Achilles' audacity was a key factor in his success, but so were the unpreparedness and the lack of training of the defenders.[18] In effect, his and Hartenstein's successes symbolized the battle for the Caribbean at this stage of the war – U-boats with daring commanders and a highly flexible command structure challenging the poorly organized, untrained and under-equipped Allies and creating great slaughter as a result.

Following their foray into the Gulf of Paria, Achilles and Bender hatched yet another daring scheme: why not return to Trinidad and resume the hunt off its northwest peninsula? Surely, no one would expect the intruder to return so soon after its miraculous escape.

They were quickly rewarded.[19] At 7:20 a.m. on February 21, the sounding room reported: "Screw noises at 350 degrees. Tanker, 3000 tons." Achilles closed range. He decided to repeat his first attack and fire from the stern tubes. Tube V: "Miss!" He had overestimated the tanker's speed and underestimated the distance to target. Tube VI. "Miss!" The target had suddenly veered off to port. Had its captain seen the U-boat? Or the track of the "eel"? "Torpedo probably passed in front of it," Achilles dryly recorded in the war diary. He ordered a new plot for a bow shot. But then the target veered off course again. By this time, hostile aircraft circled above the prey and two surface escorts approached at high speed. There was nothing to do but to head away from the scene of this misadventure submerged.

Just after 3 p.m., the hydrophone operator reported a contact at 345 degrees, northwest of the Boca Grande. "Ajax" brought *U-161* to periscope depth. "Tanker, 5000 tons. No neutrality marking. Flag extremely small, hard to make out, probably American." The unsuspecting target was approaching *U-161* at high speed. Achilles submerged and for an hour awaited its arrival. Range: 910 meters. Achilles fired a double spread. After 59 seconds, "Hit amidships. High column of water, as high as the masts." After 69 seconds, "Hit just ahead of the funnel. Likewise, high column of water." The first torpedo blew out the plates on the starboard side of the target, ripped through the inner bulkheads, and sliced open the plates on the port side. The second blew the stern off, taking rudder and propeller away with it. The tanker began to list heavily to starboard and to go down by the stern. But it did not sink. Its savvy skipper used his ballast pumps to shift water to the port holding tanks and thus managed to right his ship. Achilles maneuvered to deliver the *coup de grâce*.

"*Fliebo!*"[20] The bastard had signaled his position and predicament by "code SSS" to Port of Spain. A Royal Navy Fairey Albacore bomber flying out of Piarco Field had delivered two near-lethal hundred-pound bombs. "Emergency Dive!" Achilles took *U-161* down to 50 meters. Not a sound in the boat. The men took off their shoes; those off duty climbed into the bunks. At 4:23 p.m., Achilles brought *U-161* up to periscope depth. He spotted the wreck. In the fading sun he noticed three hostile aircraft directly overhead: two American B-18s from Waller Field had joined the hunt. All three had clearly made out the U-boat's dark shadow in the translucent waters.

"*Fliebos!*" Now all three enemy flyers were depth-charging him. Again, the Old Man took the boat down to 50 meters. Chief Engineer Heinrich Klaassens calmly reported, "No damage." *U-161* came up to periscope depth. For the crew, this was insanity. "Tanker must be brought under water, in any event," Achilles barked out. If not, it could probably be towed into Port of Spain. Given the pervasive air cover, Achilles decided to submerge and to await nightfall. At 6:49 p.m., he deemed the moment right and returned to periscope depth. For the next hour, he maneuvered into position. Range: a mere 370 meters. "Hit, funnel. Tanker sinks very quickly by the stern!" The crew managed to get the lifeboats out of the davits, and Achilles could clearly make them out in a long, white line. "Emergency Stations!" The flaming hulk seemed to be coming right at *U-161*, even though its stern had been blasted away. "Bow towers high out of the water in front of the periscope." Was it drifting with the current? There was not a moment to lose. "Dive! Dive! 15 meters. Hard turn!"

It had been a close call. Too close, in fact. Upon surfacing, Achilles found himself in a sea of fuel oil. He ordered a course in the general direction of the Mona Passage, southeast of Dominica. It was high time to leave the Trinidad sector – and to recharge the almost depleted batteries, to ventilate the boat, and to reload the bow tubes. His third victim had been the 8,207-ton British tanker *Circle Shell*, in ballast from the River Clyde to Curaçao, but diverted to Trinidad. It lost a single sailor in the action.

Shortly before midnight on February 23, *U-161*, standing about 250 miles west of the Martinique Passage, came across a fourth target. Since it was dark, Achilles ordered a surface attack. Range: 2,600 meters. He fired. After two minutes and six seconds, the torpedo slammed into the shadow's fore-ship. It ground to a halt and began to go down by the bow. It was furiously signaling "Freighter 'Lihue' sends U-boat warning," followed by "SOS I am hit sinking." Achilles was irate. He was not about to expend another "eel" on the cripple – he had only eight left – and yet it was going down much too slowly for his liking.

"Clear the decks! Prepare to fire artillery!" *U-161* broke the surface at 0:59 a.m. A human chain conveyed the heavy shells up on deck. Second Watch Officer Götz Roth opened fire with the 10.5-cm deck gun. The prey immediately replied with its 3-inch cannon and several machine guns. A few shells splashed into the water near *U-161*. Too close for comfort. "Alarm! Dive!" Achilles tracked the wreck, which had gotten up steam again, over the hydrophone set. At 1:24 a.m., he resurfaced. Roth's artillery crew blazed away at it again with the deck gun, setting its entire length on fire. Still, it refused to sink. Livid by now, Achilles maneuvered to deliver the *coup de grâce*. But the freighter's wily skipper continued to zigzag, and then put his wheel over hard to ram *U-161*. "Angle 0, distance 1500 meters."

The angle of the shot was more than 90 degrees – too risky – and so Achilles took the boat down, just to be safe. At 5:10 a.m., he resurfaced. "No sight of the freighter." He shaped a course for where he estimated it might have gone. "6:28 a.m. Freighter again in sight." By now, the sun was breaking over the horizon. Surely, hostile airplanes would shortly appear. He broke off his third plot.

"Dive!" Doggedly, he followed the target, hoping to deliver a final submerged attack. He was like a terrier seeking to hunt its prey to ground. He resurfaced. For four hours, the freighter continued to zigzag at 8 knots. All the while, it radioed its position and situation as well as appeals for help. Achilles plowed ahead. It was now noon, February 23. He took *U-161* down for a submerged attack. Range: 5,000 meters. The target's bow was down three meters. Achilles fired from Tube VI.

"Miss." He closed the range to deploy the deck gun. "Freighter looks abandoned, but I do not trust it." He broke off the surface plot.

"New approach submerged." Achilles decided on a double bow shot to kill this pesky adversary. Both "eels" missed the target. "Freighter must have been listening for the torpedoes." Achilles lost all composure. He plotted and ran four more underwater approaches. To no avail. "Freighter always lies at dead stop, then shortly before I reach firing position turns away hard toward the boat." He vowed not to expend another torpedo unless guaranteed success. "Freighter must be sounding the approach of the boat, even though I undertook the last two runs at great distance and at slow speed."

"Alarm! Aircraft approaching boat!" Achilles ordered "Emergency Dive! A-20." *U-161* crept along submerged and silent at 60 meters until 4:20 p.m., when the starboard hydrophone operator reported "great noise" dead ahead. "Ajax" brought the boat up to periscope depth. Behind and off the port side, he could make out the target – and two airplanes circling above it. There was also a small tanker some five kilometers behind the cripple.

Achilles ordered the boat to level off at 30 meters. The starboard hydrophone continued to pick up "noises." At 4:40 p.m., the men in *U-161* heard "a weak detonation and following echo." Achilles guessed that some "metal object" must have struck the outer fuel bunker and "clanked" along the length of the U-boat. He took *U-161* down to 50 meters. A second detonation. He ordered Klaassens to take the craft deep, 110 meters. "Third detonation, weaker and higher up." Was the enemy dropping depth charges? "Suspect that the freighter is a U-boat trap. Wants to maintain contact with the boat until escorts arrive."

There was no sense in further pursuit. "I let the freighter go since I cannot catch him submerged and since [*U-161*] can be seen on the surface due to the bright moon." And there was another reason: "Absolutely must recharge batteries." The plucky freighter, the Matson Navigation Line 7,000-ton *Lihue*, bound from New York to Suez with a load of general cargo, would sink two days later. Its crew of 37 was rescued by the tanker *British Governor.* Its master, W. G. Leithead, brazenly reported that he had sunk a German sub.

U-161 returned to its original northwest course on February 25 and headed for the waters between Trinidad and the Mona Passage. The bridge watch finally broke the seal of the hatch and glorious fresh, cool air rushed into the boat. The men were delighted. They had undergone 22 hours of harrowing plotting, attacking, firing, diving, surfacing, and then repeating the cycle over and over – no fewer than nine times in all – since first spotting the *Lihue* after midnight on February 23. The boat was a cesspool, the diesel buckets set out for the men overflowing with urine and excrement. The temperature inside the boat had reached 40° Celsius; the humidity almost 100 per cent. The men were white as sheets, their skins blotted and lacerated with sweat, diesel oil, and acid that had leaked from the batteries. In small groups, the Old Man allowed them up on deck to suck much needed air into their scorched lungs. Many had reached the limits of their physical and psychological capacities.

But they were immediately reminded that this was a war patrol. Through a dark haze the bridge watch spied a lone freighter at 230 degrees; then "suddenly a second freighter coming out of the mist." The two hostiles converged, then one shot off at a sharp angle. Achilles followed the remaining freighter. At 8:42 a.m., he took *U-161* down for a submerged attack. After an hour of blind pursuit, he was ready to pounce. Time: 9:58 a.m. Range: 900 meters. He fired a double bow shot. After 58 seconds the hydrophone operator reported that the sounds of the torpedo and the target had "merged." Another "Miss!" Achilles had set the "eels" to run at three meters, "too deep" for the half-laden freighter.

Achilles pursued the target for nearly two hours, hoping for a night surface attack. When he came back up, "Ajax" spied not only the hostile, but also a periscope at 800 meters. "Alarm! Dive!" He could not risk a torpedo shot with another U-boat in the vicinity. "Perhaps it is Hartenstein in *U-156*." But Achilles could not catch up to his prey.

Shortly before 9 a.m. on February 26, the bridge watch cried out, "Alarm! Aircraft at 37 degrees! Range 7000 meters!" It was a B-18 out of Waller Field, coming straight for *U-161*. The bomber was undoubtedly equipped with one of the new Anti-Surface Vessel (ASV) radars. In general, at that early stage of the Caribbean campaigns, B-18s and other American medium bombers such as the A-20s were armed with only machine guns and small 300-lb depth bombs; few American pilots had any idea at all as to how to attack a submarine. Achilles immediately ordered "Emergency Dive!" to 50 meters. "3 detonations (probably Wabo)." The last depth bomb severely rocked the boat amidships. The hull groaned and creaked. Glass and china shattered. Buckets rolled around the floor. The men were frozen in their tracks. "Chief! Damage Report!" Achilles screamed. Klaassens' report was a litany of woe. "Depth pressure gauge out of service, including back-up pressure gauge and Papenberg.[21] All glass water-leveling gauges sprung. Lights out in many compartments. Glass fuel-level gauges for the engine-oil reservoir burst. Hence, diesel oil in the lubricating oil collecting tank."

The Old Man took stock of his men. "The crew's composure good. Exceptions were quickly dealt with." Obviously, some of the younger and less experienced hands had cracked under the strain of staring death in the face. The enemy above was daily growing bolder.

Achilles thought about heading for Grenada to recharge the batteries and to reload the bow tubes, but the strong presence of enemy patrol planes convinced him instead to shape course due west, for Branquilla Island off neutral Venezuela. On the way, Klaassens and the technical crew repaired the gauges as best they could and drained the diesel from the lubricating oil collecting tank. It was hot, sweaty work. Later, Bender supervised the transfer of four torpedoes from under the upper deck plates into the bow torpedo room. More hot, sweaty work. *U-161* then retraced its steps due east, headed for St. Lucia in the Windward Islands. Achilles hoped to intercept oil tankers heading north from Aruba, Curaçao, and Trinidad. A week had gone by since the attack by the B-18.

"Ajax" did not have to wait long. At 9:13 a.m. on March 7, the watch spied a ship about 40 miles off St. Vincent Island. Achilles drove the Jumbos hard and positioned the boat for a submerged attack. Time: 11:59 p.m. Range: 1,500 meters. The first "eel" hit after 73 seconds. "Detonation. Boat is undercut." The second followed 15 seconds later. "High detonation column with several smaller explosions." Achilles could not resist a closer look through the periscope. "Freighter steams in circles, its engines still running. Boats have been lowered into the water but are only partly occupied." The twin explosions had rocked *U-161*, even though it was more than a mile away. The thing was big, damned big. It required another shot. After 34 seconds, the torpedo slammed into the wreck's side just abaft the funnel. Again, Achilles could not let go of the periscope. "High detonation column. Several successive strong explosions. Vessel sinks by the stern after 3 minutes."[22] It had flown no flag and shown no other identification markings. As it sank, it dragged three of the four lifeboats down with it. For several minutes, the men in *U-161* could hear underwater explosions. Some feared renewed depth-charge attacks. Achilles estimated the hostile at 6,000 tons. Given the tremendous after-hit explosions, it must have carried ammunition, perhaps dynamite.

He was wrong on both counts. He had torpedoed the 9,755-ton Canadian tanker *Uniwaleco*, carrying a full load of 8,800 tons of refined gasoline from Curaçao to Freetown. Thirteen of its crew of 51 lost their lives, mostly when the wreck sucked down the lifeboats. It would later turn out that "Ajax" had torpedoed the second-largest tanker bagged during Operation Neuland. The men in *U-161* were relieved when the *Uniwaleco* finally slammed into the seabed and the explosions in its holds ceased.

Achilles took the boat down to assess his situation. Klaassens reported that they had just slightly more than 100 cbm of fuel oil, and most of that would be needed for the nearly 4,000 miles back to Lorient. There were three "eels" left on board. What a shame to take them back to France.

Achilles brought the boat back up to the surface. *U-161* was on a course due east. Achilles and Bender pondered their options. A turn south toward Aruba or Curaçao, much less Trinidad, was out of the question due to the critical fuel situation. For the same reason, so was a sortie north to Mona Passage. As they brainstormed, a dark shadow appeared in the distance: St. Lucia. Its shores seemed to be totally blacked out. Welcome warm rain showers pelted the skipper, his executive officer, and the four-man bridge watch.

Achilles and Bender went below and for hours continued to mull over their options. Soon, they were just off St. Lucia's coast. Bender finally broke the impasse. He had sailed these waters before the war and thus knew the general layout of the island and its ports. Why not go out in a blaze of glory by repeating February 18's cavalry charge into Port of Spain and Chaguaramas? Capital idea! Achilles and Bender raced down into the *Zentrale* to study the sea charts of St. Lucia. All they had for Castries, its major port, was a dated commercial sailing chart. "Intention," Achilles quickly scribbled in the war diary, "lie off during the day; by night approach St. Lucia anew; and, if possible, force entry into Port Castries." St. Lucia was about to go to war.

WAR COMES TO ST. LUCIA

The mango-shaped island of St. Lucia, once thought to have been visited by Christopher Columbus on his fourth voyage in December 1502, is a lush, green jewel set in a turquoise and blue sea. Its interior is rough and precipitous. Its mountains climb steeply from the sea, their summits often shrouded in mist. Two in particular, the Gros Piton (798 meters) and the Petit Piton (750 meters), rise from their base at a sharp 60-degree angle, thus resembling immense, lush, volcanic pyramids. Since the days of Columbus, mariners have used them as reference points. For more than a century, the British and French clashed no fewer than 14 times for possession of the island before it finally passed into British hands in 1814.

The United States presence on St. Lucia began with the "destroyers-for- bases" deal in September 1940. Washington considered St. Lucia vital for the defense of the Panama Canal, a sort of fixed aircraft carrier. Planes based there would be able to watch over the eastern entrances to the Caribbean – especially the St. Vincent, St. Lucia, and Martinique Passages – and defend the Leeward and Windward Islands. Short-range craft flying between Puerto Rico and Trinidad could stop at St. Lucia to refuel. The United States chose four locations for army bases and airfields. The largest was at Vieux Fort on the southern tip of St. Lucia, where Minder Construction Corporation of Chicago, using molasses as a surfacing agent, was just putting the finishing touches to two 5,000-foot runways at Beane Field (today Hewanorra International Airport) in March 1942. Vieux Fort was already well guarded – by the 120-meter-high Mule à Chique headland and by Beane Field's fleet of warplanes, well-camouflaged behind horseshoe-shaped earthen mounds that served as hangars. It was some ten miles (as the crow flies) away from the main port and capital of Castries, the only viable entry point for all US men and equipment sent to St. Lucia. A small naval air station, home to 18 PBY flying boats, was established on Gros Islet just north of Castries. Another, much smaller airfield was built on the outskirts of Castries (today George F. L. Charles Airport). Before the Americans arrived, Castries had been connected to the rest of the island by a network of poorly surfaced, narrow roads unsuitable for any sort of heavy equipment. The roads had been improved, but in March 1942 the best way to get from one coast to the other was still by sea.

Originally called Carenage ("safe anchorage"), Castries had been renamed in 1785 to honor a German in the service of France, Charles Eugène Gabriel, Marquis de Castries. Much like Willemstad on Curaçao, it was virtually impregnable.[1] The small Ville Bay harbor is strewn with menacing rock formations and sand bars. Its entrance is only 200 meters wide, and it is almost land-locked by two headlands: D'Estrées Point to the north and the equally treacherous La Toc Point and Tapion Rock to the south. Its waters are but 10 to 15 meters deep. In 1942 the twisted inner navigable channel was guarded by massive Fort Charlotte on Morne Fortune, a fortress with a commanding sweep of the bay. It would be suicide for a 76-meter-long U-boat to enter Castries harbor. This, of course, was just the sort of challenge that Albrecht Achilles and his Executive Officer, Werner Bender, loved. When apprised of the Old Man's intentions to penetrate the harbor and attack ships at anchor therein, the men nicknamed Castries Harbor "Devil's Bay."

By 1 a.m.[2] on March 8, *U-161* was approaching St. Lucia. Achilles spied a freighter heading for Castries and shadowed it for hours.[3] At 2:15 p.m., he found another unescorted freighter, but it was too close to land to pursue. Four hours later, *U-161* stood two miles off the harbor entrance. Nightfall was imminent. Achilles ordered the boat to surface. The hatch was cracked. The bridge watch joyfully took in the intoxicating smell of anthurium, bougainvillea, hibiscus, frangipani, and beach morning glory. Achilles drove *U-161* ever so slowly up to the gap between D'Estrées and La Toc points, taking Atlas-echo soundings. Second Watch Officer Götz Roth, who in 1939 had entered Castries with the training ship *Gorch Fock*, was on the bridge to assist with navigation.[4] There were no coastal artillery batteries visible, Achilles correctly surmised, "perhaps merely a machine gun on either headland." At one point, *U-161* softly scraped along a sand bank. Around 9:45 p.m., the watch made out the broad contours of Castries. The city was blacked out, but by the light of several lamps burning on the piers the watch detected a shadow. Must be the freighter they had spotted earlier that day. Luck: the Americans, anxious to unload the freighter, had left the pier lights of Northern Wharf on. More luck: there was not just one ship at the wharf, but two others anchored in Ville Bay. A small flotilla of lighters with lanterns lit was transporting cargo from ship to shore. An hour before first light, Achilles spied what he called "Consolidated" and "Marlin" flying boats out of the US Naval Air Station at Gros Islet circling overhead. With little darkness left, he decided against risking a cavalry charge and took *U-161* back out to sea. He would return either at last light or under cover of darkness to deliver his attack.

* * *

St. Lucia's residents were happily unaware of the danger lurking off their coast. The war in Europe was an ocean away. That weekend, Clarke's Theatre in Castries had run a twin bill: Boris Karloff in *Before I Hang* and George O'Brien in *Racketeers of the Range*. Bob's Liquor Store had placed advertisements in *The West Indian Crusader* announcing a special cache of "Spirit of Love … it's a rum that can't be duplicated." The Department of Agriculture had let it be known that truckloads of cabbage, carrots, lima beans, potatoes, turnips, and Kentucky Wonder Pole Beans had just arrived at the local markets.[5] Not to be outdone, *The Voice of Saint Lucia* ran ads for Canadian Healing Oil and Eno's Fruit Salts, as well as others promising "Keating's Kills bugs, fleas, moths, beetles" and "Vic-Tabs Restores Manhood & Vitality."[6] While Achilles planned his attack, the BBC entertained St. Lucians with music from the "International Staff Band Salvation Army" and with news from "Britain Speaks" and "Political Commentary."

No one gave much thought to the harbor. It was safe. About a mile off its entrance, the launch *Welcome* maintained a vigilant sea patrol. On the heights of D'Estrées Point, an equally vigilant crew manned the Vigie Lighthouse, while Meadows Battery guarded the north side of the harbor entrance with a Nordenfelt Maxim 303 machine gun. Across the narrow harbor entrance on the southern shore, another battery stood guard at

Tapion Rock with a Lewis Light machine gun. Both posts were manned with local police constabulary. Blackout operations were in effect for the town of Castries.

* * *

Undaunted, Achilles approached Castries in the late afternoon of March 9. He decided against penetrating the harbor semi-submerged at dusk due to its shallow waters. At 8:03 p.m., he surfaced. Ever so slowly and running on the surface, *U-161* approached Castries' outer channel. There was no moon, Achilles noted, but the sky was "moderately clear." Bender and Roth once again assisted in navigation. By 10 p.m., *U-161* was between the two headlands. "Visibility is good, bright, starry night." So far, so good. The machine gunners on La Toc Point had to be asleep – or drinking, or playing cards – not to spot the intruder. The sub must have been easily visible, but no alarm was raised. Achilles decided to hug the northern headland off Vigie (ironically, French for a nautical "look-out"), from where he expected a more favorable "shooting position." He took a chance that neither the sentries at the Castries landing strip nor the encamped soldiers at Vigie was expecting a visitor.

Soon, *U-161* was in the anchorage, a mere 200 meters off Vigie. "Great tension in the boat, since all is happening so close to land," Achilles tartly noted. Bender maneuvered the boat just north of the shipping channel, so that Achilles would be able to avoid Tapion Rock on the race out of Castries. It was 10:49 p.m. Achilles fired two bow torpedoes. The boat shuddered slightly as the "eels" shot out of their tubes. "Ajax" had selected the two closest targets. There would be no time to reposition the boat for a shot at the third vessel.

"Both engines full ahead! Hard a-starboard!" Achilles screamed. *U-161* heeled over hard, blue-gray smoke spitting from its exhaust, heading straight for the harbor entrance – and the open sea beyond. All the while, the watch waited for the explosions from the torpedo hits. They came quickly.

> After 96 seconds (1490 meters), hit on passenger-freighter on the left, 8000 tons, high gray-white detonation column; can no

longer see the stern, only bow and superstructure still above water. Fore-ship begins to flame.

After 105 seconds (1500 meters), hit on the freighter off to the right, 5000 tons, bright, high fire-flash, very loud detonation with three subsequent explosions. A huge black smoke cloud looms over the freighter. Stern below water.[7]

By now, *U-161* had raced to a point midway between Vigie and Tapion Rock. It was the moment of greatest danger. Machine gun fire and tracer bullets from both shore batteries cracked through the moist, tropical air. A few bullets pinged harmlessly on the hull. Achilles ordered the bridge cleared. Surely, no one could fail to spot *U-161*: white bow waves in front, bright phosphorescent wake behind, blue-gray smoke blowing from the exhaust, its silhouette lit by the flaming ships. The tension in the boat could be cut with a knife. The radio room picked up distress signals from shore: "sss sss sss de vhq submarine Castries harbour attacked shipping 0240 gmt/lo." Then, just as suddenly as it had begun, the machine gun fire ended. Why? Had someone called it off for fear of hitting one of the lighters? Or had they lost the submarine in the dark? It was immaterial to Achilles: *U-161* was clear of both headlands and heading out into the Caribbean. A great shout went through the boat when the Old Man announced the stunning victory over the intercom.

Achilles' first victim was the 7,970-ton Canadian passenger-freighter *Lady Nelson*, bringing 110 passengers and general cargo to St. Lucia. Twenty of its crew died in the blast. The second victim was the 8,141-ton British freighter *Umtata*, en route from Durban, South Africa, to New York with a full load of chrome ore, asbestos, and meat – the latter for the island's American garrison. Four of its crew, all from Calcutta, died that morning. Nineteen passengers and sailors injured in the attacks were taken to Victoria Hospital. US Marines rushed down from Gros Islet to render assistance at the Northern Wharf.

Achilles' audacious attack inside Castries harbor was the stuff of legends. He was the acknowledged "ace" of Operation New Land. German propaganda celebrated him also as "the ferret of Castries," the *Frettchen* that had gone into the enemy's den and devoured its inhabitants.

Joseph Goebbels' minions could hardly wait for Achilles' return to exploit the triumphs.

* * *

For St. Lucia, the attack meant the end of innocence. The war was brought home to it by *U-161*, just as it had been to Aruba by *U-156*. Neither island would ever again be the same. But what had happened to Castries' defenses? By approaching the entrance to the harbor submerged, Achilles had avoided the launch *Welcome*. Still, around 9:50 p.m., the guards at Vigie Lighthouse had seen "something rise from under the water." It was approaching the inner harbor hard against the Vigie (or northern) side of the shipping channel. Police Constable B. Rachel recognized the intruder as a submarine. "I loaded my rifle." Next, he alerted the lighthouse keeper, who confirmed the sighting. "It's a submarine."[8] Rachel's immediate superior, Police Constable T. Phillips, ordered Rachel to telephone the information to police headquarters – only to find that the officer on duty, Assistant Superintendent Conway, "was sleeping in his Office." Frantic calls to the Northern Wharf, to the Tapion Gun Post, and to the harbor master's office went unanswered. Police headquarters finally managed to contact the electric power station to get the lights on the wharf switched off. It was too late: two explosions rang out at that very moment.

Chaos reigned inside the harbor. At Meadows Battery, Police Constable Spooner wildly fired off 208 rounds from the Nordenfelt Maxim Gun. At Tapion Rock, Lance Corporal Harris, one of only a handful of military men on guard duty that night, still refused to believe that a U-boat had penetrated the harbor. He rang up the Harbor Office "to enquire what Launch it was he saw"! He never managed to get off a single round at the intruder. The *Welcome* raced back to port to render assistance. Just as it approached the narrow harbor entrance, its signal lamp suddenly went dead. Unable to give the agreed recognition signal ("BP" for boat patrol), it withdrew to nearby Cul-de-Sac Bay – amidst a hail of machine gun fire, not from *U-161* but "undoubtedly coming from 'Meadows Battery'." As the final act in this *opéra bouffe*, Constable Rachel reported that he "saw the submarine going out stern first" under a hail of machine-gun fire from Meadows Battery.

British censors at once placed a tight lid on news of Achilles' brazen attack. Ironically, the very morning of the German raid, the island's major newspaper, *The Voice of Saint Lucia*, in a front-page leader had warned residents, "Enemy Subs Believed Operating Near Panama Canal."[9] On March 17, the paper's editors called on government authorities to abandon their studied "disinterestedness" in home defense and to create "bodies of Coast Watchers, Home Guard, Special Constables, Communication Service Red Cross Workers" – without ever mentioning the sinking of the two ships.[10] St. Lucia's other paper, *The West Indian Crusader*, only obliquely referred to what it called the "incident" of March 9 in Castries harbor.

But Achilles' action could not be covered up. On March 19, the editors of *The Voice of Saint Lucia* decided to ignore official censorship. The paper carried the front-page headline, "St. Lucia Can Take It!" In the story that followed, it gleefully announced that the sinking of the two ships had served finally to plaster over "petty" domestic disputes among the islanders. The death and destruction in the port occasioned by the German raider "will have seared across the screens of their minds the indelible impressions of mingled dread and sleepy surprise as heavy explosions rocked them from sleep to the first grim realities of this war."[11] Most islanders had surmised, "An earthquake," when they heard the initial blast – only to realize with the second explosion that the war had come home to them. The shock wave of the explosions had "wrenched off" many office doors and windows, had "upset" countless desks and shelves, and had scattered glass over three square blocks. Countless residents fled the capital for the safety of the rainforests in the interior of the island. Those who remained wondered whether more German submarines lurked off their shores. In utter defiance of official secrecy, the paper reported that 16 people had died and that 13 had been injured on the *Lady Nelson*, with another four dead and six wounded on the *Umtata*. Incredibly, it gave the names of the casualties that could be identified.

The immediate first task at Castries was to douse the fires on the two freighters and to begin salvage operations at once since Castries was the sole point of entry for the Americans and their supplies. Just as quickly, an official inquiry into the disaster was launched by St. Lucia's administrator, Alban Wright. It was a sobering report.[12] Castries' defenses had lacked both a "harbor boom" and adequate "artillery protection." There had not even been "a searchlight to light up the targets at night." The blackout had been "by no means wholly effective." And the "more or less untrained police" that manned the defense posts at Vigie Lighthouse, Tapion Rock, and Meadows Battery had not been up to the task. In short order, an antisubmarine harbor boom was installed at the entrance to the harbor, and a battery of coastal 155-mm artillery was rushed in to guard D'Estrées and La Toc points. Another battery of the "Long Toms" was hastily dispatched from the United States to Beane Field at Vieux Fort, lest another "ferret" steal in and shell the complex. Training was stepped up for the machine-gun companies scattered about St. Lucia, which, after Achilles' attack, had fired at anything that moved – much to the distress of the island's residents. By May 28, Wright reported to Governor Charles Talbot at Grenada that all "main deficiencies" of March 9 had "already been made good."[13]

* * *

While chaos reigned on St. Lucia, Achilles pointed *U-161* west to throw off the expected aerial searches. He then shaped a course north for the Mona Passage and home. He radioed his recent success to Kernével, informing Admiral Karl Dönitz that he planned to set out for Lorient on March 14. "Still one stern eel, 100 cbm, strong Trade Winds."

On March 10, Achilles spied lone freighters but had to let them go since they were too fast. In the early morning hours of March 13, he made out a tanker and raced after it for most of the day. While maneuvering for a shot, another tanker hove into sight. It was on course to run between *U-161* and the first tanker. Achilles simply waited for it to come into range. At 8:30 p.m., it was a mere 580 meters away. "Ajax" fired the single stern torpedo. After 29 seconds, it slammed into the tanker slightly ahead of the funnel. "High water column with minimal fire-flame, apparently boiler-room explosion." The target went down by the stern. It did not have time to put its lifeboats into the water. "Nothing more to be seen other than a large fuel-oil streak." He later learned that he had torpedoed the 1,940-ton Canadian freighter *Sarniadoc*. It was carrying a cargo of bauxite out of Demerara, British Guiana, for St. Thomas. All hands on board were lost. Achilles radioed news of the sinking to Kernével and informed

U-Boat Command about the target-rich environment that he had found west of Guadeloupe.

Just before daybreak on March 15, south of Hispaniola, the bridge watch spotted a blacked-out shadow at 256 degrees.[14] Neither Bender nor Roth could find its silhouette in any of the commercial shipping books on board. It was slow – eight knots – it was small – about 1,000 tons – and it altered course every few minutes. Achilles' first thought was: "U-boat trap." But he decided to observe it for an hour. Satisfied that it was not some sort of new "Q-ship" (an antisubmarine vessel disguised as a merchantman), he moved to attack it head on at 0 degrees, "the dog's curve" in German parlance. This would offer the hostile the smallest possible silhouette. It was 5:37 a.m.

Achilles brought *U-161* on a parallel course. "Free to fire artillery!" Lieutenant Roth opened up with the deck gun as well as with the smaller anti-aircraft guns. The victim at once signaled for help. "'Acacia' (Call-signal NRWP), position, artillery attack, abandoning ship." By now, Roth had unleashed a deadly hail of 68 10.5-cm, 92 3.7-cm, and 70 2-cm shells. Achilles watched the tracer shells from the bridge and concluded that half of the large shells had found the target. "Superb shooting!" But the 3.7-cm explosive shells were largely ineffective, and he made a mental note to suggest to Dönitz that he issue incendiary shells in future.

At 6:11 a.m., Achilles ordered "Cease Fire!" The wreck was listing badly and burning profusely in at least three places. It sank within 20 minutes. Its crew had taken to two lifeboats. As the ship went down, Achilles spied a "USA flag" in its fore-mast. "Departed at high speed since airplanes are to be expected on the basis of the freighter's signaling." *U-161* shaped a course for the Guadeloupe Passage, sailing past the British island of Montserrat on its way out into the Atlantic.

The "mystery" ship later turned out to have been the 1,130-ton American lighthouse tender USS *Acacia*. Originally constructed as the minelayer *General Joseph P. Story* for the US Army, it had been acquired by the Coast Guard in 1927 and rebuilt and renamed. All 31 of its crew survived. *Acacia* was the first Allied warship lost in the Caribbean theater.

* * *

The fifth submarine in the first wave of Operation Neuland was *U-129*, commanded by Kapitänleutnant Asmus Nicolai Clausen. "Niko," as he was known to his friends, had joined the navy as an ordinary seaman in 1929 and thus was the second oldest among the five commanders. He joined the U-Boat Service in September 1935 and learned his trade under Werner Hartmann, who was destined to become one of the great U-boat aces of World War II. When war broke out, Hartmann requested Clausen as his Executive Officer on *U-37*, where "Niko" completed three war patrols. After a brief interlude commissioning *U-142*, Clausen was given command of *U-37*. He sank 12 ships on three war patrols and was awarded the Iron Cross, First Class. In May 1941, he received the brand new Type IXC *U-129*, which he led on three fruitless war patrols, mostly in the Atlantic. Operation Neuland, he vowed, would be very different.

Unlike the other four boats, *U-129*, with "Westward Ho" painted on the front of its conning tower, was sent to hunt and destroy the bauxite traffic steaming up from Georgetown and Paramaribo in the Guianas. Clausen informed Dönitz that the "peculiarities of the inshore water" – read, the 100-mile-wide shallow continental shelf – made for "unfavorable operations," and thus positioned *U-129* about 50 miles east of Galera Point, on the northeastern tip of Trinidad. Just after 2 a.m. on February 20, the unescorted 2,400-ton Norwegian steamer *Nordvangen* hove into sight. It was carrying a cargo of bauxite from Paramaribo, Dutch Guiana (Suriname), to New Orleans by way of Trinidad. Clausen fired a single bow torpedo. The "eel" blew off the *Nordvangen*'s stern. It plunged to the bottom within a minute, taking all 24 of its crew with it.[15] There had been no time for its radio operator to get off a distress call. On March 6, a lifeboat and some debris from the *Nordvangen* washed ashore at Trinidad, a sure sign of the ship's fate.

After sinking the *Nordvangen*, Clausen cruised to a point about 120 miles southeast of Trinidad, right in the middle of the main bauxite shipping lane. Around 10 a.m. on February 22, just north of the Orinoco River estuary, all hell broke loose. A steamer suddenly appeared; neither the hydrophone operators nor the skipper had detected its approach. Clausen fired twice. "Missed!" He steamed away from the scene. Then another freighter suddenly hove into view. The radio operator reported an emergency signal from 35 kilometers away: "SSS SSS SSS ... Submarine seen."

Clausen suspected this pertained to three Italian subs north of him. *U-129* headed away from the site. At 4 p.m., a third steamer appeared. *"Alarm!* Aircraft at 45°, course SE, range 7,000 m[eters]." No time to lose: a single "eel" leaped out from Tube V, "Missed!" Then another shot, this time from Tube IV. The torpedo broke the back of the small ship instantly.[16] The victim was the 1,754-ton Canadian bauxite carrier *George L. Torian* out of Paramaribo. Four of its crew managed to clamber into lifeboats and were eventually rescued.

U-129 remained in the target-rich waters between the Guianas and Trinidad – which the Allies soon dubbed "Torpedo Junction." Within an hour of dispatching the *George L. Torian*, the lookouts spotted another heavily loaded freighter. A single "eel" struck the hostile amidships. It lowered two rafts, and sent out an SSS signal: "Torpeded [*sic*], torpeded [*sic*]." A *coup de grâce* torpedo at 600 meters broke the ship in half.[17] It was the 5,658-ton American ore carrier *West Zeda*, bound from Mombasa, Kenya, to Trinidad.

U-129 stood off the Orinoco for another 24 hours. Just before noon on February 23, Clausen spied yet another unescorted freighter. At periscope depth, he fired two torpedoes. "Two hits." But the target steered straight at *U-129*. Clausen ordered a hard turn to starboard – just as the freighter began to break apart, its screws hopelessly turning out of the water. He had torpedoed the 1,904-ton Canadian ore carrier *Lennox*. Once more, the blue-green waters of the Caribbean were covered with gray bauxite dust. He steered toward the lifeboats, "full of whites and niggers,"[18] and asked the fearful, weary sailors the name of their ship, its cargo, and its destination. Then he told them that Trinidad was 120 miles to the northwest and handed over some food and water. Unbeknown to Clausen, just before going under, the ship's master, Daniel Percy Nolan, managed to get off an SSS. Trinidad Naval Station was alerted anew to the presence of "gray sharks" off its waters.

Clausen decided to leave the area and move closer to the Guianas to attack the bauxite carriers at their source. Yet again, the shallow waters of the continental shelf gave him little leeway to dive, and shipping seemed to have all but disappeared. After four days of frustration, his luck returned. Just after supper on February 28, he torpedoed the 2,605-ton Panamanian ore carrier *Bayou*. Then, another week of empty seas.

At dawn on March 3, a return to good fortune: Clausen destroyed the unescorted 5,105-ton American freighter *Mary*, carrying war stores from New York to Suez. And at dusk on March 6, a final strike: the unescorted 6,188-ton American ore carrier *Steel Age*, caught off Vichy French Guiana (Guyane).[19] In all, Clausen chalked up seven ships at 25,600 tons.[20] On March 13, Admiral Dönitz awarded Clausen the coveted Knight's Cross. Clausen was more methodical, perhaps, than the daring Hartenstein or Achilles, but with no less dramatic results as the Allies suddenly discovered that their major source for bauxite was also endangered by this new U-boat offensive.

* * *

Albrecht Achilles was still not done. At dawn on March 21, he came across a tanker in Quadrant DE 9772, mid-Atlantic. He plotted a surface artillery attack. At a range of 3,500 meters, Lieutenant Roth and his gun crew opened fire with all three cannons. Then all three guns jammed. As well, the distance to target had been too great. Too bad, for it was a fat prize: the 6,000-ton tanker *Empire Gold*. It ran off at high speed, showing *U-161* only its slender stern. Then it opened fire with a deck gun – shells with timed fuses splashed 150 meters off *U-161*. Next came three smoke bombs. All the while, the tanker was signaling its position.

Within half an hour, Roth fired 30 shells from the 10.5-cm deck gun at the hostile. Achilles made out two hits, one amidships and one on the stern. Then the shells from the tanker's gun ranged in on *U-161* and so he took it down to avoid being hit. But the thought that the shell Roth had fired had landed on the tanker's stern and might have damaged its rudder gear nagged Achilles. At 9:11 a.m., he was back up on top. It was hazy and rain began to fall. The target seemed to be turning circles – and then disappeared into the rain. Achilles passed its position on to U-Boat Command in the hope that another sub might be in the area.

U-161 tied up in Lorient at 9:30 a.m. on April 2. From its extended periscope tube flew eight pennants – for five ships of 27,997 tons sunk and three ships damaged – including two black pennants for tankers and a red one for the warship. A vast crowd was at dockside to celebrate the "ferret." Goebbels' camera crews were on hand to record the glorious scene for that week's propaganda newsreel, *Die Wochenschau*. Second Flotilla Chief Viktor Schütze, proudly wearing his Knight's Cross, welcomed Achilles home. A young girl handed "Ajax" a huge bouquet of fresh flowers. Three shouts of "Hurrah!" thundered across the harbor.[21] *U-161* headed straight for the cavernous Kéroman bunkers for repairs. Achilles and his crew left the boat for the usual round of banquets and much needed shore leave.

Admiral Dönitz was delighted with the war patrol. Not content with the nickname "ferret" for Achilles, he devised one of his own: *Lochkriecher*, or "borer." His official evaluation gushed with praise:

> Superbly executed first operation by a young commander with a new boat.
>
> Especially to be praised are the penetrations of the Gulf of Paria and the harbor of Port Castries on Santa Lucia, executed with daring and cunning.[22]

On April 5, the "Great Lion" awarded Achilles the Iron Cross, First Class. There was a man to be closely watched for future awards – and future war patrols.

* * *

The first wave of Operation Neuland created panic and chaos in the Caribbean and in Allied capitals. In just 28 days, the five Type IX U-boats sank 41 ships, 18 of which were tankers, for a total of 222,657 tons; they damaged a further 11 ships. "Diplomats Blame Hull For New Sub Activities; Expect Cabinet Ouster" the *Miami Herald* declared in a front-page story on February 24. The newspaper's Washington correspondent claimed that British and Russian diplomats were "indignant" that Secretary of State Cordell Hull had allowed a situation to develop wherein the Caribbean was "swarming with Nazi submarines based on the French islands."[23] This was an allusion to Hull's careful approach to Vichy France and its colonies in the Americas. None of that was true, of course, which made the real implications of the disaster even more serious. The U-boats were steaming 3,000 miles across the Atlantic, gliding easily through the gaps in the island chains, and striking at will from the Florida Strait to the waters east of Trinidad. The Allies were unready, divided, disorganized, untrained, under equipped, and terrified.

On March 12, Prime Minister Winston S. Churchill wrote President Franklin D. Roosevelt's close advisor Harry Hopkins: "I am most deeply concerned at the immense sinkings of tankers west of the 40th meridian and in the Caribbean Sea.... The situation is so serious that drastic action of some kind is necessary." Churchill urged the Americans to pull some of their destroyers out of the Pacific and to put them to work escorting convoys off the US coast, in the Caribbean Sea, and in the Gulf of Mexico. Britain had promised to give the US Navy ten Flower-class corvettes to bolster the defense of shipping off the east coast, where the U-boats had just completed another great slaughter in Operation Drumbeat, and Churchill hoped that these escorts, bolstered by American destroyers, would hold off the U-boat offensive in the Caribbean. He pointed out that, unless an effective form of convoy protection was worked out, the Allies faced two stark alternatives – temporarily stop the sailings of tankers, which would "gravely jeopardize our operational supplies," or diminish the number of convoys crossing the North Atlantic in order to release sufficient escorts to cover the Caribbean.[24] Either move was fraught with danger. But with the Imperial Japanese Navy romping over the Pacific, and US destroyer production just two years into a long-term expansion program, Churchill's suggestion was ignored.

* * *

By late April 1942, tensions between some 50 Chinese stokers and the Curaçaose Shipping Firm Maatschappij (CSM), caused by the sudden loss of dozens of tankers since mid-February, exploded into what the Curaçao historian Junnes Sint Jago has called "one of the greatest mysteries of our nation's history." In a tragic series of events "fifteen Chinese sailors [were] killed and dozens more wounded" by police bullets at a camp just outside Willemstad.[25] The so-called *bloedbad*, or "blood bath," was brought about by the shipping company's failure to address the growing fear of the Chinese stokers.

Although CSM had immediately halted further transports of oil from Venezuela after *U-67* had torpedoed the tanker *Rafaela* on February 16, news slowly seeped into Willemstad via United Press bulletins of *U-156*'s sinking of *Oranjestad* and *Pedernales* at San Nicolas as well as of *U-502*'s dramatic destruction of *Tia Juana*, *Monagas*, and *San Nicolas* off the coast of Venezuela. Three days later, word filtered through that another raider, *U-161*, had torpedoed *British Consul* and *Mokihana* off Port of Spain, Trinidad. Not surprisingly, these additional sinkings greatly alarmed the Chinese engine crews of the lake tankers.

Some 500 Chinese indentured sailors lived in squalor in four large "lodgments" in Punda, the old part of Willemstad. They were non-union and without full citizenship rights. Most were men in their early to mid-forties. Many had come from Guangdong province to work for Dutch shipping firms before the war – hence, their common nickname "Rotterdam-Chinese." Many had accepted long-term contracts with the Dutch fleet of small tankers that in endless rhythm hauled crude oil from Lake Maracaibo to the Royal Dutch Shell Santa Anna refinery in Curaçao for processing. None had bargained for war, or for U-boats.

The Chinese stokers pleaded with their nearest consul, Hing King in Trinidad, and through him with the Chinese ambassador in London, Dr. Wellington Koo, to put pressure on the Dutch government-in-exile to mediate the dispute with CSM on Curaçao. The stokers demanded a rise in wages from their current 50 florins ($450 in 2012 US dollars)[26] per month; a 10 per cent cost of living allowance; a war bonus for dangerous work; repatriation to China after the expiration of their contracts with CSM; and, above all, the convoying and screening of the tanker fleet between Willemstad and Maracaibo by Allied warships.

To no avail. Neither CSM nor the Dutch authorities on Curaçao or in London would budge. Exasperated, on March 14, the stokers mounted a peaceful demonstration near the Governor's Palace in Punda. The Dutch General Military Commissioner, Baron Carel van Asbeck, was not amused by what he and his staff termed a "mass strike" of an "aggressive" nature. They rolled out military trucks and instructed the Chinese: "We go camp!" The police took them "over the hills" to Camp Suffisant, which had served as British barracks from June 1940 to February 1942. As numerous other Chinese stokers returned from Maracaibo on board lake tankers, they voluntarily interned themselves, in an act of solidarity, at what was now called "concentration camp" Suffisant. The inmate population quickly swelled to 420. News of additional sinking of tankers by U-boats, coupled with tight official censorship, further fanned the flames of unrest.

The action by the "Rotterdam-Chinese" alarmed Dutch authorities. What if other, non-Chinese sailors joined their protest? Would the vital flow of oil out of Maracaibo be curtailed? And how would Curaçao, which produced virtually no food and had no major artesian wells, survive if general cargo shippers also were crippled by strikes? An example had to be made. On April 18, Dutch civilian and military police as well as CSM company guards, under the command of Willem van der Kroef, ordered 58 putative "ringleaders" to muster in the barracks square for a peremptory roll call – and to receive instructions on how they were to be removed to another camp. Several Chinese sailors stepped forward and shouted some incomprehensible commands, most likely in Chinese. Thereupon, armed with pipes, rocks, and sticks, they stormed the entrance gate. Panicked by this act of defiance, Van der Kroef ordered the police, carrying carbines with bayonets fixed, to draw up in a line. Shots rang out. Thirteen stokers were dead and 40 wounded; two died later of wounds inflicted that day.

The Dutch police seized the rocks, sticks, and pipes and ordered the Chinese to return to their barracks, and eventually to work. Most did – once they received promises that the lake tankers would, indeed, be convoyed across the Caribbean Sea – but 52 hard-core "strikers" refused to return to work for CSM under any circumstances. They were sentenced to isolation arrest in police barracks at Camp Suffisant. Subsequent attempts to dispatch what Dutch authorities now called "unwilling Chinese" to the United States or to send them to serve with the Chinese Expeditionary Army in India failed. No record of their eventual fate has ever been found. Twelve of the 15 stokers killed at Camp Suffisant are buried in a neglected cemetery at Kolebra Bèrdè (Papiamento dialect for "green moray") at Kas Chikitu on Bonaire Island.

* * *

On the morning of April 20, 1942, Adolf Hitler exited his bunker at the Wolf's Lair near Rastenburg, East Prussia. His paladins stood at attention in two parallel lines: Field Marshals Wilhelm Keitel and Erhard Milch, General Alfred Jodl, Grand Admiral Erich Raeder, Reichsführer-SS

Heinrich Himmler, Foreign Minister Joachim von Ribbentrop, Armaments Minister Albert Speer, and Chief of the Party Chancery Martin Bormann, among others.[27] A selected group of local children bounced up to the Führer and handed him bouquets of fresh flowers. It was Hitler's 53rd birthday.

The day's festivities had started shortly after midnight. For the first time since the invasion of the Soviet Union on June 22, 1941, Hitler's entourage had broken out champagne and heartily toasted their (albeit at the time absent) Führer. Lunch – cutlets, red cabbage, potatoes, and fruit salad – was served with Rhine wine on white table linens, as was supper – ham with home fries and asparagus salad. Hitler, as usual, touched neither meat nor wine. He entertained the birthday well-wishers with tales of how Deputy Führer Rudolf Hess, who had mysteriously piloted a Bf-110D fighter-bomber[28] to Scotland in May 1941, would immediately be locked up in an insane asylum or summarily executed if he ever returned to Germany. That night Hitler watched the first newsreels featuring the new steel-reinforced concrete U-boat bunkers built by the Organisation Todt along Bay of Biscay ports in France. He had good reason to celebrate the opening of these behemoth bunkers. The submarine war was going well. Surely, the Allies could not withstand this onslaught much longer.

For the Allies, in fact, things might have been far worse. Admiral Dönitz apparently never realized that the shallow-draft lake tankers bringing Venezuelan crude from Lake Maracaibo to the refineries on Curaçao and Aruba were purpose-built and limited in number. A few of these ships had been sunk at the very beginning of the operation, but they were never specifically targeted. If they had been, the flow of oil from Venezuela could have been stopped altogether, at least until sufficient escorts were available and new tankers built.[29] It remains a mystery why this vital weak link was not cut. It may well be that Dönitz's almost religious belief that every Allied ship sunk constituted a loss to the Allied war effort, and thus that all ships were to be attacked whenever and wherever they might be found, is at the heart of the mystery. This was *Tonnagekrieg* (tonnage war), a struggle that did not distinguish between a lake tanker and an ocean tanker, or even a dry cargo ship. In other words, a ship was a ship was a ship, and any effort to target particular classes of ships would result in opportunities lost to sink other types of ships. But whatever the reason for this strategic mistake, Dönitz certainly realized that the Caribbean was a very important weak point in the Allied war effort. He quickly dispatched the next wave of U-boats to the Caribbean as Werner Hartenstein, Albrecht Achilles, "Niko" Clausen, Jürgen von Rosenstiel, Günther Müller-Stöckheim, and others, began to arrive back at Lorient. In the months that followed, the Germans would add significantly to the toll they had already taken.

7

TORPEDO JUNCTION

Admiral Karl Dönitz launched the second wave of the Caribbean offensive five days before the first boat returned from the opening attack. The first three Type IXC submarines to sortie from Lorient at the end of March were *U-154*, to patrol the Mona and Windward passages; *U-66*, bound for Trinidad; and *U-130*, headed to Curaçao. *U-108* followed in short order, directed to Puerto Rico. These boats, together with *U-123*, eventually destroyed 29 ships, 13 tankers among them, for a total of 164,000 tons.[1] *U-130*, commanded by Ernst Kals, surfaced to shell the Curaçao refinery in the early morning hours of April 19, but the island's defenders were not caught napping. The 155-mm "Long Toms" of 252nd Garrison Artillery were emplaced and ready to fire, and fire they did. Kals was forced to retreat to deeper water.

Dönitz used radio reports from the Neuland captains to paint a picture of Allied defenses in the Caribbean and the prospects for future operations.[2] While Allied air cover over Aruba was sufficient numerically, Dönitz concluded, it was "inexperienced and <u>bad</u> compared to English air surveillance." Above all, the skippers had experienced no "crisp, well-thought out" antisubmarine operations; at best, only "spur-of-the-moment panic reactions" to the sinking of the tankers. Commander U-Boats was "surprised" that so much tanker traffic continued to operate in the Caribbean, which to him only revealed how desperate the United States was for the oil, especially given that much of it had to be shared with Britain. As far as surface antisubmarine warfare was concerned, Dönitz surmised that due to lack of available escorts, there would be no "long-term real, effective protection" against the U-boats in the Caribbean.

Yet again, a heated war of memoranda had raged behind the scenes between Dönitz and Grand Admiral Erich Raeder concerning Neuland. From Berlin, Commander in Chief Navy on March 26 telegraphed Kernével that he wanted the U-boats to mount a "continuous occupation" of the Caribbean, with boats constantly spelling each other in the area in rotating "waves."[3] Dönitz replied two days later with what amounted to a lecture on submarine operations.[4] First, U-boats simply could not "occupy" any area of sea. Second, it took three to four weeks to reach the operations area. Third, to stagger departures from the Bay of Biscay, when boats were provisioned and ready to sail, would have "a very negative psychological effect on crews ready for war patrol," as well as an "unwanted congestion of the [Biscay] bases and docks." Fourth, there were only five boats available at any time for Caribbean operations. When they departed, there naturally had to be a "hole" in further sailings. To sweeten the message, Dönitz promised greater activity in the Caribbean once U-tankers were available to resupply the boats on station.

Raeder took a week to respond. On April 2, he had his staff send Kernével an acid one-sentence telegram: "Commander-in-Chief wishes that his dispatched order [of March 26] will be carried out through deployment of all suitable units."[5] Dönitz chose not to respond. But, ever the consummate bureaucrat, he knew that his actions needed to be documented. Thus, on April 14, he penned a lengthy justification of his "tonnage war."[6] It was simple mathematics:

1. The shipping of the enemy powers forms one great whole.
    Thus, in this context it is immaterial where a ship is
    sunk; in the final analysis, it has to be replaced by a new
    construction.

2. The decisive question in the long run is the race between
    sinking and new construction.

The real enemy in this area was the United States, not Britain. "Thus I will strike the evil at its root by tackling the supply, especially oil, at this center of gravity." Every ship sunk translated not just into a lost bottom, but into a further diminution of the American shipbuilding and armaments industries. Every ship sunk translated into delaying a possible British attack on Nazi Germany. Therefore, the U-boats had to attack enemy shipping where it was "most rational" and "cheapest" in terms of potential U-boat losses.

Dönitz once more reassessed American ASW. While it was improving in quantity, its quality ("its attentiveness, its will to attack and to destroy") left much to be desired. "Soldiers do not fight [for America]," he philosophized, "but rather people who are paid for their presence in areas endangered by the U-boats."[7] The will to win would decide the war's outcome. And that "will" was with Germany.

The "Great Lion" had also come up with a technological innovation: the so-called "milk cows" (*Milchkühe*).[8] These deep, broad-beamed 1,700-ton Type XIV craft were basic Type VIIC boats converted to oil tankers; each held 432 tons of diesel. By resupplying the subs in the Caribbean, the "milk cows" could extend the war patrols of twelve Type VIIC boats for an additional four weeks, or five Type IXC boats for an extra eight weeks.

Kapitänleutnant Georg von Wilamowitz-Moellendorf's *U-459* was the first operational *Milchkuh* and was immediately assigned to the Caribbean boats. Known throughout the service as "wild Moritz" for his antics both on shore and at sea, Wilamowitz-Moellendorf, a veteran of the Great War, was the right man for the job. Calm under the most violent of actions, he could be relied upon by Dönitz to undertake the arduous trip to the Caribbean. *U-459* could refuel Type IXC boats at the rate of 35 tons per hour; it also carried 34 tons of lubricating oil, 10.5 tons of fresh- and three tons of distilled water, spare parts, four torpedoes, extra rations, a modest medical service, and a bakery that could produce 80 loaves of bread per hour. Supplies and spare parts would be transferred in calm seas on a six-meter rubber dinghy; and in heavy seas on a "dead-man's cradle." Diesel transfers would be undertaken by way of a main fuel hose with several manila lines wrapped tightly around it for strength and to protect against chafing (and sparks) on the steel hulls.

Positioning *U-459* in the Atlantic south of Bermuda gave U-Boat Command the opportunity to send the smaller Type VII boats into the Caribbean as well. Dönitz actually preferred the smaller subs because they were more nimble and maneuverable, though with decidedly shorter range. The first of the Type VIIs to venture into the Caribbean was

Dietrich Hoffmann's *U-594*, which had been patrolling the Atlantic sea lanes off the United States since March 1. Hoffmann replenished fuel and supplies from *U-459* and then headed into the Caribbean. It was followed by *U-69*, *U-558*, and *U-741*. Hoffmann's sortie was a complete flop, and he was relieved of command when he returned to France at the beginning of June; the other three sank more than 30,000 tons in total.[9]

* * *

In the first half of 1942, Royal Air Force raids on Brest, only 60 miles north of Lorient, and a daring British commando raid against the St. Nazaire dry dock prompted Hitler and Raeder to order Commander U-Boats to leave Kernével for a safer location. Dönitz resisted, but Hitler and Raeder insisted. The new headquarters were established on the Avenue Maréchal Maunoury in Paris, and, at 11:00 a.m. on March 29, 1942, control passed from Kernével to Paris. Orders to the U-boats emanated from the powerful transmitter at the former French Colonial Office in Saint Assise, southeast of the capital. And just to be safe, the Führer grounded Dönitz's private Junkers Ju 52 aircraft; he could not afford to lose his most dedicated naval commander.

The last two weeks of April were slim pickings for the German subs. The Americans, along with the British, the Dutch, and the Venezuelans, tried to halt tanker traffic temporarily in order to mount an interlinked convoy system over the major Caribbean and South Atlantic shipping lanes. Their main obstacle, as Prime Minister Winston S. Churchill had anticipated, was the lack of escort vessels. There were fewer than five destroyers in the entire Caribbean basin and all of these were World War I ships. Other vessels were on hand – sub-chasers, patrol craft, converted yachts, motor torpedo boats – but these did not have the range, armament, or submarine detection equipment needed to battle the U-boats. As a prominent historian of the campaign in the Caribbean put it: "In the first five months of the Caribbean offensive the U-Boats' only worry in the area, was the threat of air attack by [US Army] Air Corps aircraft."[10]

But lack of escort vessels was not the only obstacle. Although both the US Army Air Forces and the US Navy had a not-insignificant air presence along the island chains, the Americans had almost no experience in the use of aircraft to escort convoys. The air crews were untrained in ASW, and a reliable system of relieving covering aircraft by other aircraft so as to provide air cover 24 hours and seven days a week had not been worked out. As well, some captains, answering primarily to their own ship owners and – if registered in neutral countries – outside the purview of the Royal or US navies, chose to continue steaming on their own. Advantage Dönitz.

* * *

On April 22, 1942, Hartenstein left Lorient shortly after dusk on his second war patrol to the Caribbean. To his distress, *U-156* encountered numerous French fishing boats in the Bay of Biscay the next day. "I do not like this gathering of fishing boats; it opens the door to collaboration with the enemy."[11] Shortly before midnight on April 23, Dönitz sent a long, convoluted radio message. Hartenstein quickly guessed its essence: "Means: Panama Canal." He decided to enter the Caribbean through the Mona Passage and to "graze" off the northern coast of Puerto Rico, where he expected to encounter traffic from the canal to San Juan and St. Thomas Island. Many of the crew entertained themselves with tales of Caribbean pirates, Spanish galleons, and gold.

Hartenstein and the other four boats of Operation Neuland had pioneered the Caribbean campaign, but by the end of April the waters in the Caribbean basin, the Gulf of Mexico, and immediately outside the island chains were swarming with submarines. At any given time that month, at least 13 boats, mostly Type IXs but with a handful of Type VIIs, were either in the area or en route to it. As Dönitz had told Raeder, in practical terms, the only way to keep the pressure up in the Caribbean was to mount continuous sorties. Thus, a conveyor belt process fed subs into the Caribbean as soon as they were ready from a previous war patrol, or as soon as they were commissioned and had had their first shake-out patrol. In the spring months of 1942, most of these boats concentrated on the waters of the Windward Passage between Cuba and Hispaniola, around the Dutch Islands, and in the waters surrounding Trinidad. Allied seamen came to call that area Torpedo Junction.[12]

On this second passage out, Hartenstein's "garbage tour" was sheer misery. Day after day, unending rain showers and fog. In the Bay of

Biscay, British aircraft forced *U-156* to submerge for 16 hours. On April 30, *U-156* passed San Marina Island in the Azores – without ever sighting its soaring cliffs due to unabated rain showers. Morale plummeted. This time, there were no swims in the warm waters, no fishing off the deck. The sea continued to roil. Four more days of howling wind and rough seas. Below decks, the first cases of pubic lice and crabs demanded attention with a special "Kuprex"[13] ointment before the affliction spread to the rest of the crew by way of the shared bunks and blankets. The only cheer came by news from Paris that Midshipman Max Fischer had just been promoted lieutenant. Schnaps for every man on board!

At noon on May 5, the Enigma lit up: "Hartenstein area of 300 nautical miles in Grid Quadrant ED 99.... Previous order rescinded ... proceed ... to north corner of Barbados." Richard Zapp in *U-66* had reported "moderate traffic" just outside the chain of the Lesser Antilles. At midnight, Chief Engineer Wilhelm Polchau reported that two steering racks and a fairlead bushing in the diesel compressor had broken down. A return to Lorient would mean four weeks of lost time. Hartenstein remembered that Zapp and *U-66* were on their way home, and hence he radioed U-Boat Headquarters to ascertain whether Zapp could spare the parts. On May 7, Zapp replied that *U-66* was desperately low on fuel and requested the transfer of five tons of oil.

Three days later, the two boats met in Grid Quadrant DQ 7937. "Cloudy, misty, occasional rain." Hartenstein steered a course parallel to *U-66* until he was abreast of it at a distance of 25 meters. The two steering racks were floated across with the aid of a buoy, and then *U-156*'s rubber dinghy paddled the oil hose over to *U-66*. As always, the war diary was terse: "5 cbm in 26 min." Zapp reported stray freighters in Quadrants EE 36 and 39 as well as in EF 1190. Hartenstein was thrilled. "That seems to be the golden vein New York – Cape Roque [Brazil]. Am heading for it." The northeasterly trade winds kicked in and *U-156* made good time. The sun finally appeared.

Zapp had been right on the money. At 2:05 p.m.[14] on May 12, the bridge watch called out, "Steam freighter! 17 Degrees!" Quadrant EE 39. For hours, Hartenstein worked *U-156* ahead of the target. It was a clear, star-filled evening. No moon to give the U-boat away. Strangely, no smoke from the target's funnel.[15] At 7:20 p.m., Hartenstein was ready for the kill:

he fired two bow torpedoes. Anxiously, the boatswain counted down the seconds. Nothing. "Both misses!" Hartenstein approached the shadow in its wake – and discovered that it had drastically reduced speed at the moment of attack. "That explains the miss. Reloaded." A little before 10 p.m., he fired again. The electric torpedo was a surface runner! It veered off target, but then steered toward it. "Hit machine. Steamer is putting lifeboats over the side. Steamer sinks by the stern." The Old Man approached the lifeboats to ascertain nationality and displacement of the victim, but could not make out a mumble that sounded like "Ouney." He could not find the name in any of his shipping registers. He would later discover that he had torpedoed the Dutch 4,551-ton motor freighter *Koenjit*, in transit from Halifax to Egypt with 8,000 tons of general cargo. The crew of 37 was rescued.

At 9:31 a.m. on May 13, the watch spotted a smoke smudge on the horizon off Barbados. The target was running on an erratic zigzag course, but in the general direction of the U-boat. Given that it was daylight, Hartenstein opted for a submerged shot. At 4 p.m., the hydrophone operator warned, "He is turning to run at us!" Calmly, Hartenstein counted down the range: 1,500 – 1,000 – 700 – 450 meters. He fired a stern shot from Tube V. The G7e headed straight for the target. "Hit just in front of the bridge. Stops. Swings lifeboats out. List of 2 degrees." Still submerged, Hartenstein circled his victim. On the stern he could make out "*City of Melbourne*. Liverpool." *Lloyds Register* listed it as a British 6,630-ton general cargo steam freighter.

But the victim refused to go under. "Surface! Ready the Artillery!" It was time to test the newly installed 10.5-cm deck gun. Second Watch Officer Fischer and his gun crew pumped 24 shells into the fore-ship. It broke off and the stern lifted up out of the sea. "Steamer still refuses to sink!" Fischer fired another five shells into the wreck. "Slowly sinks!" Hartenstein was beside himself. Many of the shells' nose fuses failed to detonate on impact and the missiles harmlessly passed out the other side of the ship. "Behavior of the 10.5cm ammunition unsatisfactory," he laconically noted in the war diary. The *City of Melbourne* lost only one of its crew of 78 that day.

To preserve precious fuel, Hartenstein let the current take *U-156*. The men took turns coming up on deck to shower, to swim, and to wash their sweaty shorts and neck rags. Shortly after noon on May 14, the watch spotted yet another "smokeless" steamer. It seemed to be in ballast, and it mounted a heavy gun on the stern. A few minutes before 3 p.m., *U-156* attacked. No detonation. "Inexplicable Miss! Probably ran under the ship!"

For four hours, the two MAN diesels roared on full speed to get *U-156* ahead of the target again. No moon. A star-studded clear night. Just before 8 p.m., Hartenstein fired from Tube II. Another terse entry in the war diary: "Inexplicable Miss!" And another hour to plot yet another attack. This time Hartenstein let loose from Tube III. The "Eto" broke the surface, but this time there was no mistake. "Hit amidships. 20 m[eter] high dark explosive cloud. Swings lifeboats out, begins to list. Steamer sinks!" The crew in the lifeboats revealed it to be the Norwegian 4,301-ton motor freighter *Siljestad* out of Oslo. It was carrying general cargo and war material from New York to Alexandria, Egypt. Two of the crew of 33 died.

Hartenstein ordered four of the old "Ato" torpedoes to be moved from below the upper deck plates to the bow tubes. *U-156* once more drifted with the current. At 8:26 a.m. on May 15, the watch screamed "Steam Freighter in sight!" Hartenstein began his approach. At that moment, the target blew off steam and stopped – to pick up survivors from the *Siljestad*. "They have a surprise in store for them!" the Old Man chuckled. The freighter resumed its course, zigzagging wildly. At 3 p.m., Hartenstein was in position. Range: 1,200 meters. He fired a single bow torpedo. "Hit just in front of bridge and cargo room 2. Lists 2 degrees to port. Swings lifeboats out." On its deck, Hartenstein could make out large wooden crates. The victim turned aimlessly in circles for 15 minutes, then the sea swallowed it. Hartenstein surfaced. A dozen lifeboats bobbed up and down on the gentle sea. They carried the survivors of both ships. The water was littered with wooden boxes revealing airplane and automobile parts. Hartenstein had the men fish 14 automobile tires and about 100 inner tubes as well as packs of Chesterfield cigarettes out of the water. The survivors (39 out of a complement of 41) informed him that he had sunk the Yugoslavian 4,382-ton freighter *Kupa*, bound from New York to Egypt.

For two days, *U-156* encountered no new targets. Again, halcyon days for showers and swims up on deck as Hartenstein let the boat drift with the trade winds. Then, just before noon on May 17, the welcome shout, "Steam Freighter in sight!" The target mounted two guns on the stern and was laden down with wooden crates on deck: "Probably automobiles or airplanes." For more than three hours, the Jumbos drove *U-156* ahead of the target. At 3:04 p.m., Hartenstein fired a stern shot. The torpedo ran true. After 25 seconds, "Hit front of funnel. 40 m[eter] high black-brown column of fire. Steamer sinks."

"Surface!" *U-156* broke the sea in an immense field of crated airplane parts. The watch spotted a figure floating amidst the smashed wooden crates. It was a young American sailor. He informed Hartenstein that he had sunk the British 5,072-ton freighter *Barrdale*, en route from New York to the Persian Gulf with "airplanes, tanks, automobile tires, and general cargo." A good loss for Joseph Stalin and the Red Army, Hartenstein must have noted. He fished several of the large airplane tires out of the water as a "trophy." Then he drove the American over to the lifeboats.

There was no time to rest. At 3:39 a.m. on May 18, the watch was at it again: "Dark shadow to starboard!" Hartenstein was up on the bridge in a flash. He decided to position *U-156* west of the target to silhouette it against the first light of dawn. It would still be sufficiently dark to risk a surface shot. At 4:18 a.m., he fired. After one minute and 22 seconds, "Hit stern superstructure. 40 m[eter] high black column of fire and smoke. Steamer sinks." The blast killed 11 of its crew of 41. Once again, Hartenstein approached the lifeboats. He was informed that he had torpedoed the American 4,961-ton freighter *Quaker City*, en route from Bombay to Norfolk with a full load of manganese ore. Junior Third Mate Charles Stevens recalled Hartenstein as being "very courteous" and giving the survivors the coordinates for the nearest landfall, Barbados.[16] The men in the lifeboats declined the skipper's offer of water and food but requested playing cards to while away the time. These boys have a sense of humor, Hartenstein thought, and passed them three decks.

Still, the crew of *U-156* got no rest. At 8:07 a.m., the by now familiar cry "Steamer in sight!" rang down from the bridge. For five hours, Hartenstein drove *U-156* hard to get ahead of what he took to be a tanker in ballast. Just before 1 p.m., he decided on a double bow shot. The two electric "eels" ran for just over one minute. "Hit under the bridge. 2nd hit amidships." The tanker began to list to port, but its wily skipper quickly ran his bilge pumps and managed to right the vessel. He continued on course at 11 knots and fired his deck guns at *U-156*. Hartenstein was furious. He pursued under water, hoping that the adversary would "stop or show a sign of weakness. Nothing of the sort." Was it a U-boat trap? Was the "tanker" a decoy, a Q-ship?

"Surface!" Hartenstein was determined to hunt this one down. At that moment, 5:18 p.m., the Enigma lit up:

> To Hartenstein. Proceed at once to Grid Quadrant ED 66. Task: 1. attack American warships suspected operating off the harbor. 2. Scout harbor and anchorages, if this can be done without being seen. 3. Destroy departing French warships and merchant ships so that they will not fall into American hands. [Aircraft carrier] "Béarn" especially important.

Vichy France had cautioned the Germans that the Allies were patrolling Fort-de-France with one cruiser and four destroyers.

What was taking place at Martinique was, in fact, a classic game of tit-for-tat. The Allies worried lest Admiral Georges Robert's tidy fleet of 70,000 tons of warships and treasury of 12 billion francs in gold would join the U-boats in their assault on the vital Caribbean oil supply. In the near-panic atmosphere of 1942, J. Edgar Hoover at the Federal Bureau of Investigation warned the Administration that "1400 airplanes and 50 submarines are near readiness at Martinique for an attack on the Panama Canal, Puerto Rico, Florida or the Florida Keys and Cuba."[17] The Germans, for their part, were equally panicked that Robert, far away from Vichy, might have had a change of heart and joined the Allies. Whatever the case, both stepped up their surveillance of Martinique.

Hartenstein did not have to consult the navy's grid chart to know that Martinique was in Quadrant ED 66 – precisely where he had dropped off his Second Watch Officer during the last war patrol. But he was less than pleased. This was just the sort of micromanaging by the "Great Lion" that the U-boat skippers hated. "First the tanker must be disposed of. Half-finished work should not be allowed to languish. Course for Martinique will have to be shaped without me." This bordered on insubordination. It had better result in a major success.

An hour before noon on May 18, Hartenstein fired another torpedo at the tanker. It struck the hostile abaft the bridge after 37 seconds, causing a 20-meter-high column of dark smoke to rise. Incredibly, the tanker continued on course at seven knots as if the torpedo had missed! Hartenstein pursued. Three hours later he fired yet another torpedo. After one minute and 16 seconds, it hit near the engine room. Still, the damned thing continued on course at seven knots. In a towering rage, Hartenstein leaped ahead of the target yet again. Since it had taken four torpedoes in the starboard side, he decided to "break it in half" with a shot in the port side. And since the torpedo gang had not had enough time to reload the bow tubes, he was forced to make a stern shot. At 3:17 a.m., the electric torpedo sped on its way – and missed! What did he have to do to sink this character? Doggedly, Hartenstein ordered "Pursue!" But the tanker outran him on an erratic zigzag course, swinging wildly from 120 to 330 degrees south and west.

Finally, "Crazy Dog" gave in. He let it go. He had spent 20 hours pursuing the target, had plotted five attacks, and had torpedoed it four times. And nothing to show for it. Then, reality set in like a cold shower via the Enigma machine: "Shape course for Martinique at once."

Some time later, Hartenstein would learn that the tanker he had chased through the night was the 8,042-ton *San Eliseo*, in ballast out of Liverpool. Ironically, it belonged to the Eagle Oil and Shipping Company of San Nicolas, Aruba, the scene of his first Neuland war patrol triumphs. The *San Eliseo* had been severely damaged but managed to make it to Aruba.

At 9:25 a.m. on May 19, the watch screamed, "Alarm! Aircraft! 270 degrees! Course 0!" Executive Officer Paul Just had the watch. "Emergency Dive!" The lookouts tumbled down the hatch. The crew heard the air blowing out of the dive tanks. The diesels cut out; the electric motors began to whir. Thirty seconds and *U-156* was below the surface, on a downward angle of 20 degrees. Within two minutes, two depth charges exploded near the boat. Glass broke. The lights went out. Dim emergency bulbs came on in the *Zentrale*.

"Damage Control, report!" It was the Old Man. Chief Engineer Polchau was ready. "Damage to both hydroplane motors, vertical rudder motor, gyro indicator, lights, starboard electrical motor. Major damage:

leak in ballast tank I, 2 batteries torn and slowly draining, echo sounder." Emergency teams began their repair work. At 110 meters, just beyond recommended maximum depth, two more depth charges rocked the boat. Polchau ordered "Both hydroplanes up!" and finally leveled the boat.

"Periscope depth!" Hartenstein surveyed the scene through the sky periscope. "Alarm! Airplane at 240 degrees!" *U-156* dove again. This time there were no depth charges. The boat remained submerged until dusk. At 5 p.m., it resurfaced. Course: Martinique. For hours, the torpedo gang muscled four "Ato" torpedoes from under the upper deck boards into the bow torpedo room. At 9:40 p.m., Martinique came in sight. Lieutenant Just noted the strain on the crew. "30th day. 16 hours submerged with 45-degree [Celsius] heat and 90 percent humidity in the boat."[18] Mildew had spread everywhere. The men were covered with heat sores.

Cautiously, *U-156* circumnavigated the island. Officers and men were on edge. Headlands appeared menacingly in the distance; the watch mistook "La Perle" cliff for a hostile craft. Hartenstein submerged off Fort-de-France to reconnoiter the harbor. It revealed five tankers as well as the aircraft carrier *Béarn*. Off to the side in Flammand Roads, he spotted a modern passenger liner, *Agittaire*. Dead ahead was a warship: perhaps the cruiser *Émile Bertin*? Two American flying boats buzzed around Fort-de-France. *U-156* remained on station. Suddenly, at 12:06 p.m. on May 21, a dark shadow with two masts appeared. A signboard on the bridge revealed its name: *President Trujillo*. It flew a Dominican flag, "thus enemy." Hartenstein wasted no time, firing an "eel" from Tube II. After 29 seconds the "Ato" ripped into the ship's stern, sending up a 30-meter-high column of fire and smoke. The 40-year-old Dominican 1,668-ton freighter sank within a minute, taking 27 of the crew of 39 as well as beer-making machinery and forage down with it.

Three airplanes appeared at once and dropped depth charges randomly. *U-156* submerged. Every half hour, Hartenstein brought it to the surface to reconnoiter Fort-de-France. It was sheer hell for the crew. Sixteen hours submerged. Then 20 hours. Then 14 hours. Heat and humidity were almost unbearable. Finally, on May 25, *U-156* surfaced in an isolated bay to recharge the batteries. Rain, glorious rain! It came down in sheets. The watch could hardly see beyond one meter. Hartenstein called the men up in shifts to take in the tropical air and the sweet-cold sea spray. And then the rain stopped and the sun broke through. Steam rose from bay and boat. It was a sauna.

Shortly after 1 a.m. on May 25, Dönitz was back on the airwaves. "Danger exists that ships in the main harbor will be turned over to the USA. Thus main task is attack on USA warships and other ships leaving port. Attack incoming ships only if identified as hostile." *U-156* was off Cape Salomon. The watch reported "Shadow ahead!" It was a fellow traveler, *U-69*. Paris had ordered it to join *U-156* in patrolling Fort-de-France. Hartenstein returned to Fort-de-France. At 7:43 a.m. on May 25, the hydrophone operator reported, "Screw noises at 235 degrees." Hartenstein ordered periscope depth. He could hardly believe his eyes. "American 4 stack destroyer." Would he finally get a crack at one of the "destroyers-for-bases" craft that Dönitz had lectured the Kaleus about at Lorient?

The destroyer was the 1,154-ton USS *Blakeley*. It had seven survivors from the *Quaker City* on board. Launched in Philadelphia in 1918, the "flush-decker" had seen no action in World War I and then had been decommissioned at Philadelphia from 1922 to 1939. It escorted troop convoys to Curaçao in February 1942. On May 25, *Blakeley* was assigned to patrol a base course roughly north to south off the west coast of Martinique. It was steaming at 15 knots and zigzagging with the galley deck guns and the .30- and .50-caliber machine guns manned.

But it was too far off to attack. "Perhaps he will return," Hartenstein wrote in the war diary. As per his wish, the destroyer reappeared two hours later off Precheur Light, zigzagging and making 15 knots. "He is coming! Battle stations!" Clear sky, calm sea. Range: 800 – 700 – 600 – 500 meters. On board the destroyer, a fatal mistake: the sonar had been turned off at 1:45 p.m. while a maintenance man went to the tracking room to lubricate the equipment. But the Officer of the Deck wasn't notified. When the maintenance work was completed, the sound-detector gear was turned back on; it started to sweep off the starboard beam, but too late to deter Hartenstein.

Time: 10:52 a.m. Hartenstein fired two bow shots. After 25 seconds, he gleefully recorded: "Hit in fore-ship. High column of fire. Fo'c'sle torn off. Hit must have been 2nd torpedo." It was a strange sight: bow and fo'c'sle shot off and listing 15 degrees to starboard, the destroyer's rump continued to move ahead. The oil from the forward tanks shot 1,000 feet

USS *Blakeley* after a direct torpedo hit to the fo'c'sle from *U-156*. Source: Ken Macpherson Photographic Archives, Library and Archives at The Military Museums, Libraries and Cultural Resources, University of Calgary.

into the air and then showered the ship with "torrents of oil, water and debris." The bow was lifted clear out of the water and the fantail was "set to whipping" by the explosion. Several sailors had seen the torpedo's telltale "bubbles" at the last moment, but it had been too late to alert the bridge. The explosion was so powerful that radio tubes and resistors as well as the gyro compass on the U-boat were heavily damaged. But there was no time to deliver the *coup de grâce* as enemy aircraft were already overflying the Bay de Fort-de-France and dropping depth charges all about. *U-156* headed back out to the open sea.

Blakeley's skipper, Lieutenant-Commander M. D. Matthews, at first did not know how badly damaged his ship was. As soon as the debris cleared, it became obvious that the *Blakeley*'s bow had been blown off and that it had developed a 15-degree list to starboard. He gave the order to prepare to abandon ship. But the damage-control party sprang into action. It pumped oil from the starboard tanks into the port tanks, righting the ship. Matthews concluded that *Blakeley* would stay afloat. He canceled the order to prepare to abandon ship and lowered a whaler to pick up several men who were in the water. He tried to back the ship into Fort-de-France harbor, seven miles away. The distance was too great. Matthews then ordered the *Blakeley* turned about and, with 20 meters off its bow, steamed ahead into the harbor. The destroyer docked alongside the *Béarn* some three hours after the torpedo hit. French surgeons treated the 21 wounded sailors. After the legal stay of two days, *Blakeley* was escorted to Castries, St. Lucia. Muster revealed that six sailors had died or were still missing.[19] The loss of the *Blakeley* came as a severe shock to Washington. The Navy Department hastily dispatched the destroyers *Breckenridge*, *Greer*, and *Tarbell* as well as two Catalina flying boats to the Caribbean to deal with the marauding "gray sharks."

* * *

For Hartenstein and *U-156*, the coming days brought only a succession of emergency dives to avoid attacks by land aircraft and flying boats. It was hell for the crew. Every time the boat surfaced to recharge the batteries and to take in fresh air, hostile aircraft forced it to dive. Day after day, aerial depth charges rained down all about the craft. Executive Officer Just again expressed concern about the state of the crew:

> We look like cellar wood lice. The skin is a greenish white; shriveled and wrinkled due to the constant sweating. Some of us are tortured by rashes and abscesses. Others have ear infections from the temperature changes [caused by the] dives. When we surface at night, the rush of air into the compartments is ice cold. The seawater shower in the diesel room brings no refreshment, but still stimulates a bit.[20]

Around midnight on May 26, Hartenstein surfaced off Pointe des Négres. "Alarm! Flying boat at 250 degrees!" The Catalina was flying 30 to 50 meters above the water and coming straight out of a bright moon. "Engine full speed ahead! Hard-a-starboard!" As the boat heeled over to the right, three depth charges exploded in its wake. Close call. After surfacing, the routine set in anew. And then a chilling report from the radio room: "Destroyer noises at 60 degrees!" The men could hear the sickening "pings" of the destroyer's ASDIC bounce off the hull. Hartenstein ordered "Silent Running!" and took the boat down to 120 meters. Six depth charges burst around *U-156*. Glass broke, lights shattered, two of the heavy batteries shorted out, and both hydrophones broke down. Total darkness. For hours, *U-156* crept along in the deep, listening to the occasional rumble of depth charges off in the distance. The heavy, humid air burned the men's lungs.

By 3 a.m. on May 27, Hartenstein had no choice but to surface, for both the men and the electric batteries were drained. Forty minutes later came the dreaded cry, "Alarm! Flying Boat 160 degrees." Down again. Then up again. After 20 minutes, "Alarm! Airplane!" To hell with this harbor patrol! Dönitz may have claimed that American air reconnaissance was "inexperienced and *bad*" and that there existed no "crisp, well thought-out antisubmarine operations," but that was the picture back at Lorient and not here in the Caribbean. The Old Man ordered a course for the open sea with the last juice left in the electric batteries. At 7:24 p.m., he brought *U-156* to the surface. Coast clear. "Both engines full ahead!" The Jumbos roared up to power and Hartenstein shaped a course for St. Lucia.

At 11 a.m. on May 28, the watch reported, "Smoke cloud at 280 degrees!" Hartenstein at once gave chase, but in the excitement of the moment he brought *U-156* too close to the target. A torpedo detonation would have damaged both vessels.

He followed the hostile, which was heading back to Martinique. Precious hours wasted. By 7 p.m., *U-156* had caught up to the shadow. Hartenstein fired a single "eel" from Tube IV. The old "Ato" ran true. After 30 seconds, "Hit amidships, down by the stern. Lists to starboard, but does not sink." Afraid that he might already have drawn enemy aircraft, Hartenstein delivered the *coup de grâce* from Tube II. After 45 seconds, "Hit forward hatch." Then he discovered that the target's stern deck gun was manned. Too late! The ship slipped beneath the waves with a last tremendous rattle of detonations. He had torpedoed the British 1,913-ton freighter *Norman Prince*, in ballast en route from Liverpool to St. Lucia. *U-156* was down to its last three torpedoes.

Shortly after 3 a.m. on May 30, Hartenstein fired off a unique radio-gram to U-Boat Command in Paris. After reporting on enemy traffic off Fort-de-France and his latest "kills," he pressed on Dönitz the toll that the war patrol was taking on the crew. "In 7 days in the tropics, 121 hours submerged. Limit of capacity reached." The Old Man had taken careful measure of his young crew. They needed relief. He shaped a course for the Atlantic, planning to pass Vincent Channel in the Lesser Antilles between Barbados and St. Lucia.

The first of June brought a fat target in Grid Quadrant ED 5329. The Old Man noted in the war diary: "Flies an indiscernible flag. Name painted over." No time for niceties. He fired from Tube V. "Hit abaft mast. Sinks down by the stern." He circled the victim. From its stern flew a small Brazilian flag. The smudged plate revealed the name *Alegrete*. The boatswain snatched up *Lloyds Register*. "Has 5,970-tons, Herr Kaleu!" Owner: Lloyd Brasileiro. Home port: Rio de Janeiro. Damn, it was a neu-tral! Hartenstein decided that he could not just leave the freighter to sink by itself. He surfaced and ordered Lieutenant Fischer to pump 20 10.5-cm rounds into the wreck. It sank by the stern, bow high out of the water. This would take some explaining back in Lorient.

At 2:40 a.m. on June 3, Hartenstein spied a darkened schooner off Cape Moule à Chique, the southernmost tip of St. Lucia. He ordered it to strike sails. It refused. He sent a 3.7-cm shell across its bow. The schooner set 18 inter-island passengers off in a lighter and continued its course. From the abandoned passengers, Hartenstein learned that it was the Venezuelan sloop *Lilian*, loaded with rum out of Jamaica. He could read the thoughts on the crew's collective face. But his temper broke at the cheek of its captain: Fischer fired 52 light rounds into the *Lilian*. A terrible waste of good Jamaican rum.

Paris ordered *U-156* to shape a course for Lorient – 12 days away. But "Crazy Dog" still had supplies for 41 days, more than a hundred 10.5-cm shells below decks, and one torpedo in the tubes. At 10 a.m. on June 23, the watch detected a smoke smudge on the horizon. Hartenstein pursued. He wanted this one badly. The freighter ran a wild zigzag course. Time and again, the Old Man approached for a shot, only to see the target dash off at high speed in another direction. He was finally in position at 2:20 a.m. on June 24. The "eel" fired from Tube I was a "hot runner": it stuck in the tube, its small compressed-air motor running wildly. The danger of a premature explosion of its warhead was high. Executive Officer Just ordered double air pressure for Tube I and the torpedo finally left the bore. It veered erratically off target and then sank.

"Clear the decks to engage with artillery!" Fischer fired 65 rounds from all three guns. The target's captain sent out a distress signal, from which Hartenstein learned that he was shelling the British 4,587-ton freighter *Willimantic*, in ballast from Cape Town to Charleston. Hartenstein took the ship's captain prisoner. He learned from Master Leon Everett that the crew consisted of elderly men ("well over 60") taken from an existing "pool" of sailors. The Allies could build ships at great speed, but experienced skippers were hard to come by. Another 20 rounds from the deck gun and the *Willimantic* sank.

U-156 glided through the Kernével Narrows and docked in Lorient at 2:06 a.m. on July 7, 1942. Ten pennants flew from its periscope tubes. The last entry in the war diary was terse, as ever: "Total distance 10,465.4 nautical miles, of this 546.9 underwater." For the first time, Hartenstein signed the KTB with his new rank: Korvettenkapitän (lieutenant-commander). He was especially pleased that he had offered water, food, and directions to every lifeboat from the ships that he had torpedoed. For July, the entire crew of *U-156* was invited to be feted by their "sponsor," the city of Plauen in Saxony, Hartenstein's birthplace. Paul Just received his own command, *U-6*, and Lieutenant Gert-Fritjof Mannesmann took his place as Executive Officer.

Chief Engineer Polchau used his final report on the war patrol to underscore the Old Man's radio signal to Dönitz that the crew had reached the "limits" of their "capacity" off Martinique. "The boat remained underwater for a long time in tropical waters, once seven days in a row and on average 18 hours per day. Water temperature 30 degrees [Celsius], air temperature in the boat on average 34 degrees." The capacity of the batteries had been reduced once the acid mix reached 42 to 44 degrees. And at 47 degrees, it had proved impossible to recharge the cells.[21] It was a sobering report.

Dönitz was pleased with the war patrol. "The commander exploited well the numerous chances for success and thus scored a very nice success. Especially to be praised is the special assignment off Martinique, conducted with tenacity."[22] The new technological innovation, the "milk cow," had proved its mettle: *U-459* had resupplied two outbound boats, four returning boats, and five boats on station off the Caribbean Sea.

En route to Plauen, Hartenstein paid the customary call on Dönitz in the Avenue Maréchal Maunoury.[23] The "Great Lion" was in a particularly good mood and offered Hartenstein (as well as Karl Thurmann of *U-535*) his Mercedes limousine to do some sightseeing. Hour after hour passed. At 8 p.m., Dönitz had to borrow a smaller car to take him to a meeting with the city commandant. "It is always the same story," he growled to an aide, "If you offer these types your little finger, they'll grab the whole hand." When he returned from the meeting, the Mercedes was still not back. He ordered his adjutant to have both skippers report to him immediately upon their return. Well after midnight, Hartenstein and Thurmann finally returned – after a night of barhopping and sampling what Paris had to offer. Seeing that both officers were three sheets to the wind, the adjutant suggested they wait until morning to report.

Not Hartenstein. He put on his dress uniform and insisted on being taken to Dönitz. A workaholic, the admiral was still at his desk. He let loose with a tirade concerning the ingratitude of the two skippers. Hartenstein took it all in, saluted, and recited from the arch-rascal Baron Karl von Münchhausen:[24] "On many a flag have I laid my hand swearing loyalty in this wicked war, many an admiral have I served." As his voice trailed off, Hartenstein simply turned around and left the room. Dönitz recounted the story the next morning at breakfast in great mirth. Things were going very well for the "Great Lion."

8

HUNTING OFF THE ORINOCO

The struggle in the Caribbean between Dönitz's submarines and the Allies was not confined to the Caribbean basin alone. Tankers, bauxite carriers, and other Allied merchantmen were just as easily torpedoed outside the island chain as they were within. And when the Allies really turned the heat on in the basin itself, Dönitz merely sent his U-boats to the east and south, along the coast of Venezuela and Brazil. *Tonnagekrieg*, after all, knew no boundaries. Cargoes that would aid the Allies were sunk off the coastal bulge of Brazil, or even off the beaches of Rio de Janeiro, as they were near Trinidad or Cape Race. It did not take long before the Americans realized that the defense of the Caribbean was closely tied in with the defense of mid- and South Atlantic waters. Even before the summer of 1942, the war began to spread beyond the Caribbean to the waters of the Torrid Zone and into the South Atlantic.

Virtually all the men who commanded Karl Dönitz's "gray sharks" during Operation Neuland began their careers in the surface navy; Fregattenkapitän[1] Jürgen Wattenberg, skipper of the Type IXC *U-162*, was one of them. He had entered the navy in 1921, and at the outbreak of World War II served as navigation officer on the "pocket" battleship *Admiral Graf Spee*. After Captain Hans Langsdorff, trapped by British Hunting Group G, scuttled his ship in the Rio de la Plata off Montevideo in December 1939, Wattenberg escaped from the temporary internment camp at the Naval Arsenal in Montevideo. Friends provided civilian clothes, and the adventurous Wattenberg walked into Argentina, hiked across the Andes to Santiago, Chile, and coolly boarded a commercial flight to Germany in May 1940.[2] He at once volunteered for the U-Boat Service. *U-162* was his first command. He thirsted for revenge.

Wattenberg's first war patrol was disappointingly unsuccessful. In 40 days at sea he sank only one ship, the British 4,300-ton freighter *White Crest*, out of homebound North Atlantic convoy ONS-67. Wattenberg returned to Lorient in March 1942 to a sharp rebuke from Dönitz. He had not kept in constant touch with Kernével because he feared that the Allies were intercepting those signals. He had misjudged the course of a convoy near the Azores and had lost contact. He had been hesitant in attacking an escort destroyer and a lone freighter because he believed that the escort was part of a larger convoy, for which he had searched in vain. As far as U-Boat Command was concerned, he had "balked" at the chance for a "kill." Dönitz had been unsparing in his critique: "A commander never knows how a situation will develop, therefore always attack at first opportunity and never undertake experiments."[3]

Wattenberg's war patrol to the Caribbean began on April 7, 1942. He headed for the waters off Venezuela, the Guianas, and Trinidad and spent a good part of the last week of April scouting enemy traffic. He did not see much, but each day dutifully sent back long reports on what he did see. On April 25, he received orders to proceed to Grid Quadrant EE, the suspected shipping lane for West Indies–Gibraltar sailings. En route, he put the artillery crew under Second Watch Officer Berndt von Walther und Cronex through their paces, firing at jettisoned wooden egg crates by day and by night.

At 5 p.m. on April 29, the bridge watch spied the mastheads of a tanker. Wattenberg drove his boat hard for two hours to get in position to attack. Just before 9 p.m., he fired a bow shot at a range of 600 meters. "Surface runner!" He well remembered Dönitz's reprimand, and ordered *"Second bow shot at once."* Somehow, both torpedoes sliced into the shadow. It stopped and began to list. Its crew took to the lifeboats. The survivors informed Wattenberg that he had torpedoed the British 8,941-ton tanker *Athelempress*, in ballast out of Liverpool. Since the wreck refused to sink, Wattenberg had it riddled with shells from the 10.5-cm deck gun, whereupon it quickly slid beneath the waves. It was the Old Man's first major success.

Shortly after 1 p.m. on May 1, standing off the mouth of the Orinoco River, the watch reported first a smoke smudge and then masts on the horizon. The diesels roared to full power. For two hours, *U-162* pursued.

Wattenberg took careful measure of the target. "Freighter has stern deck gun, also a gray-green camouflage stripe and no flag, thus hostile." At 2:46 p.m., he fired a single bow torpedo at 600 meters. It struck the freighter abaft the funnel. It stopped at once, down by the stern, and swung out four lifeboats. Wattenberg surfaced and had Walther-Cronex fire several heavy shells into its superstructure. No reply from the ship's gun. Wattenberg circled his victim. On both bow and stern, he could make out a name in small letters: "Pernahyba" out of "Rio." He still believed it to be a "hostile." Since it refused to sink, he pumped 56 rounds from the deck gun into the hulk. As it went down, he made out barrels of lubricating oil and intestine skins on the deck. He had destroyed the German-built 6,692-ton *Parnahyba*, bound for New York with a cargo of coffee, cotton, and cocoa.

Like Werner Hartenstein in *U-156*, Wattenberg had sunk a neutral vessel. He at once radioed the news to Paris. This undoubtedly would not sit well with Dönitz. Among the floating jetsam of oil and wood and guts, the watch spied a fat turkey perched on a barrel, chickens fluttering on the waves, and two black pigs swimming furiously among the wooden staves. They were all brought on board as "welcome booty" – then promptly butchered to feast the crew.[4] Thereafter, Wattenberg shaped a course for the estuary of the Demerara River at Georgetown, British Guiana, to hunt bauxite carriers.

The sinking of the *Parnahyba* was another blow to Brazilian neutrality. When the war began, Brazil was closer to the Axis than the Allies. There were a significant number of German expatriates in Brazil, and Lufthansa, the German airline, had pioneered air routes connecting several South American countries. Brazil was a potential buyer of German military equipment, especially anti-aircraft artillery from Krupp. But Brazil was also a potentially important partner for the Allies because of the raw materials, especially rubber, produced there, and also because of its geography. The country's eastern bulge – U-boat skippers referred to it as the "mid-Atlantic" in their war diaries – was a key refueling point for aircraft flying from the United States to Africa and the Middle East, and Brazil's proximity to the sources of bauxite in the Guianas made it an excellent place to put antisubmarine aircraft.

At the end of 1942 Brazil was swayed to break diplomatic relations with the Axis after receiving promises of US military aid and American airlines (especially Pan American World Airways) to fly Brazilian routes. But it was still officially neutral when Wattenberg and other U-boat commanders began to sink Brazilian ships. As the U-boats penetrated further into the South Atlantic, Brazil's shift to the Allied cause became the key to antisubmarine defenses in the area – especially given the pro-German "neutrality" of Argentina. Each sinking brought that day closer.

Wattenberg's decision to head for the waters off the mouth of the Demerara River was a good one. At 10 p.m. on May 3, the watch spied two shadows at two miles. Wattenberg approached them submerged, but decided that they were too small to risk losing the element of surprise. At 1:27 a.m., the next day the watch again reported a shadow, dead ahead in the moonlight at 10 degrees. An hour later, *U-162* loosed a bow shot at 2,800 meters. The boatswain counted off the seconds to 120. Nothing! The steam-driven torpedo apparently had not been fully charged and sank below the target.

Wattenberg renewed the attack. Range: 3,800 meters. "Hit! The freighter's stern sinks down to the base of the funnel." He surfaced and circled his prey. "No name discernible." It was 6 a.m. The first light of day was beginning to break over the horizon. He assumed that the stricken freighter would sink and left the area before hostile aircraft appeared. His victim was the American 3,785-ton bauxite carrier *Eastern Sword*, en route from New York to Georgetown. Eleven of its crew of 29 went down with the ship.

Around noon on May 4, *U-162* came across a three-masted schooner. It was the *Florence M. Douglas* out of Demerara, "thus hostile." After ordering its crew into lifeboats, Wattenberg sank the schooner with 18 rounds from the deck gun. He saw something floating in the debris. "Yet again, a small black piglet comes on board. It is too small to be butchered and so it will eat our scraps and leftovers. It is quickly named: 'Douglas'." No sooner had it been stowed in the diesel room than an aircraft appeared out of the sun. "Alarm! Aircraft dead ahead! Emergency Dive!" Somehow, the enemy pilot failed to spot *U-162*. "Douglas" survived the steep dive unharmed and came squealing into the control room. Wattenberg decreed it to be the boat's "lucky pig."

Wattenberg headed out to sea. Shortly before midnight on May 6, the watch reported "Shadow with heavy smoke cloud 165 degrees!" Since there was a bright moon, he opted for a submerged attack with a stern torpedo. "After running for 10.4 sec. bright, yellow fire-flash against hull." Then, a metallic, hollow "clank." He had miscalculated, coming too close to the target – 180 meters – and thus the "eel" had not had ample time to arm the trigger mechanism. *U-162* came round for a bow shot. "Miss! Miscalculated target's speed!" It was now 2:33 a.m. The Old Man expended a third torpedo on the target. "Hit amidships! High detonation column. Freighter breaks in half and sinks in 7 min." The men cheered the hit and "Douglas" ran through the boat squealing with delight.[5] The vessel's master stated that his ship was the 7,000-ton bauxite carrier *Runciman*. But Wattenberg could not find it in any of his shipping registers. In fact, the wily skipper, Ingvald Hegerbeg, had given Wattenberg a false name. The victim was the Norwegian 4,271-ton freighter *Frank Seamans*, bound for Trinidad with a load of bauxite. The crew of 27 all took to the lifeboats; Wattenberg offered provisions and directions.

Just before 10 a.m. on May 8, *U-162*'s watch sighted a new target. Wattenberg plotted a submerged attack with the stern tubes. But after the destruction of the neutral *Parnahyba*, he had grown cautious. "Cannot absolutely discern origin of the vessel. No zigzagging, no deck gun, no camouflage stripe, but also nowhere anything indicating marker as a neutral." He could not make out the flag fluttering from its stern. On the bow, he saw what he thought was the name "Louis" preceded by the letter "M." Was it a Vichy French carrier? He tracked the target for nine hours. It never set evening lights and so he guessed it to be "hostile," possibly a Canadian bauxite carrier. At 10:12 p.m., he fired a bow torpedo at it. "Hit in the after-ship. Mighty explosion with resulting dark-black smoke, which immediately envelops the entire ship." The vessel disappeared so quickly that he thought it loaded with explosives. His hunch had been right. It was the Canadian 1,905-ton freighter *Mont Louis*, en route from Paramaribo, Dutch Guiana, to Trinidad.

Thereafter, the sea was empty. The radio room picked up urgent long-wave calls from "Navy Commander Bermuda" to Georgetown and Paramaribo warning ships to avoid "area seven." The Old Man remembered from his days on the *Graf Spee* that British merchantmen used numbered codes for specific areas of the ocean. He decided to head for the more promising coast of Guiana, where he would hook up with Hartenstein in *U-156*.

Once again, his decision proved to be right. Just before sunrise on May 12, still well off Barbados, the watch reported "Mast tops at 253 degrees." He pursued. Time for another submerged attack using the stern tubes. At 1:49 p.m., he fired two torpedoes; both missed the target. Wattenberg refused to give up. At 8 p.m., he was again in position to fire. But the target veered wildly off course, apparently hoping to ram *U-162*. Wattenberg coolly dived under the hostile at full power and came up on its other side. "Stern shot hits amidships, immediately causes oil bunkers to explode, so that a column of fire lights the entire ship for one minute." The victim stopped and put its lifeboats in the water. Its captain reported it to be the 7,699-ton Standard Oil tanker *Esso Houston*, with a full load of crude from Aruba to Montevideo – scene of the *Graf Spee* disaster in 1939. Wattenberg delivered the *coup de grâce* with another "eel."

U-162 stayed in the target-rich environment of Grid Quadrant EE 73. At 4 p.m. on May 13, it came across another tanker. After a three-hour chase, Wattenberg ran a nighttime surface attack: a single torpedo struck the target and caused a single column of flame to rise, but no further detonation. It blew alarm whistles. Its crew manned the stern gun and began to fire at *U-162*. "Immediately set out to renew attack from the East." Another bow shot: the tanker saw the telltale bubbles of the "Ato," heeled hard to starboard and avoided a hit. It continued to fire. Furious, Wattenberg plotted a new approach. At 1:39 a.m. on May 14, he fired a double spread. "Hit astern. High detonation columns combined with bright red fire flames, caused by a fuel bunker explosion on the target." The crew took to the lifeboats.

Still, the tanker remained afloat. Another torpedo failed to send it down. Wattenberg fired yet another torpedo at the target. "Huge fire flames combined with pervading black smoke clouds." The tanker burned furiously and finally slid beneath the waves. Its captain informed Wattenberg that he had destroyed the 6,917-ton tanker *British Colony*, en route from Trinidad to Gibraltar. The kill had cost ten hours and six torpedoes. Later that night, Wattenberg had the last four "eels" brought below decks. Kernével radioed that a Wehrmacht communiqué reported 21 ships of 113,000 tons sunk in the Caribbean and singled out *U-162* and *U-156* for praise. "Jubilation and pride" throughout the boat, the Old Man wrote in the war diary. "Douglas" squealed joyfully as ever.

* * *

The initial Allied reaction to the spring U-boat offensive in the Caribbean was to put temporary halts to shipping, especially of tankers and bauxite carriers. Tanker traffic under Allied control was stopped in mid-February and mid-April; bauxite traffic was stopped after Nicolai Clausen's first predations east of Trinidad with *U-129*. But clearly, this was no long-term answer to the Caribbean attacks. At a sequence of meetings of the US, Royal, and Royal Canadian Navies in Washington and Ottawa in late spring 1942, several key decisions were made regarding the Caribbean. British escort group B5, consisting of the Royal Navy destroyer HMS *Havelock* and four Flower-class corvettes, were to be taken off the North Atlantic run and sent to the Caribbean. The Royal Navy escort group was to receive air cover from No. 63 Squadron Royal Air Force (Coastal Command), consisting of 20 Lockheed Hudson twin-engine bombers based at Trinidad.

A Canadian escort group of four corvettes and an occasional destroyer would escort tankers from the Caribbean to the east coast of Canada. Admiral Ernest J. King also redoubled his efforts to convince President Roosevelt to support the building of destroyer escorts, which eventually began to appear in the Caribbean – but not for another year at least. The withdrawal of the British and Canadian escorts from North Atlantic convoy duty forced Britain to "open out" the convoy cycle – stretching the time of departure between convoys – while shortening the North Atlantic voyage by sending convoys on a more northerly route.[6] Both moves resulted in a decrease of tonnage reaching Britain. To this extent at least, Dönitz's tonnage war in the Caribbean directly affected the Battle of the Atlantic.

The first convoys to sail in the Caribbean appeared in the first half of May; within the next two months, a complicated network of convoys was put in place that locked into the new coastal convoys off the US coast. Some convoys proceeded directly to Britain from the Caribbean via Bermuda or Gibraltar, while others – mostly bauxite carriers – traveled between the waters off eastern Trinidad along the east coast of South America as far as the bulge on the Brazilian coast. Some of the prime routes were Halifax-Aruba-Halifax, Trinidad-Aruba/Curaçao-Trinidad, Aruba-Guantánamo-Aruba, Aruba-Colón-Aruba, Guantánamo-New York-Guantánamo, Key West-New York-Key West, and Key West-Texas-Key West.[7]

On May 20, *U-155* Kapitänleutnant Adolf Piening in *U-155* spotted the first large convoy sailing from New York to Trinidad. The 31-year-old Piening had taken command of the brand new Type IXC boat in August 1941; this was only its second war patrol, but Piening had already sunk six ships of 33,500 tons. The Kaleu approached the convoy off Venezuela's Testigos Islands, 110 miles northwest of Port of Spain. He was spotted by the US Navy four-stack destroyer *Upshur* and dove to trail the convoy as it sailed toward the Dragon's Mouth entrance to the Gulf of Paria. At dawn, about 40 miles out, Piening sank the Panamanian-registry 7,800-ton *Sylvan Arrow*. At Curaçao the tanker had taken on 125,000 tons of bunker oil, which quickly blew all over the ship and burst into flames. *U-155* received the usual plastering by depth-charges from *Upshur* and a patrol craft but escaped without damage.[8]

The introduction of a convoy system into the Caribbean certainly did the "gray sharks" no service, but it did not hinder them much either. Sinkings for the balance of May and June continued at a rate of from one to two ships per day. The convoy system in the Caribbean could never be as effective in fighting off, or avoiding, submarine attack as it was in the North Atlantic. There, the space was so vast that a convoy could shift course and avoid a U-boat concentration if intelligence indicated that subs were gathering on its track. Once a convoy left the east coast, whether Halifax or Sydney in the early days of the war or New York later, it was in the open sea, and even the largest convoys could be difficult for the U-boats to find. Aside from the Strait of Belle Isle, which was only infrequently used, there were no narrow seas, nor "chokepoints" where subs could lie in wait for traffic. But the Caribbean was much smaller, meaning that convoys that attempted to take alternate routings to avoid U-boats were usually easily found; and the Caribbean was ringed with chokepoints. From the Florida Straits east and south along the arc of the island chain, the Windward Passage, the Mona Passage, and the rest of the narrow waters between the islands offered excellent ambush points. After all, at some point in time all the ships moving in the Caribbean Sea, except for inter-island traffic, had to enter or exit the Caribbean basin. And when they did, the "gray sharks" were often waiting for them.

* * *

With Hartenstein's *U-156* moving into Barbados waters, Wattenberg took *U-162* into Bridgetown Roads. The harbor was devoid of targets, but the shore was lit up as in peacetime. He could clearly make out brightly lit fishing boats, houses, and cars. He returned to the open sea. At 3:22 a.m. on May 17, the watch spied a tanker, but given that dawn was not far away and that he was well within range of land-based aircraft, Wattenberg decided to track it until he was in optimum position to attack. At 9 p.m., he fired two "eels" at the tanker. One hit. "The 1st torpedo exploded with a high black-gray detonation column. The fo'c'sle broke in half." The tanker sent its crew off in lifeboats and Wattenberg fired the *coup de grâce* at the flaming hulk. "Hit amidships and caused the ship to sink in 10 minutes after showing a gigantic fireball and heavy smoke." Its captain revealed it to be the British 6,852-ton tanker *Beth*, bound from Trinidad to Freetown. A single crew member lost his life.

On May 18, Wattenberg dispatched a lengthy radiogram to Kernével, summarizing his experiences off Venezuela and Barbados. Enemy antisubmarine activity from the air was "minimal," that from surface vessels "absent." He deemed Quadrant EE 71 to be the "chokepoint" for all West Indies traffic bound for Gibraltar, Freetown, and South America. His experience with *Athelempress* convinced him that single "eels" could not sink crude-oil tankers and that the deck gun was inadequate to do the job. Bauxite carriers, on the other hand, were easy game. His bag to date: nine ships of 47,162 tons.[9]

Wattenberg had only a single old "Ato" torpedo left in the tubes and 67 rounds for the 10.5-cm deck gun. He decided to head for Quadrants EE 30 to 60 in hopes of sinking a "capital South American" target as a "Whitsuntide roast."[10] But the sea remained empty for the next five days. The radio room intercepted a news flash from the Transocean Press Service stating that the United States had ordered a halt to all shipping in the area to save bottoms for the vital supply line to Britain. Wattenberg decided to return to the coast of Guiana, to fire his last torpedo at a tanker and perhaps to destroy some sailboats with the deck gun.

"Whitsuntide roast at 183 degrees!" came the call from the bridge at 6 p.m. on May 23. "Another tanker. Jubilation throughout the boat." Wattenberg decided to "dine" on the target under the cover of darkness. At 1:28 a.m., he fired his last torpedo. It harmlessly raced by the bow of the tanker, which for some reason had suddenly reduced speed. Had it spied the boat, or the torpedo's bubbles? The sea churned up to Force 5. No weather for an artillery attack.

Wattenberg shaped a course for Lorient. Ahead lay 12 days of the return "garbage tour." *U-162* tied up to the hulk *Isére* in Lorient at 7:10 a.m. on June 8, 1942. It proudly flew ten sinking pennants, including three black ones for tankers, from its periscope tubes. There was the customary military band, the welcoming *Blitzmädchen*, the post-patrol feast at Flotilla Headquarters, and crew rotations. "Total distance covered 9657.3 nautical miles" was the last clinical entry in the war diary. "Douglas" was ceremoniously handed over to Commander Viktor Schütze of 2nd U-Boat Flotilla for "safekeeping."

Admiral Dönitz read *U-162*'s war diary with great interest. He was pleased: "Good and superbly conducted operation. The commander exploited well the numerous chances for success and achieved a very good success. Very good firing technique." Wattenberg had atoned for his first war patrol.

The second wave of Neuland boats had enjoyed a "merry month of May," sinking 78 per cent of the total bag of 109 Allied ships destroyed that month. The Caribbean boats alone destroyed more tonnage than was coming down US slipways. The simple mathematics of Dönitz's "tonnage war" was proving to be on target and the admiral looked toward summer 1942 with high expectations. With a few notable exceptions, neither Allied surface patrols nor Allied aircraft had caused German commanders much concern. Despite the introduction of convoys, merchant ships – and especially tankers – continued to move as single units rather than in the convoys. The deployment of the U-boat tankers, the so-called "milk cows," promised even greater operational time on station. Dönitz was determined to dispatch a third wave of New Land boats to the Caribbean at the earliest possible moment. This time, he would send them out in groups of three or four to maintain steady pressure in that theater of the war.

* * *

As the Lorient boats in staggered formations began to head out once more for the Golden West, Dönitz used the transit time to reassess Operation Neuland with Adolf Hitler and Grand Admiral Erich Raeder. He remained ever the optimist. On May 14, 1942, he reassured Hitler that the "race between enemy new construction and U-boat sinkings" was "in no way hopeless."[11] The U-boat war was a simple matter of "combat against enemy merchant tonnage." He regarded Britain and the United States "as *one!*" The trick was to deploy the U-boats where they could do the most damage with the lowest risk. They were sending some 700,000 tons of enemy shipping to the bottom of the seas every month – more than enough to stay ahead of US shipbuilding. Thanks to the U-boat tankers, the Type IXC boats could operate in the Caribbean and the Gulf of Mexico for four to five weeks. Finally, Dönitz informed his Führer that US antisubmarine warfare remained woefully incompetent. "American flyers cannot see anything; destroyers and their air escorts operate at too high speeds ... and they are not sufficiently tough in pursuing depth charge attacks after detecting U-boats." In short, there was every reason to expect summer 1942 to bring another rich harvest in the Caribbean.[12]

As soon as the third wave of New Land boats approached the Caribbean, Dönitz in a "secret assessment" reminded his commanders that Allied ASW efforts were largely ineffective. "Clumsy, questionable conduct. Overall impression: security forces lack training and toughness; U-boats superior. Little ASDIC." While there was moderate air cover in the Windward Islands and the Mona Passage, and "somewhat stronger" air cover off Curaçao and Trinidad, Allied surface ASW forces remained "inconsequential."[13]

* * *

Albrecht Achilles' *U-161* had barely spent three weeks at Lorient for repairs, refitting, and resupply before it, along with *U-126* and *U-128*, was ordered out at the end of April 1942. Destination: Fernando de Noronha Island, off the eastern "bulge" of Brazil. Several old friends from the "first wave" (*U-156*, *U-162*, and *U-502*) would soon follow.

On May 11, off the Cape Verde Islands, Achilles ran across a large, well-protected convoy. The 12 ships of SL-109 were arranged in three columns and screened by what the Kaleu took to be a dozen escorts. He realized that the enemy had learned some hard lessons. For three days and three nights, *U-161* was repeatedly forced to crash dive by energetic depth-charge attacks by the escorts. Several times, the boat shook violently as the bombs exploded close by. China and glass gauges shattered. Fresh food rolled on the floor. Lights blew out. Nerves were on edge. At night, a barrage of star shell illuminated the ocean and betrayed the U-boat's whereabouts; during daytime, the hostiles had an uncanny ability to pinpoint its position. Achilles informed U-Boat Command: "Enemy operates very effectively with listening device."[14] In fact, it was the new ship-borne "High-Frequency Direction-Finding" (HF/DF or "Huff-Duff") that was causing *U-161* such grief.

Dönitz ordered the boats to abandon their chase of SL-109 and to shape a course for their designated operations area off Brazil. But Hitler was uneasy about declaring Brazil a war zone and hence, on May 15, Dönitz issued new orders. While the boats were still to proceed to Fernando de Noronha Island, thereafter they were immediately to work their way northwest, up the coast of Brazil toward British Guiana and the Lesser Antilles. There, they were to disperse. Unsurprisingly, "the ferret of Port of Spain" headed back to his old hunting grounds off Trinidad.

The site of his former triumphs had been radically transformed: almost every day during the first week of June, Achilles ruefully entered terse, telling comments into the war diary: "Alarm! Aircraft," "Alarm! Flying boat!" Every time *U-161* surfaced, it was met by an aircraft, sometimes by two or more. They came out of the sun or through heavy rain clouds. They came by day and they came by night. They were bombers or flying boats. They delivered a constant hail of aerial bombs, star shell, machine-gun, and rocket fire.

Shortly after surfacing around midnight on June 9, Able Seaman Otto Tietz had the bridge watch. "Alarm! An aircraft mounting lanterns is approaching the boat over the stern." The rest of the watch split their sides laughing. "Dear Able Seaman," one of the lookouts chided Tietz, "have you ever heard of an aircraft having identified a U-boat by night on the surface?" In seconds, the drone of the plane's engines could be heard. "Alarm! Emergency Dive!" Three bursts of star shell illuminated the dark waters around the boat. Then, three aerial bombs detonated in the water, the last one "severely rattling" the slender craft as it began to slip beneath the waves.

"Damage report. At once!" Chief Engineer Heinrich Klaassens was as efficient as ever. "Damaged: depth-regulator motor, gyrostabilizer, magnetic compass, depth-pressure gauge, all water-gauge glasses." For the rest of the night, U-161 stood out to sea while Klaassens and his technical crew undertook what repairs they could. U-161 had just survived its first encounter with the Leigh Light, the suspected "lanterns" swinging from under the plane's fuselage.[15]

After endless days of tropical rain showers and almost daily attacks from the air, Achilles surfaced off the Dragon's Mouth, the northern exit of the Gulf of Paria. The heat and humidity in the boat were unbearable. The men desperately needed fresh air and a few hours of rest. They got neither. As soon as U-161 broke the surface around 5 p.m. on June 13, the watch spied a tanker, then a freighter. For most of the night, Achilles shadowed the freighter. In fact, he had discovered a large convoy on an easterly course.

"Ajax" awaited the convoy's arrival at periscope depth just off the Los Testigos Islands. At 3:34 p.m., the merchantmen and their escorts hove in sight.

"4 columns of 4 freighters each; one destroyer each on the port and the starboard side." He took the boat down to position it between the two inner columns, "where the larger freighters are." The escort off the starboard side was already launching depth charges. How had it ascertained the U-boat's position? Time: 6:12 p.m. Range: 1,250 meters. He fired two bow shots. The usual slight jolt and the G7a torpedoes were on their way to the targets. After one minute and 20 seconds, Achilles heard a muffled detonation. Through the attack periscope he could see two columns of water rising from the side of a large freighter.

"Alarm! Alarm!" The hydrophone operator was screaming that, apart from the high-pitched whine of the torpedoes, he could hear the dull thrashing of a freighter's propeller blades. In his excitement, Achilles had brought the boat too close to the foremost ship on the starboard middle column. "Hard a-port! Both engines full ahead! Down periscope!" It was too late. With a sickening feeling in his stomach, Achilles witnessed the horrendous screech and rending of metal as one of the freighters scraped along the entire starboard side of the boat. *U-161* heeled hard to port. The steamer ground away part of the conning tower and the "shark's teeth" net cutter on the bow. As *U-161* righted itself, water trickled into the boat: the hostile had also sheared off several of the radio dipoles and jumping wire on the conning tower. But the pressure hull remained intact. Achilles at once took the boat down to 150 meters. The men in *U-161* could hear the distant thunder of depth charges exploding.

The Old Man apprised the crew of what had taken place over the intercom. "We just barely managed to escape that one. At the last moment, with rudder hard to port and full power on both engines, we managed to avoid a fat freighter and to bring in the periscope.... Perhaps they saw the periscope when I ran it out for the attack."[16] He was always straight with his men.

After an hour, *U-161* came up to periscope depth. The convoy had broken up, ships heading in all directions. The two destroyers were frantically working the strays like sheepdogs, trying to regroup the lost vessels. "Ajax" could make out only eight remaining targets. Since it was still light, he decided to stay down. At 10 p.m., he brought *U-161* to the surface to survey the scene and to inspect the damage to the boat. He was barely able to squeeze through the damaged bridge tower hatch. An awful sight greeted him. The collision had bent the starboard shielding of the tower inward. On the outer tower, some radio dipoles and jumping wire were gone, as was the starboard antenna. The spray deflector had been ripped from the bridge. Both the bow and the stern net cutter on the starboard side had been cleanly sheared off. But neither of the periscopes had been damaged.[17] It had been a close call. Too damned close.

The convoy was nowhere to be seen. The detonation of ammunition aboard *U-161*'s target had caused chaos within the convoy, and its captains had taken evasive action in every conceivable direction to avoid plowing into the burning wreck. Achilles had alerted Jürgen von Rosenstiel to the whereabouts of the convoy, and it is likely that *U-502* torpedoed the

American 8,001-ton freighter *Scottsburg* out of New York, bound for the Persian Gulf with 10,500 tons of general cargo and war material, including five tanks and seven bombers, as well as the 5,010-ton freighter *Cold Harbor*, en route from New York to Basra, Iraq, with seven aircraft and 28 tanks on board.[18] The unrequited slaughter of Allied shipping continued. So far, not a single Neuland submarine had been lost. But that was about to change.

9

WAR BENEATH
THE SOUTHERN CROSS

On June 1, 1942, the brand new Type IXC *U-157* departed Lorient under the command of Korvettenkapitän Wolf Henne, bound for the waters between the Bahamas and Cuba. Born in 1905 at Fuzhou, China, Henne had joined the navy in 1924 and had served mainly on torpedo-boats until he was seconded to the Submarine Service in March 1941. *U-157* set course for the mouth of the Mississippi River via the Old Bahama Channel and the Florida Straits. It arrived off the north coast of Cuba less than two weeks later. Just before midnight on June 10, Henne spied the lone American 6,400-ton tanker *Hagan* running northeast at about ten knots. It was carrying 22,000 barrels of blackstrap molasses from New Orleans and Antilla to Havana. At 2 a.m. the next morning, the Kaleu fired two torpedoes at *Hagan*; one struck its starboard quarter below the waterline in the engine room; the other the port side fuel bunkers, spraying burning oil over the ship. The tanker sank rapidly by the stern, but 38 crew members managed to scramble into two life boats and make their way to shore.[1]

The Commander of the Gulf Sea Frontier, Rear Admiral James L. Kauffman, ordered "all available forces, both air and surface ... to hunt this submarine to exhaustion and destroy it."[2] A radar-equipped Army Air Forces B-18 out of Key West spotted *U-157* on the surface at first light on June 11. It swooped in for the attack. But the bomber's bay doors were not fully open as it passed low over *U-157*: a startled Henne executed an emergency dive as the B-18 swung around for a second pass at 300 feet. This time it dropped four depth bombs, but all missed their mark. Henne apparently surfaced as soon as the hostile disappeared because *U-157* was spotted about four miles further west very shortly after by a Pan American World Airways passenger plane.

Admiral Kauffman dispatched more B-18s to search the area while a small flotilla of 14 antisubmarine warfare vessels sortied from the ASW schools at Miami and Key West to the waters east of the Florida Straits between Key West and Havana. The force included several old destroyers – *Noa*, *Dahlgren*, and *Greer* – as well as the Coast Guard cutters *Thetis* and *Triton*. The surface vessels failed to make contact during daylight on the 11th, but a B-18 flying from Key West spotted *U-157* on its radar after nightfall the next day. Henne was running doggedly for the Gulf of Mexico under the cover of darkness. Kauffman recalled the Miami ships but ordered the force from Key West to search the area about 90 miles southeast of there.

USS *Thetis* was a 165-foot "B" Class cutter designed and built to enforce prohibition. It and its *Thetis*-class sister ships were created to patrol the offshore waters of the United States to detain smugglers' mother ships that offloaded illegal alcohol to smaller, speedier vessels, which took their cargoes ashore. But as *Thetis* patrolled the waters between Florida and Cuba on June 13, 1942, booze was the last thing on the mind of its commander officer, Lieutenant Nelson C. McCormick. His academic career was far from distinguished – he had graduated near the bottom of his Coast Guard Academy class in 1935 and, as a consequence, only received a temporary commission until late 1937. But by 1942, he had served on three different cutters and in command of *Dione* had gained a good deal of experience hunting submarines off Cape Hatteras, North Carolina.

Thetis began its search for *U-157* at about 2:00 p.m. on June 13. Two hours later, the cutter's soundman picked up a solid contact. McCormick ordered an immediate depth-charge attack, heeling *Thetis* about at 14 knots and charging straight over the target. He dropped seven depth charges from the stern racks at five-second intervals and two more off both sides by the cutter's "Y" guns. They were set for 200 and 300 feet, and it was not long before debris and fuel oil roiled to the surface. The crew fished a couple of pairs of leather pants, some wood planking and an empty grease tube with "Made in Düsseldorf" from the water. Five other ships made runs on the target, but *Thetis* was given credit for the kill. *U-157* was the first German casualty of the Caribbean campaign. The second was to be *U-158*.

Like *U-157*, *U-158* was on its second war patrol, having departed Lorient for the Gulf of Mexico on May 4 under the command of Kapitän-leutnant Erwin Rostin. He had joined the navy in 1933 and commanded the minesweepers *M-98* and *M-21* before joining the Submarine Service in March 1941. On his first war patrol off the US east coast, Rostin had sunk or damaged seven ships of 54,049 tons. He began his second war patrol, en route to the Gulf of Mexico, on May 20, by torpedoing the British 8,113-ton tanker *Darina*. His next sinking came two days later – the Canadian 1,748-ton cargo ship *Frank B. Bair*. But Rostin's greatest successes were scored after U-Boat Command ordered him to pass the Florida Straits and to patrol the coast of Mexico from the Yucatán Channel to Tampico, the country's primary oil port. In what was to become the most successful war patrol by any U-boat in the Americas, between May 20 and June 29, Rostin sank twelve ships of 62,500 tons.[3] On June 28, Admiral Karl Dönitz radioed the Kaleu the news that he had been awarded the Knight's Cross. Rostin then headed home, again via the Florida Straits.[4]

Rostin was a skilled U-boat commander who had chalked up a total of 19 ships of 116,500 tons sunk or damaged on his two war patrols. But, like his fellow skippers, he had a bad habit (as ordered) of constantly reporting to Dönitz in Paris. Even though the Allies could not read his signals, they could follow his call signs and triangulate his positions with High-Frequency Direction-Finding. As a result, Rostin was tracked almost daily as he dodged American ASW patrols in the Florida Straits and headed for Bermuda on his way back to Lorient. On June 29, he stopped the Latvian 3,950-ton freighter *Everalda* and forced its crew to scuttle it under threat of shelling from his 10.5-cm deck gun. This, too, he reported to Paris. And this signal was also picked up and an estimate made of his speed and course.[5]

At 3:45 p.m. on June 30, a new US Navy Martin Mariner from Patrol Squadron 74 (VP-74) flown by Lieutenant Richard E. Schreder, a naval reservist based at Hamilton, Bermuda, searched for *U-158* in the waters near that island. The Mariner was a large twin-engine flying boat, which had first entered US Navy service in the fall of 1940. With a range in excess of 3,500 miles, radar-equipped, and a large carrying capacity for

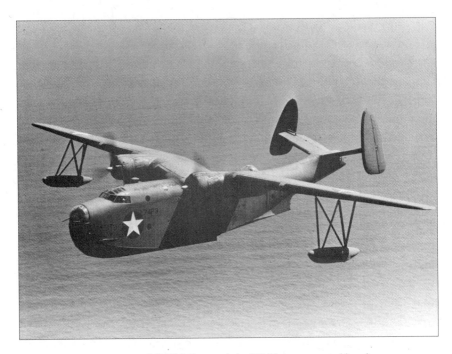

The Consolidated "Mariner" (PBM-3) served the US Navy as a patrol bomber or transport. The crew of nine men was armed with 50-calibre machine guns and bombs. Source: Library of Congress, Prints & Photographs Division, FSA/OWI Collection, [reproduction number, LC-DIG-fsa-8b08013 (digital file from original neg.)].

ASW weapons, it was an ideal sub hunter. While flying above scattered cumulous clouds at 3,500 feet, Schreder's radar man detected a target three miles off the starboard beam.[6] It was carelessly running on the surface. Schreder, flying at 175 knots, swung the Mariner to the right while descending rapidly through the clouds before steadying on a course right up the U-boat's stern, 75 to 100 feet above the water. It took a minute-and-a-half before the large flying boat was on *U-158*'s wake.

The Mariner's bow machine guns opened fire and jammed almost immediately. Schreder was now so close to the sub that he could see "about fifteen men" lounging on the bridge tower and looking up at his aircraft. Two demolition bombs and two Mark XVII depth charges were ready in the plane's open bomb bay, but the demolition bombs failed to release. Of the two Mark XVIIs dropped, one exploded under the submarine's stern and the other crashed through the wooden deck slats about 15 feet aft of the tower and lodged there. At about the same moment as the first charge exploded under the stern, *U-158* began a crash dive. The Mariner flew over the boat and as Schreder set up to make another attack, a large explosion rent the sea and a very large dark oil slick began to spread on the surface. Schreder later concluded that the second depth charge, wedged against the U-boat's hull, must have blown up when the submarine descended to the depth at which the charge was set to explode. It was the end of Rostin and *U-158*.

* * *

The cruel nature of the U-boat war for Allied merchantmen was an individual affair – survival or death – from master down to seaman. When Albrecht Achilles' *U-161* torpedoed the 8,000-ton steamer *Scottsburg* in the engine room and the number one hold off Grenada around dusk on June 14, Merchant Seaman Archie C. Gibbs joined 43 of his mates in two lifeboats.[7] The next morning they were spotted by two patrol planes, as was a lifeboat with survivors from the 5,000-ton steamer *Cold Harbor*, which Jürgen von Rosenstiel had sunk northwest of Trinidad that same day. By afternoon, the 6,062-ton Matson Line freighter *Kahuku* had rescued the 112 men in the three lifeboats. Since *Kahuku* had only two lifeboats, Master Eric Johanson decided to tow a third boat behind his ship – just in case. He shaped a course for Trinidad.

Kahuku never made it. Some time after 9 p.m., about 90 miles west of Grenada, Ernst Bauer's *U-126* dispatched Johanson's vessel with a single torpedo. The men on board panicked and lowered the two lifeboats while *Kahuku* was still moving; 17 drowned as a result of their rash action. Gibbs made for the ship's stern and slid down a rope to his old lifeboat. He missed the raft as it glided by in the darkness. He retrieved a flashlight from a small bag around his neck and frantically began to signal for help. It came quickly – in the form of *U-126*. Fished out of the water by Bauer's crew, Gibbs was taken to the conning tower and for an hour watched helplessly as the Kaleu fired two more torpedoes into the *Kahuku*, then finished it off with 30 rounds from the deck gun. Later, Gibbs was fed what he called a "poor" diet of canned stew, vegetables, apricots, and peaches.

On the morning of June 18, Kapitänleutnant Bauer brought his captive up on deck. A small motor skiff, the Venezuelan *Minataora*, stood off to the side. The skipper ordered Gibbs to swim for it. He was hauled out of the water. By way of pigdin English and hand signs, he was informed that the vessel was carrying food and six cases of rum – as well as two Venezuelan prostitutes en route from LaGuaira. Some 40 miles off Willemstad, an unidentified U-boat fired five rounds from a machine gun at the *Minataora* and hailed it to halt. The U-boat's captain ordered yet another Allied sailor to swim toward his rescue. Later that day, Gibbs and his companion were safely landed in the Curaçao capital.

Saving survivors almost never occurred in the cold seas of the North Atlantic. There, bad weather, rough seas, darkness, the need for the U-boats to keep moving under the growing threat of Allied escorts and ASW aircraft – not to mention official policy which forbade rescuing survivors – all militated against such efforts. But the warm waters of the Caribbean and the more leisurely pace of life on board the U-boats seemed to create a more permissive attitude toward the rescue of one or two individuals. Besides, the U-boat commanders liked to surface alongside boatloads of survivors, question the crew as to the name, course, and cargo of their ship, and even pass over food and water to ease their passage somewhat. These very real events, combined with the ever-present rumors of U-boats re-provisioning at Vichy French Martinique, led to bizarre tales of what German submariners did when they were not sinking merchantmen.

Uninhabited islands off the Yucatán Peninsula, in the Bahamas, and in the Virgin Islands were havens for submarine crews who could air out their boats, swim in the warm Gulf Stream waters and catch fresh fish. One prominent historian of the Caribbean campaign writes: "Nearly all the boats in those golden days [the early period of the campaign] took an occasional day off to rest and recuperate." He tells of boats stopping at small obscure ports to buy provisions and even "female company." The most intriguing tale – never substantiated – tells of a U-boat commander stopping a small inter-island steamer and, after warning the captain not to continue on his journey, producing ticket stubs to a movie then being shown in Port of Spain's leading theater and highly recommending the movie.[8]

After he had torpedoed the *Scottsburg* at dusk on June 14, Achilles decided to head west to scour the eastern terminus of the Panama Canal for prey. En route, he came across the small Dominican sailing vessel *Nueva Alta-gracia*. It was a wonderful bag, full of fresh fruit and chickens. "Ajax" took its crew of eight on board and dispatched the 30-tonner with a single shot from the deck gun. Once underway again, *U-161*'s cook, Helmut Baier, decided to relieve the monotony of the war patrol. As the eight Dominic-ans huddled on the foredeck, Baier suddenly appeared from the forward hatch – scruffy beard, flowing blond wig, bare upper torso smeared with cooking oil, and a long butcher knife in his hand. Menacingly, he strode up and down the line of captives before selecting a frightened young cabin boy – to pluck the dead chickens still lying on the deck.[9] After a delightful dinner that night, Achilles stopped another Dominican sailboat, *Ciudad Trujillo*, "liberated" more fresh fruit, threw its cargo of corn overboard, and placed his eight captives aboard for transport to Curaçao. On June 20, *U-161* effected a prearranged rendezvous with Helmut Witte's *U-159*. Lying under the protection of the Venezuelan shore, the two boats tied up alongside one another. Achilles took on 20 cbm of fuel oil and rations for seven weeks.

Day after day, tropical rain showers pelted the boat and made life below decks a humid, moldy hell. By June 22, *U-161* stood off Colón, Panama. Achilles was quickly reminded yet again of how the balance in the U-war was shifting. The port was defended by a small armada of de-stroyers. Tethered observation balloons kept a sharp lookout for intrud-ers while flying boats and wheeled aircraft constantly buzzed overhead. *U-161* was forced to remain submerged for most of the rest of June. On the few occasions when it dared surface, tropical showers and the heavy spray produced by the strong trade winds reduced visibility to a few meters.

After being driven below the surface four times by aircraft on June 30, Achilles shaped a course for Puerto Limón, Costa Rica. From offshore, he spied smoke rising from the port's piers. It was almost too good to be true: a perfect repeat of his earlier attacks on Port of Spain and Castries. In broad daylight, just before 2 p.m. on July 2, he took the submerged *U-161* into the harbor, carefully weaving his way around rock formations and sandbars. Tied up at the pier was a single steamer of the United Fruit Company. Achilles retraced his steps out to sea, determined to deal with the vessel later that night.[10]

At 10 p.m., *U-161* was back inside Puerto Limón harbor. The piers were well lit and the city only partially darkened. Range: 1,650 meters. Achilles fired two "eels" from the stern tubes. Both ran true. "After 105 seconds, first hit under the bridge at 1,650 meters. High fire column, steamer goes down by the bow." The second torpedo hit amidships. "Strong, bright detonation." Then the freighter disappeared from view as the lights in the port were doused. Still, Achilles had been able to make a positive identification: "Steamer was American 'San Pablo,' 3,305 tons." The torpedoes had struck as the freighter was unloading its cargo; one crew member and 23 stevedores died in its hold. As confusion and fear gripped the port, "Ajax" once again made good his escape.

The sinking of the *San Pablo* greatly alarmed US authorities. If a single U-boat could penetrate Puerto Limón with impunity, then Colón, just 100 miles away guarding the Panama Canal, surely was equally vulnerable. Within days, work began on a massive minefield across the mouth of the port; coastal defense guns and infantry arrived; US Navy vessels were routinely stationed at Limón; and construction had begun on an airfield. Yet again, with a single dashing entry into a harbor, Achilles had set in motion a gigantic Allied effort to defend Caribbean ports against the feared "ferret."

By now down to 45 cbm of fuel oil, Achilles decided to return to Lorient, more than 4,000 nautical miles to the east. He drove *U-161* past Jamaica and left the Caribbean via the Windward Passage, off the coast of Haiti. To keep the crew alert, shortly after 9 a.m. on July 16, he ordered a practice dive some 500 miles north of the Virgin Islands. "Alarm! Convoy at 300 degrees!" The bridge watch had spotted the fast convoy AS-4 headed for Sierra Leone. Its freighters were arranged in four columns and seemed to be escorted by only two Gleaves-class destroyers. Achilles decided at once to position *U-161* inside the convoy, between the escorts and the biggest steamers. The sea was smooth as glass. But this time "Ajax" had put his hand in the proverbial wasps' nest. AS-4 was a special fast military convoy out of New York on personal orders from President Franklin D. Roosevelt to resupply British Eighth Army after it had been driven out of

Tobruk and back to Egypt by Erwin Rommel's Afrika Korps. Hence, it was especially well escorted by two cruisers and seven destroyers.

U-161's hydrophone operator picked up the deadly ASDIC "pings" at 3,000 meters.[11] Undaunted, Achilles maneuvered the boat to loose a full bow spread at the nearest large vessels. He was about to fire, when at 9:42 a.m. the entire convoy sharply wheeled ten degrees to starboard. There was not a moment to lose. Four torpedoes raced out of their tubes. After 215 seconds, the Kaleu heard what he called a "short, weak, cracking detonation" at 3,300 meters. No visible confirmation. Thirty-four seconds later, he saw two detonations on what he took to be a 9,000-ton freighter. No confirmation of a hit from the fourth shot. "Emergency Dive!" Achilles took the boat down to 160 meters. He heard another explosion, perhaps the victim's engines, bulkheads, or cargo. Then depth charges went off all around *U-161* "in no discernible pattern." The blasts from the detonations ruptured the outer door of Tube I and the balance valve in Tube II; sea water seeped into the boat through both tubes; the bilge-pump filter in the control room began to leak. Nothing serious, Klaassens assured him.

After an hour, Achilles brought *U-161* up to periscope depth. A destroyer was off in the distance at 5,000 meters; another closer still at 2,000 meters. Cool as ever, "Ajax" had the technical crew replace the bilge-pump filter. Suddenly, the hydrophone operator reported rapidly approaching fast propeller noises. The Old Man took *U-161* down to 120 meters. Just in time: four depth charges rocked the boat severely. For nine hours, the destroyers USS *Kearny* and *Wilkes* savaged the boat. *U-161*'s ever-terse war diary gave some idea of the constant pounding:

> 11:58 a.m. Out of action: all glass depth gauges; depth indicator outside control tower; repeater compasses; internal depth indicator, and lesser outages.
>
> 5:52 p.m. Outage of port [electric] motor due to burnout of thrust bearing caused by increasing water intake in the aft-ship.
>
> 6 p.m. to 9 p.m. Constant detecting noises from 3 destroyers.... Go down to 140 meters.... Irregular depth charges off in the distance.

Shortly before midnight, the hammer blows suddenly stopped. Achilles took *U-161* back to the surface. Though severely mauled, the boat resumed the homeward trek to Lorient.

Achilles' one confirmed "kill" out of Convoy AS-4 was the American 6,200-ton freighter *Fairport*. It was a grievous loss. In the rush to resupply the British in North Africa, the Americans had loaded 300 brand new Sherman tanks into fast cargo ships – without their engines. These were later loaded into a single ship, the *Fairport*. All 300 engines went down with it. The *Kearney* had broken off its pursuit of *U-161* to rescue 133 seamen and US Army personnel.

At 4 p.m. on July 22, *U-161*'s bridge watch joyfully sighted Lieutenant Wolf-Harro Stiebler's "milk cow" *U-461*, appropriately adorned with a nursing she-wolf on its tower. *U-161* took on board 40 cbm fuel oil, 1 cbm lubricating oil, and provisions for eight days. Most welcome were the fresh bread, fresh meat, and fresh lemons. The doctor on board *U-461* tended to *U-161's* numerous cases of boils, heat sores, stomach disorders, and diarrhea. In heavy seas, *U-161* then pointed for Lorient and at 3:37 p.m. on August 6, it docked beside the hulk *Isére*.

Achilles received yet another hero's welcome. He had completed a record war patrol lasting 102 days and had destroyed three freighters of 17,500 tons. Admiral Dönitz in his evaluation took due note of the difficulty of the patrol: "Very long patrol by the boat, which, without blame to the commander, brought only minimal success despite determined, proper procedures."[12] *U-161* would require five weeks of repairs at the Kéroman bunkers.

In a special report to Dönitz, Achilles took stock of the effect of such long war patrols in tropical waters. Overall, morale had remained high – despite the constant attacks by aircraft. But fatigue had set in by the tail end of the patrol. Especially the technical crew manning the electric motors when the boat was submerged often worked for hours in 39-degree Celsius heat. The men suffered from skin sores, boils, digestive disorders, and exhaustion. The canned bread spoiled far too quickly and malnutrition added to the sailors' woes.[13]

The three boats of the special Brazil group had torpedoed 96,000 tons of Allied shipping. Overall, the 13 boats operating in the Caribbean from April to July 1942 had shattered all existing records: 95 confirmed "kills"

The German Navy's grid chart (Quadratkarte) of the Caribbean basin, showing the quadrants to which the boats were directed by U-Boat Command in France. Source: Federal Military Archive, Freiburg, Germany.

for a total of 483,000 tons. A single boat, Rosenstiel's *U-502*, was lost in the Bay of Biscay on its way home. On his wall charts at U-Boat Command in Paris, Dönitz cheerfully noted the record "kill ratio" of 95 to 1. He could only surmise that *U-157* and *U-158* would not be coming home.

On May 16, 1942, all U-boats received instructions from Dönitz to attack without warning all armed merchant ships belonging to Central and South American states that had broken off diplomatic relations with Germany, "that means, all states with the exception of Argentina and Chile."[14] The order drove Brazil closer to a declaration of war and to an active role in fighting the "gray sharks" in the South Atlantic. Rio de Janeiro had watched anxiously in the late 1930s as Adolf Hitler brought Europe closer to war. In November 1938, the Brazilian ambassador to the United States, Mario de Pimentel Brandáo, advised Foreign Minister Oswaldo Aranha that Brazil would inevitably have to decide between Germany and the United States. Cyro de Freitas Valle, Rio's ambassador to Germany, echoed these concerns when he reported that Hitler's plans for the future included global spheres of influence based on a German-dominated Europe, a US-dominated western hemisphere, and a Japanese-dominated East Asia. Under such circumstances, he asked, would it not be better for Brazil to ensure that it was tightly within the American orbit?[15]

This notwithstanding, President Getúlio Vargas played it cagey with the United States; his primary interest was, quite naturally, to wring as many American economic and trade concessions as possible while at the same time building up Brazil's defenses. He may have underestimated the fear growing in Washington that Axis influence in Brazil was rising to unacceptable levels; in late May 1940, Roosevelt approved Operation Pot of Gold, whereby 100,000 American troops would seize key spots along Brazil's coast from Belém to Rio de Janeiro. Fear of such action and important US concessions on building a major steel mill in Brazil changed matters very quickly. At the end of September 1940, the Brazilian government decided that in the event of a German attack, all of the nation's resources would be placed at the disposal of the United States. Shortly after, the Vargas government granted Pan American World Airways the right to fly directly from Belém to Rio, thus cutting three days off the trip, and quietly allowed the United States to begin setting up bases in Brazil's northeast, particularly on Fernando de Noronha Island.

The small bases in the northeast were just the start of a major American presence in Brazil. In November 1940, the US Army negotiated a secret agreement with Pan American to set up a chain of airfields from the United States to the Brazilian northeast. The American base at Parnamirim, Rio Grande do Norte, as well as others, especially at Belém, Natal, and Recife, became major jumping-off places for flights ferrying US aircraft to British forces in the Middle East even before either Washington or Rio entered the war. After Pearl Harbor, the US Navy began operating in Brazilian territorial waters with Vargas' permission, even though Brazil was still neutral. At the Rio Conference of January 15 to 28, 1942, Rio supported Washington's efforts to persuade all of the Latin American nations (with the notable exceptions of Argentina and Chile) to cut diplomatic relations with Germany. Thereafter, U-boat attacks on Brazilian shipping mounted, with four vessels lost in February and March. In April, Vargas demanded that the United States provide convoy escorts and arms for Brazilian merchant ships, or he would embargo them; but a month later, he met with Vice Admiral Jonas H. Ingram, Commander, US Navy South Atlantic Force, and took the extraordinary step of secretly opening Brazil's ports, repair facilities, and military airfields to the US Navy. And ordering the Brazilian military to cooperate fully with Ingram's command. In May, Brazil lost another four freighters to the U-boats.

Hitler, outraged over what he considered to be Brazil's blatantly pro-American basing policy, demanded on June 15, 1942, that the Kriegsmarine launch a sudden strike against Brazilian ports by ten submarines.[16] Dönitz reluctantly agreed; Foreign Minister Joachim von Ribbentrop did not. He vetoed the scheme for fear that it would alienate the pro-German governments in Buenos Aires and Santiago de Chile. The operation was canceled on June 26.[17]

Still, Dönitz was not displeased when late in July Korvettenkapitän Harro Schacht, who had scored no "kills" off the coast of Freetown, Africa, requested permission to take his *U-507* across the Atlantic to Brazil. He readily approved the request – provided that Schacht was "extremely careful" not to sink Argentinean and Chilean ships and that he not attack Brazilian harbors. A long-time veteran – he had joined the navy in the spring of 1926 – Schacht had served on the light cruisers *Emden* and *Nürnberg* before joining the U-Boat Service in June 1941. The newly commissioned *U-507* was his first command.

Schacht's one-boat foray became a veritable slaughter for Brazilian shipping: beginning on August 16, and operating so close to shore off Arajacu and Sergipe that he could watch the locals playing tennis, he sank five ships for 14,822 tons in just two days.[18] One of his victims that first day, the 4,800-ton *Baependy*, went down with 250 soldiers, seven officers, two artillery batteries, and other equipment; two others (the 4,800-ton *Araraquara* and the 1,900-ton *Annibal Benévolo*) sank with the loss of 131 and 150 passengers and crew, respectively. Schacht's actions caused an uproar in Brazil, with the result that Foreign Minister von Ribbentrop, again fearing negative repercussions in Argentina and Chile, demanded that the U-boats remain 30 miles offshore.[19]

Schacht assured Dönitz that he had operated well outside Brazil's territorial waters and that none of the ships attacked had displayed either national flags or "neutrality signs." Then, on August 19, he brutally shelled the 90-ton yacht *Jacyra*. It was the proverbial last straw: three days later President Varga's government declared war on Germany. Just for good measure, Schacht that same day sank the Swedish 3,220-ton freighter *Hammaren* with five torpedoes and eight rounds from the 10.5-cm deck gun.

In September 1942, Vargas gave Admiral Ingram full authority over Brazilian air and naval forces and complete responsibility for the defense of the long Brazilian coastline. Equipped with new US-built sub-chasers and a growing fleet of eight former American destroyer escorts and Brazilian-built, British-designed Marcilio Diaz–class destroyers (adding to its existing small and mostly obsolescent fleet), Brazil's navy helped immensely to throttle German blockade runners and U-boats crossing the Atlantic narrows between French West Africa and the Brazilian "bulge." In the words of the official history of the Royal Navy in World War II: "The Allied shipping control organisation could now be extended almost to the great focal area off the River Plate … but an even greater advantage was the stronger strategic control of the whole South Atlantic gained from the use of Brazilian bases."[20] American flying boats heading out to sea became a daily sight over Rio and São Paulo. As one expert on Brazil's role in World War II wrote: "It is incorrect to say that unwarranted German aggression compelled Brazil to become a belligerent. Vargas's policies were unfolding to their logical conclusion." However, the U-boat attacks in August 1942 "stimulated public support for mobilization and for un-reserved alignment with the Allies."[21]

* * *

In late summer 1942, Commander U-Boats reassessed the operational effectiveness of the three waves of boats that he had sent out to the Caribbean. The balance sheet was not entirely positive. Vastly enhanced Allied ASW – through surface escorts with "Huff Duff" and centimetric radar, but especially through bombers equipped with new air-to-surface-vessel (ASV Mk.II) radar with a range of between one and 36 miles as well as 22-million candlepower Leigh Lights installed under their fuselages – was taking a heavy toll on the U-boats. Dönitz and his staff faced a cruel dilemma: the large Type IX boats could operate in the Caribbean for up to four weeks thanks to the "milk cows," but they were unwieldy and slow in evading hostile air attacks; the smaller Type VII boats were much more maneuverable, but they lacked the necessary sea legs for Caribbean operations. Moreover, Hitler, fearing an imminent Allied invasion of North Africa, ordered Dönitz to transfer six Type VII submarines from the Atlantic to the Mediterranean, where four of the boats were soon lost.

On September 28, 1942, Grand Admiral Erich Raeder, Commander in Chief, Navy, and Admiral Karl Dönitz, Commander U-Boats, met with Hitler at the Reich Chancery in Berlin for two hours to discuss the overall situation. Raeder, in fact, had not wanted the meeting and as late as September 13 had refused to invite Dönitz to meet with Hitler. As always, Hitler got his way. The Führer began by heaping "great praise" on the "Volunteer Corps Dönitz." It, of all the service branches, seemed to appreciate fully his notion of *Kampf*, of struggle first and foremost. The "moral impact" of sinking merchant tonnage alone, Hitler allowed, greatly affected the Allied war effort. American new construction, despite the most blatant Rooseveltian "propaganda," could never balance losses at sea. And the mere production of "hulls" could not overcome the dearth of men and machines to sail them.[22] He suggested two enhancements of the U-boat campaign: "new technical developments" to allay the Allies' ASW campaign needed to be rushed to the front as quickly as possible so that they could be "practically deployed"; and something needed to be done about the "large number of sunken ships' crews" that somehow managed to survive torpedoing and to "return to the sea on newly-built ships."

Dönitz replied in his usual cold, rational way. The cost-effectiveness of U-boat warfare along the eastern seaboard of the United States had been vastly reduced by enhanced ASW efforts; thus, he was moving the war patrols back to the mid-Atlantic and to the South Atlantic, off the coast of Africa. He maintained that morale among his skippers remained high; that the enemy would not be able to offset tonnage torpedoed with tonnage built; and that greater deployment of airpower in the form of long-range Heinkel-177 bombers as well as U-boats with faster speed while submerged could help Germany regain the initiative in the Battle of the Atlantic.

Left unspoken was that the U-Boat Service still enjoyed superiority in the area of codebreaking.[23] In February 1942, Dönitz's engineers had introduced a fourth rotor (*alpha*) to the Enigma machine. Called "Triton," the new M4 cipher raised the number of possible variations, or "cribs," from 16,900 to 44,000. It blinded British codebreakers at Bletchley Park for most of 1942. Conversely, Naval Intelligence (*B-Dienst*) was now reading the Allied Cypher Number 3 with sufficient speed to allow U-Boat Command to use the information tactically against Allied convoys.

Concerning Hitler's suggestion to rush "new technical developments" to the front, Dönitz had several innovations ready. His commanders had reported in detail on the Allies' deadly combination of ASV radar and the Leigh Light. The answer seemed to lie in the area of radar detection. The French firms Metox-Grandin and Sadir had developed a VHF-heterodyne receiver that, in combination with a primitive wooden aerial (Biscay Cross), could pick up incoming radar beams and alert a U-boat's crew by a loud "pinging" in the receiver. The Metox receivers were installed in U-boats beginning in August 1942.

As well, German boats were outfitted with *Bold* sonar devices. Short for *Kobold* (goblin), these 15-cm diameter capsules filled with a calcium and zinc compound could be released from the stern of a submarine by way of a special ejector. They were designed to maintain neutral buoyancy at 30 meters depth and, upon contact with seawater, to produce hydrogen gas that in turn created bubbles. To Allied ASDIC operators, these bubbles would resemble the "echo" produced by true submarine contacts.

Interestingly, when Hitler had demanded production of such a device on September 28, Dönitz had been cool to the idea as it "might mean the regrettable loss of a torpedo tube."[24]

Finally, the "gray sharks" received anti-radar decoys designed to create more false radar echoes. Codenamed *Aphrodite*, the system consisted of eight-meter-long aluminum foil strips or dipoles attached to a 120-meter-long wire that connected a large hydrogen-filled balloon to a sheet anchor. Crews were instructed on how to inflate the balloon on deck and then simply to toss anchor, wire, aluminum strips, and balloon overboard. *Aphrodite*, approved by Hitler in summer 1942, was operational by September. The Battle of the Atlantic was becoming a chess game between white-smocked engineers while the growing number of sunken tankers and bauxite carriers continued to foul the azure waters of the Caribbean.

10

THE ALLIES REGROUP

In the U-boat war in the Caribbean, the German force was united and operated at the direction of one commander, Admiral Karl Dönitz. He daily sized up the course of the campaign, evaluated Allied successes and failures, estimated Allied strengths and weaknesses, and deployed his forces accordingly. By contrast, the Allied defense was about as disunited as it could be. In the United States, army-navy rivalry was as old as the Republic and had contributed, at least in part, to the debacle at Pearl Harbor. The United States and its Caribbean Allies – the United Kingdom and the Netherlands government-in-exile – agreed on the overall aim of the war and the Caribbean campaign, but disagreed on details. Central and South American nations drawn into the fray controlled all of the western and southern shore of the Caribbean; they had a long and troublesome history with their powerful northern neighbor. Unity of command would not be easy to achieve, but victory would not be possible without the help and cooperation of all the anti-Axis forces in the theater.

Unity of command among US forces was the first obstacle to be overcome. Secretary of War Henry Stimson discovered on December 12, 1941, that no scheme for establishing unity of command existed for Panama, which was then considered the most likely target for attack by Japan. He asked Army Chief of Staff General George C. Marshall to draw up a directive placing all US forces in the Panama Coastal Frontier, except for fleet units, under army command. When Stimson laid the plan before the Cabinet later in the day, President Franklin D. Roosevelt approved the idea by the simple expedient of taking a map and scrawling "army" over the Panama Canal Zone and "navy" over the Caribbean Coastal Frontier, and then adding "O.K. – F.D.R." When General Leonard T. Gerow later took the papers to a meeting of high-ranking naval officers including Admiral Ernest J. King, Commander in Chief Atlantic Fleet, only one man – Admiral Richmond K. Turner – opposed the unity scheme. He was overridden by the argument that if the army and the navy did not work together effectively, the president might establish a new Department of National Defense to which both the army and the navy would be subordinate. By December 18, Roosevelt's directive had been instituted. The army assumed command of the defense of Panama, the navy that of the Caribbean; units of each service in either area would be at the disposal of the service commander in that area.[1]

In the weeks following the Pearl Harbor attack, reinforcements were sent to both Panama and the Caribbean as quickly as possible. The Canal Zone came first with two infantry regiments, two barrage balloon units, one field artillery battalion, and 1,800 artillery replacements. Anti-aircraft guns as well as fighter and bomber reinforcements were also dispatched, raising the total garrison to 47,600 combatants by the end of January 1942. The buildup on Puerto Rico was impeded both by the flow of supplies to Panama and because it seemed to lie far beyond any enemy's grasp. Thus, only about 800 men were added to the island in the two months following America's entry into the war. US naval defenses in the Caribbean were wholly inadequate for the job. There, Rear Admiral John H. Hoover's force consisted of two old "four stacker" destroyers, three small "S" class submarines, two World War I vintage sub-chasers and twelve patrol planes. Trinidad, which was to play a key role in the battle, was covered by two converted yachts, two small patrol craft, and four patrol planes.[2]

The overall defense plan for the area, RAINBOW 5, called for armed assistance to "recognized governments" in Latin America and for the "protective occupation" of colonies belonging to allied European powers to alleviate those Allies of the burden of defending their colonies – and also, no doubt, to shore up the defenses of the Panama Canal. Washington shortly entered into discussions with the Dutch government-in-exile concerning the defenses of the Netherlands Antilles. The discussions were somewhat sticky – the Dutch were appalled at the prospect, wholly theoretical, that Venezuelan troops might defend Aruba and Curaçao – but were concluded on January 26, 1942. Eventually, a combined Dutch-American headquarters was established. In the meantime, six A-20 attack bombers flew to the islands in mid-January. A British garrison of 1,400 troops was pulled off and sent back to the United Kingdom; 2,300 American soldiers equipped with 155-mm coastal guns were sent from New Orleans in early February. The U-boats struck before those troops were ready.

Local Combined Defense Committees were set up to coordinate between the military and civil authorities. They consisted of the colonial governor as chair and convener, the senior officers of the local military (American and British in most cases), and other local bureaucrats. Only on Trinidad was there significant trouble, as the governor, Sir Hubert Young, insisted on his authority as local commander, a position the senior American officer refused to accept. Eventually, the Americans replaced the local commander with a man of higher rank; London fired its governor and dispatched a more reasonable man.[3]

* * *

The attack on Aruba brought home to all Caribbean nations that German submarines were suddenly a clear and present danger to the lives, property, and livelihood of any whose existence depended on the commerce of the Caribbean Sea. Even so, the independent Latin American republics of the region had a long and troublesome relationship with the United States and, in some important cases, the mistrust generated thereby did not disappear overnight.

Between 1890 and 1932, US military forces intervened in the Latin American republics 19 times. They occupied Cuba in 1906–8, and then returned twice after that. They intervened in Haiti in 1891 and in 1914 returned for nine years. They operated in Nicaragua five times before virtually occupying that country for 20 years starting in 1914. The United States also used troops to protect its interests in Argentina, Chile, Panama, Honduras, Mexico, the Dominican Republic, Guatemala, Costa Rica, and El Salvador. As well, Washington played a heavy hand in enforcing debt collection, backing major American-owned enterprises such as the United Fruit Company, intervening in elections, protecting allied dictators, or installing friendly regimes. Roosevelt declared a "Good Neighbor" policy toward these republics shortly after he was inaugurated in 1933, and three years later relinquished the "right" of the United States to intervene in Panama – a "right" which the United States had declared in 1903 when its intervention against Colombia had "allowed" Panama to declare independence.

In 1936, Anastasio Somoza García seized power in Nicaragua. On the night of December 7, 1941, the pro-American Somoza threw his country's support on the side of the Allies and placed Nicaragua's entire territory, including sea, air, and land, at the service of the United States until the Axis was defeated. The airports at Managua and Puerto Cabezas were operated by US Army personnel and the US Navy built a sea plane base at Corinto. In 1943, a medium-range radio loop station was opened at Managua airport and the runway was extended to facilitate long-range bombers flying antisubmarine sweeps between Managua and Guatemala City. Nicaraguan cooperation, combined with the considerable US air and naval contingents in the Panama Canal Zone, allowed the Americans to establish a formidable air presence over the southwestern stretches of the Colombian Basin and the northern approaches to the Panama Canal.

Honduras began to mount its own antisubmarine sweeps on the Caribbean coast in July 1942 but did not notify the Americans. It also authorized the use of Puerto Castillo as an advanced naval air base for sea planes near the Strait of Yucatán.

Mexico had cooperated with the United States in counterespionage and intelligence since before Pearl Harbor. President Manuel Ávila Camacho, elected in 1940, took an increasingly pro-American position as Axis military triumphs mounted. On April 13, 1941, Mexico and the United States signed an agreement calling for reciprocal use of air bases for mutual defense, and nine days later Mexico closed all German consulates and expelled Berlin's diplomatic corps in response to its perceived interference in Mexican politics.[4] The Mexican government had little to offer in air and sea forces at the start of the Caribbean campaign, but its army and air force gladly accepted American weapons and training. Mexico granted permission to the United States to build an airfield on the island of Cozumel from which sweeps were conducted across the Yucatán Channel. On May 25, 1942, Mexico declared war on the Axis in the wake of the sinking of the Mexican-flagged tankers *Potrero del Llano* (by *U-564*) and *Faja de Oro* (by *U-106*).

Cuban cooperation was vital to guarding the entrances to the Gulf of Mexico – the Yucatán Channel and the Straits of Florida. Both were easily patrolled by air from western Cuba, southern Florida, and Mexico. Cuba, in the words of the official US Army history of World War II, "promised a great deal as a cobelligerent," but its "participation in definite activities was small." The Cubans took little interest in protecting their coastline from submarine threats. After James L. Collins, Commanding General of the Puerto Rican Department, pressed Cuba in late September 1942, Havana promised "the fullest Cuban cooperation." Nonetheless, on a visit to the island a month later, Collins came away convinced that the Cubans were still dragging their heels in patrolling "certain designated deep water shorelines" to deny their use by German submarines.[5] Nevertheless, Havana granted Washington permission to build airbases near San Antonio de los Baños and at Camagüey, both of which allowed US aircraft to patrol the Bahamas Channel as well as the Strait of Yucatán. The US Navy also operated airship bases at San Julian on the western tip of the island and at Caibarién and the Isle of Pines.

Cuba's initial reluctance vanished quickly after signing a naval and military cooperation agreement with the United States on September 9, 1942. The old cruiser *Cuba* was modernized, along with the school ship *Patria*. Cuba's small flotilla of gunboats and patrol craft were also modernized, while 12 patrol craft were transferred from the US Navy to the Cuban Navy. In May 1943, a flotilla of Cuban patrol aircraft in conjunction with a US Navy Kingfisher aircraft sank *U-176*, which had destroyed 11 Allied ships of 53,307 tons. By the end of the war, Cuban air and naval units had escorted close to 500 ships and rescued 221 shipwrecked sailors.[6]

Colombian President Eduardo Santos Montejo in August 1938 had campaigned on a foreign policy that favored improved relations with the United States. His position stemmed partly from his distaste for the racism of the Nazi government and partly because protecting the approaches to the Panama Canal was a fundamental Colombian national interest. Thus, throughout 1939 and 1940, US air and naval missions began to work more closely with Colombian armed forces to concert defensive arrangements for the canal. But Colombia insisted on limiting just what Americans would be allowed to do from its territory or airspace. It extended no invitations to the Americans to establish bases prior to Pearl Harbor. On

March 17, 1942, after the first wave of submarine attacks in the Caribbean began, Bogotá and Washington signed a lend-lease agreement providing for some $16.5 million of military assistance.[7]

When a Colombian schooner plying the waters between the mainland and the San Andrés Archipelago was sunk by a U-boat in August 1942, and four of its survivors were machine-gunned in the water, Bogotá granted the Americans permission to establish a base at Cartagena. Subsequently, the US Navy set up a PBY base near Cartagena and used the civil airport at Barranquilla as a base for army patrol bombers.[8]

Venezuela was, without question, the most strategically important of the Latin American republics bordering the Caribbean. Its northern coastline constituted the longest shoreline of any Caribbean nation; the key islands of Trinidad and the Netherlands Antilles lay within easy flying reach of its shores; and the oil fields of Lake Maracaibo were a major petroleum asset to the Allies. When war broke out in Europe in September 1939, Venezuela declared strict neutrality and continued to trade with both sides. Right after Pearl Harbor, Venezuela's new president, General Isaías Medina Angarita, reaffirmed its neutrality, but then froze Axis accounts, impounded Axis ships, and severed diplomatic relations with the Axis countries.[9]

Following the Third Conference of the foreign ministers of the American Republics held at Riò de Janeiro beginning on January 15, 1942, Venezuela and the United States agreed to exchange information concerning western hemisphere defense, but little progress was made. Venezuela's hesitant approach to Caribbean defense continued even after the February 16 U-boat attacks on Aruba and the lake tankers. On the one hand, Caracas and Washington concluded arrangements regarding blackouts of tankers, shutting off Venezuelan coastal navigation aids at times of danger, and placing the movements of the lake tanker fleet under the voluntary control of the Royal Navy on Curaçao. On the other hand, a US initiative to work collectively in defense planning essentially led nowhere. In March, military representatives from Venezuela and Colombia arrived at General Frank M. Andrews' Caribbean Defense Command headquarters in the Canal Zone to coordinate defense planning with the United States. The Colombian officers appeared to be operating with a considerable degree of executive authority, but not those from Venezuela, who referred almost everything to President Medina – which was virtually useless for planning purposes. In mid-May, two of the three Venezuelan officers returned home; Colonel Centano, the leader of the delegation, stayed in Panama until early 1943 but "performed hardly any service in his liaison capacity."[10]

Venezuela's armed forces – like those of all the Central and most of the South American states – were obsolete, under-equipped and poorly trained. The country eventually acquired a handful of gunboats and aircraft from the United States under Lend-Lease, but its greatest defense potential for the Allied war against the U-boats was its ports and airfields. Until the submarine attacks of February 16, President Medina refused to allow US military aircraft to fly over Venezuelan territory without 24 hours' advance notice; eight days later he relented. The Americans constantly pushed for greater cooperation but with little success. As one US Army account of the period concluded, there was considerable discrepancy "between [Venezuela's] expressed desire to cooperate with the United States on mutual military measures and her reluctance to take the necessary steps therefor."[11] In June, Medina allowed the US Sixth Air Force to station a small detachment at Maracaibo to service and refuel American antisubmarine aircraft, and gave US warships permission to enter Venezuelan waters. In mid-March, a detachment of 283 US soldiers was admitted to set up coast artillery units at Puerto de la Cruz and Las Piedras, about 150 miles east of Caracas.[12]

The attack on Aruba galvanized the British, the Dutch, and the Central American Republics to cooperate far more fully with the United States in mounting a defense of the Caribbean. What mattered most was both the willingness of these Allies to give US forces virtually unlimited access to their air space and coastal waters and to allow the United States to set up bases for either land or sea-based antisubmarine aircraft. Once sufficient antisubmarine resources – particularly aircraft – were marshaled, the ring of bases around the Caribbean promised to turn it into an Allied lake.

* * *

The first layer of defenses against the U-boats consisted of antisubmarine nets, searchlights, sea mines, and coastal artillery. The entrances to the

Panama Canal were well guarded by all three by mid-February 1942, but little else was in the Caribbean or Gulf of Mexico. The US Army laid sea mines in the approaches to the Panama Canal, but none in the Puerto Rico or Trinidad sectors. The US Navy mined the entrances to the Gulf of Paria in Trinidad – over the objections of the Royal Navy, which maintained that the fast flowing currents through the Serpent's Mouth and the Dragon's Mouth would render them ineffective. The Royal Navy's predictions proved all too true. The water flow pulled many of the 350 mines well below the surface and ships drawing up to 22 feet were able to pass safely over them. That included surfaced U-boats running the passages by night. Later, some of the mines broke from their moorings and began to drift, some as far as the waters around Curaçao and Aruba, posing a severe hazard to shipping.[13] After the war, a US study of antisubmarine measures in the Caribbean theater concluded that "the effectiveness of the minefields was extremely doubtful."[14]

The buildup of coastal artillery around the Caribbean began shortly after the fall of France in June 1940. The gun batteries had two purposes – to protect minefields in those few places where they were laid, and to deny surfaced submarines (or other vessels) the ability to enter or even approach important harbors and chokepoints. Many US coastal artillery units were built around the 155-mm "Long Tom" howitzer, which threw a 95-pound shell some 15 miles with high accuracy, or the 90-mm anti-aircraft or anti-torpedo boat guns, which fired high-velocity rounds with a flat trajectory up to 12,600 yards. Dutch and British artillery were also emplaced.

In March 1941, the 252nd Coast Artillery Regiment, training at Fort Moultrie, South Carolina, was ordered to send its headquarters and "C" and "D" batteries to Trinidad, where it set up on Chacachacare Island. The unit was directed to guard the entrance to the Serpent's Mouth. The unit's "A" and "F" batteries were sent to Aruba. They began to arrive in early February 1942 but were not properly emplaced or ranged when Werner Hartenstein and U-156 attacked and thus were unable to return fire.[15] In July 1942, a composite field artillery battery of 121 men and four 155-mm guns was sent to Puerto Barrios on the Caribbean coast of Guatemala. The men languished there. There was constant friction between the Americans and the locals. The GIs suffered high malaria and venereal disease rates and had to be relieved after six months. Eventually, the battery was turned over to the Guatemalan government.[16]

This was the pattern throughout the islands. American troops were sent in small garrisons to Haiti, Antigua, St. Lucia, St. Croix, St. Thomas, Jamaica, Trinidad, Aruba, Curaçao, Venezuela, and Puerto Rico, where they saw virtually no action – the vast majority of their guns were never fired in anger. The islands may at first have seemed like small Gardens of Eden, but everything was different from the life the men had known and they were far from homes and families. The opportunity for bored and pissed-off soldiers to get into trouble was everywhere, from too much island rum to too many island girls. In many cases, the Americans could hardly wait to turn these garrisons over to local forces.

Nowhere was the problem more intractable than on Trinidad. In the fall of 1941, Rear Admiral Jesse B. Oldendorf took command of the Trinidad Sector and turned Chaguaramas into a formidable naval station for six US Navy and three Royal Navy destroyers, six gunboats, one submarine, 14 patrol boats, and a host of smaller craft. Over the coming year, "Oley" Oldendorf and his staff greatly built up their forces to face the expected third wave of U-boats: some 50 US Navy and Coast Guard, 26 Royal Navy and two Royal Canadian Navy escort vessels; 13 sea-gray Martin PBM Mariners (VP 74); and 12 PBY Catalinas (VP 53).[17] The first American advance ground elements consisted of the 11th Infantry and the 252nd Coast Artillery, whose 155-mm guns were sited on Chacachacare and Monos islands at the northern entrance to the Gulf of Paria. It was these units Albrecht Achilles and *U-161* had outwitted after their daring raid on Port of Spain in February 1942.

The pace of base-building in the interior of the island was frenetic. The Walsh Construction Company and the George F. Driscoll Company first removed a thick canopy of jungle vegetation over a 17,000-acre tract and then built a temporary gravel runway – followed in January and June 1942 by two mile-long concrete runways. At a cost of $52.4 million ($717 million in 2012 dollars), the most expensive Atlantic base ever built by the Corps of Engineers, Waller Field (and the adjacent Army post of Fort Reid) eventually housed 8,500 men and 51 aviation-fuel storage tanks. But even this gigantic complex was inadequate for the massive air assault being organized against the U-boats, and in December 1941 work began on new 5,000-foot runways at adjoining Xeres Field and Edin-burgh Field, about 12 miles southwest. Seasonal summer rains brought much delay, with the result that Waller Field's temporary runway was not ready for operation until October 1941. The first American airmen – 432 officers and men of the 1st Bomber Squadron based on Panama – arrived at the end of April and had to be housed in a tent camp at the commercial airport, Piarco Field.

Over the next year, fleets of new aircraft such as the PV-1 Lockheed Venturas, the North American B-25 Mitchells and the Consolidated B-24 Liberators arrived. Edinburgh Field alone received the PV-1 Venturas of US Navy bomber squadron VB-130, the B-25s of the US Army 7th An-ti-Submarine Squadron, the B-25 Mitchells of the 59th Bomber Squad-ron, the Douglas B-18 "Bolos" of US Army 10th Bomber Squadron, the 23rd Anti-Submarine Squadron, and a host of reconnaissance dirigibles. It was a far cry from the days of the first German wave of submarine attacks, when only the 1st Bomber Squadron had been stood up on Trinidad.[18]

* * *

The dramatic scale of construction virtually wiped out Trinidad's chronic unemployment overnight. But it also upset the delicate wage scales on the island and became a major source of trouble. While President Roosevelt had expressed a desire that local workers be paid "top-scale prevailing rates rather than average-scale prevailing rates,"[19] in reality wages paid to local workers – especially the unskilled – remained based on low local rates. Overall, the war brought little enhancement to wage scales – 17 to 57 per cent in skilled industries, 3 to 7 per cent in the skilled agricultural sector, 1 to 11 per cent in sugar mills, and none for stevedores and lighter men.[20] The concurrent 70 per cent increase in the cost of living more than wiped out all of these minuscule gains.

American labor, on the other hand, was paid United States union scale – plus an added "differential" for increased cost of living and overseas ser-vice. The sight of many American construction workers setting up black mistresses in fine homes, supplied with electric stoves and refrigerators "liberated" from base stocks, infuriated black Trinidadians.[21] Still, the prospect of work on the American bases brought about an uncontrolled influx of workers from Caribbean islands without US bases as well as an unwanted exodus of workers from local sugar estates and oilfields to the bases. All this caused further anxiety and ill-feeling.

Perhaps most offensive for many Trinidadians was the vast swath of destruction that the American contractors brought to their island. Armies of workers armed with chainsaws razed entire stands of poui and palm and cocoa trees. Thereafter, fleets of bulldozers leveled country homes and shanties alike. The novelist Ralph de Boissière left a vivid picture of the beehive of activity swarming around Chaguaramas:

> Endless streams of military trucks, long trailers carrying bulldozers or tanks, moved between Docksite and Cumuto; planes roared overhead in such numbers that it seemed they bred like mosquitoes in the swamps of Caroni.... Out of the mud of the foreshore, out of the inland forests, arose complete American towns.

He chronicled the effect of this feverish construction activity on one family estate: "The land lay naked, cut up like a corpse, and black workers swarmed all over it, like flies. The hill had disappeared, pushed into the swamp.... Huge and unfamiliar machines were everywhere at work, wiping out the past."[22] Yet, thousands of West Indians, he noted, continued to flock to Trinidad "as barnyard fowls who rush for the corn that is scattered by a lavish hand at sunrise," attracted by the lure of high pay.

Trinidad's laborers reacted to pay inequity in various ways. Many worked excessive hours of overtime. Others overstated their qualifications. One of the characters in Samuel Selvon's short story "Wartime Activities," when asked by an American female clerk ("a good-looking sport sitting behind a desk") what his "line of work" was, blurted out: "The first thing that come in my head is mechanic, so I say that." He signed up at "twenty bucks a week."[23] Still others resorted to making off with base equipment – from flats of beer to food to construction tools – and selling it on the black market. Peremptory firing of these culprits brought a sharp rise in labor unrest and violence, and it soured relations between American military personnel and contractors, on the one hand, and local black workers, on the other. In time, even the "better-class" houses on Trinidad closed their doors to American soldiers, sailors, and construction workers.

* * *

In May 1942, 2,484 black soldiers of the 99th Anti-Aircraft Artillery Regiment landed on the island. While the regiment would eventually constitute but 12 per cent of the total US force of 20,000 military personnel on Trinidad, its presence caused significant problems for Trinidad's British colonial regime and for the United States. From the start of the US presence, Governor Young had expected Washington to deploy only white troops to Trinidad; when the 99th arrived, he expressed his "very indignant" reaction to the Secretary of State for the Colonies.[24] He told London that the men of the 99th were "recruited from the lowest types of American Negro and are having an extremely bad influence ... debauching [local black] people and their family life." The American black soldiers had caused a crime wave to break out on the island, the Governor claimed, and he suggested they be replaced with Puerto Rican troops as the lesser of two evils. "Though neither are wanted here, Puerto Ricans would be preferable to United States Negroes." Whatever London thought of the governor's remarks, it could not afford to alienate the United States on such a sensitive matter.

Official Washington had, of course, been alert to concerns about transplanting American black troops onto the heterogeneous Trinidadian society. The State Department as well as the War Department had been well informed by both the British colonial authorities and the Island's local white and white Creole populations that they opposed such a move. They feared labor unrest among the local blacks due to the higher rates of pay that American black troops received. In particular, Governor Young had been apprehensive lest the delicate balance of colonial administration (wage scales, taxation rates, and customs tariffs) be disturbed by the arrival of well-paid and well-clothed American "Negro" soldiers. The quickly ensuing social unrest on Trinidad only confirmed such fears.

The black soldiers of the 99th Anti-Aircraft Regiment manned the batteries on the Laventille Hills overlooking Port of Spain as well as the hills protecting Waller Field and at the Army Air Forces complex, Fort

Reid. A rickety railroad linked Fort Reid to Port of Spain; later on, these two points were connected by the new Churchill-Roosevelt Highway that cut across the base of Trinidad's Northern Range. The men of the 99th Anti-Aircraft Regiment thus had ample opportunity to seek out the pleasure spots in the capital.

The US Army did almost nothing to prepare American troops of either color for duty on Trinidad. A seven-page mimeograph entitled "The American Soldier's Guide Book to Trinidad" explained almost nothing of the Island's history, its social structure, or its multicultural population and their local taboos. Instead, there was a map of Port of Spain, a recreation of a poster showing a drowning sailor with the exhortation "Somebody Blabbed … Button Your Lip," and very brief sections on security, dangerous insects, the basic geography and weather of Trinidad, shopping, sports, recreation facilities, and diseases, especially VD. American soldiers were told that the Island's economy depended heavily on the export of pitch, but not a word about the oil fields or the refineries. It was a half-hearted effort, to say the least, and probably had little or no impact on the GIs.[25]

The worst fears of the critics of the policy to garrison Trinidad with black soldiers were soon realized. Race relations, both between American white and black soldiers and between American black soldiers and local blacks, quickly broke down. US black troops resented their segregation from white troops, their deployment in what they deemed to be "less desirable" sites, and the Army's unstated assumption that only whites could command black troops. Their work often was boring and the hours long. They resented food rationing due to the rapacious activities of the U-boats. The only cheer seemed to come from readily available cheap liquor – and from indigenous black females.

The inevitable clashes between American and Trinidadian blacks were not long in coming. In the words of historian Annette Palmer, "American soldiers were accused of manslaughter, indiscriminate shootings, armed robbery and assaults against the members of the local population."[26] Few, if any, of the soldiers accused of such crimes were tried in local courts, and US Army military courts proved hesitant to convict American soldiers of crimes against local black civilians. As well, the "sight of white Americans engaged in manual labour or drunken white sailors shattered the image of white racial supremacy."[27]

Port of Spain became the focal point of much of the racial unrest. Already overpopulated at the start of the war, the situation was exacerbated by the 20,000 new American military personnel and by the uncontrolled migration of black labor from the other West Indian islands, especially Barbados, seeking work on the bases. The nightly arrival of the black soldiers of the 99th Anti-Aircraft Regiment based in Queen's Park barracks, just outside Port of Spain, and from Waller Field, set the stage for confrontation. Transportation was not a problem: buses charged three cents for a ride, trams four cents.

The capital offered the soldiers an exotic alternative to the drab (and segregated) conditions that prevailed on the bases. Again, journalist Albert Gomes provided a rich portrait.[28] Especially on weekends, the city's streets bore witness to a "prosperity of swank American cars, traffic strangulation and neon signs." The culinary odors of such local delicacies as black pudding, *souse*, *acras*, and *floats* wafted through the streets, as did the smell of warehoused sugar and nutmeg. Fresh markets bustled with vendors hawking bananas, plantains, pawpaws, green vegetables, and pork as well as shark meat. Peddlers spread before the foreigners their notions – pencils, sunglasses, lace, ribbons, combs.

Away from Port of Spain's commercial and administrative core, the air was filled with "the sulphurous stink from the nearby mangrove swamps, the flies and the incessant, suffocating, eye-smarting smoke from the burning mounds of fresh refuse," as well as from the "foetid and suffocating stink" of the city's open cesspits. In Shanty Town, dead dogs, cats, and birds added to the pungent aroma. So-called "pharmacies" doled out exotic potions such as "Spirit of Love," "Confusion Powder," "Man You Must," and "*Vinaigre Sept Voleurs*" as good "magic" to encourage romantic encounters. The halls run by the Shouter Baptists and the Shango Dancers, as well as the magic parlors of the Scarlet Sisters and Mother Holy Ghost, offered the soldiers a heady concoction of chants, drumbeats, prayer, ritual dances, and superstitions. Chinese opium dens vied with Portuguese rum shops for the "Yankee dollar." In the colorful language of the novelist Robert Antoni, "half the whores in Venezuela" crossed the Gulf of Paria "in salt-fish crates" to get their piece of the action.[29]

Unsurprisingly, fights, both among black and white sailors and soldiers and among them and the local blacks, became the weekend norm.

Crime and corruption were commonplace. Natives cheekily spoke of their island as "Trickydad." Common diseases carried by the anopheles mosquito, by the small vampire bat, and by hookworm as well as venereal disease from unprotected sexual encounters ran rampant. Soon, the US Army had no choice but to build the Caribbean Medical Center to control the "alarming" spread of VD.

The spark needed to trigger a racial fire was struck at Arima, near Fort Reid, in April 1942, when an American soldier was charged with the murder of a Trinidad civilian.[30] The local black population poured out into the streets to demand justice. The police intervened. Some 33 persons were arrested, most of them Barbadian migrant workers. As one of his last acts as governor, Sir Hubert Young immediately raised the matter of jurisdiction, demanding that the accused American soldier be handed over to the British colonial administration. While Young and the State Department in Washington exchanged lengthy diplomatic memoranda on this thorny issue, General Ralph Talbot, Jr., commanding US troops in Trinidad, ordered a military court-martial to proceed with the case. The soldier was acquitted of the charges against him. Anglo-American relations on Trinidad hit their nadir.

All of this was grist for the mills of local satirists, the calypso singers. For, in the competition for female company, the American black soldiers were the clear winners. They had hard currency, the glamour of a crisp uniform, and the attraction of being outsiders. One calypso singer, complaining as a victim, commented: "I was living with me decent and contended wife/Until the soldiers came and broke up my life." Another noted that while local blacks offered only "love and misery," American black soldiers held out "romance and luxury." Yet another lamented that the local young girls had become "frisky frisky" upon the arrival of the Yanks. "They say the soldiers treat them nice/They give them a better price!"[31] Of course, the classic social comment rested with "Lord Invader":

> Since the Yankee come to Trinidad
> They got the young girls all goin' mad
> Young girls say they treat 'em nice
> Make Trinidad like paradise.

Drinkin' rum and Coca-Cola
Go down Point Koomahnah
Both mother and daughter
Workin' for the Yankee dollar.[32]

Behind the musical satire existed a harsh climate of racial tensions and open hostility. For, in the words of the US Caribbean Defense Command, American troops "conducted themselves in a manner more in keeping with the occupation of a captured country."[33]

The incident at Arima forced official Washington to take action. Under-Secretary of State Sumner Wells appreciated that the entire issue was "most explosive," and Army Chief of Staff George C. Marshall considered it "too dangerous to be handled on paper."[34] He therefore suggested that closed discussions be held to resolve the issue. These apparently took place both within the Washington establishment and with the British Colonial Office. The upshot was that the War Department ordered the soldiers of the 99th Anti-Aircraft Regiment home – "at night on secret orders." But nothing was ever secret in Trinidad. The railway line from Arima to Port of Spain, used for the regiment's embarkation to the United States, was "thronged with women waving goodbye." By the end of 1943, "white Puerto Ricans with a knowledge of English and high educational standards" garrisoned the Trinidad bases.[35] All the while Admiral Karl Dönitz's "gray sharks" continued to ravage the Caribbean sea lanes.

The US descent onto Trinidad had brought much unrest. The hopes of well-educated Trinidadians that President Roosevelt would extend the Four Freedoms to their island never materialized. Nor did those that Prime Minister Churchill would apply to Trinidad the Atlantic Charter's provision for peoples to freely choose their form of government. Moreover, wages rose but modestly and the sharp rise in the cost of living erased what few gains were made. An Anglo-American Caribbean Commission, established jointly by Roosevelt and Churchill in March 1942, brought about some improvement in methods of agriculture, housing, education, and public works, but it was largely viewed by local activists as another colonial administration in new garb.

The hard reality was that Trinidad was but one small part of a global struggle. The US military and contractors had arrived in spring 1941 to throw up army, navy, and air bases to meet the mounting German submarine threat; they had little interest in rearranging the island's labor, political, or social relations. The Germans had pierced the Caribbean basin with but one aim, to disrupt the Allies' vital flow of oil and bauxite out of the region. They had no interest in native populations; their racial doctrines held no appeal to Trinidadians; and their amateurish spy network operating out of the Panama Canal Zone found no fertile soil on Trinidad. The character "Cassie" in Ralph de Boissière's novel *Rum and Coca-Cola* perhaps best captured the great-power reality: "So now you hear, Friends, the English not givin' us anything, the Americans not goin' to give us anything, nor the Germans."[36] Albert Gomes echoed the feelings of many fellow political activists during the war when he stated, "Whenever we pass into other hands, both hands must be our own."[37]

WHITE CHRISTMAS

The Hollywood blockbuster for the fall of 1942 was the musical *Holiday Inn* with Bing Crosby, Fred Astaire, Marjorie Reynolds, Virginia Dale, and Walter Abel. Released in August, the picture became a national phenomenon in just a few months. As summer gave way to fall the movie's big number, "White Christmas," written by Irving Berlin and sung by Crosby, became the unofficial anthem of the holiday season for the rest of the war. By the end of November 1942, more than 600,000 records as well as more than a million copies of the sheet music had been sold. "White Christmas" became the longest-running song ever played on the weekly radio show "Your Hit Parade." In that dreary Christmas of 1942 – the second Christmas coming in the midst of war in the United States – Berlin's melancholy words gave voice to the heartfelt wishes of millions of Americans, at home and overseas, for a Christmas more joyous and brighter than the one they were about to celebrate.

A modern American Christmas is a holiday of light – the colorful Christmas lights of homes and businesses, shop windows all ablaze, every city and town putting a tableau of the manger in Bethlehem by the old court house or town square, home fireplaces and Yule logs, and the national Christmas tree on the White House grounds. But that year the deepening energy crisis in the United States, and especially in the country's most populous region and the heartland of its industry – the northeast and the Ohio River Valley – made Christmas a lot darker. Outdoor holiday lights were dimmer where they were found at all. Posters everywhere exhorted Americans to save energy for the fight at the front and for the production of the machines of war that would ensure victory. Coal, oil, and gasoline for personal civilian use were tightly restricted. Car trips to visit family were virtually out of the question. On top of all that, it was a cold and dry winter with little snow. Well did Americans wish for a white Christmas that year.[1]

The severe energy and particularly oil shortage in the American northeast (and eastern Canada) was the inevitable result of the war that Admiral Karl Dönitz's "gray sharks" had been waging against all shipping since September 1939, but especially against tankers. Tanker losses had started to mount over the winter of 1939–40, but the Anglo-French allies had initially been able to counter those losses by leasing neutral tankers, rationalizing tanker traffic, convoying, and other means. The constant struggle to preserve tanker capacity, and possibly to build more tankers than the enemy was destroying, suffered a severe blow with the entry of the United States into the war. Suddenly, every tanker in the Atlantic, the Gulf of Mexico, and the Caribbean was at risk.

Shortly after his American offensive began, Dönitz told the German people over the radio: "Our U-boats are operating close in shore along the coast of the United States ... so that bathers and sometimes entire coastal cities are witnesses to the drama of war, whose visual climaxes are constituted by the red glorioles of blazing tankers."[2] Those "red glorioles" erupted 64 times between mid-January and mid-June 1942, when as many tankers (along with 62 other ships) were sent into Davey Jones' Locker off the US east coast.[3] But these were not the only tankers lost in American waters or in proximity to American waters; from the opening of the submarine offensive in the Caribbean on February 16, 1942, until the end of that year, another 75 tankers were destroyed in the Caribbean, along with 310 other ships.

Every tanker loss meant that nearly 100,000 passenger cars on the east coast would be devoid of gasoline; 35,000 homes or small businesses would suffer with little or no heating oil for up to a year. The average tanker on an east coast run carried 80,000 barrels of oil up from the Gulf or the Caribbean every 20 days, thus delivering 4,000 barrels a day. This translated into 1.5 million barrels per year of carrying capacity.[4] When multiplied by the 222 US, British, and other Allied or Allied-chartered tankers lost in 1942,[5] the impact was staggering – 330 million barrels of carrying capacity lost. The Germans did almost no damage to the drilling, lifting, or refining capacity of the United States, Venezuela, or Trinidad.

But they were making it very difficult to get any of that oil to points vital to the Allied war effort.

The result of the great damage to the transporting capability of the tanker fleet was felt up and down the line. In the first six months of 1942, Venezuela was forced to curtail production by 12.5 million, Mexico by 2.95 million, Colombia by 1.15 million, and Trinidad by 0.2 million barrels. The oil could not be shipped, so it had to be left in the ground – "shut in" – or stored somewhere near the drilling site.[6] No one had foreseen the need for major oil storage facilities, so the oil was left in place. Local and even national economies suffered. United States production destined for the northeast was drastically cut back. No pipelines existed to carry oil from Texas or Louisiana to New York or New Jersey.[7] Daily tanker shipments from the Gulf of Mexico to the eastern seaboard dropped from the 1941 average of 1.42 million barrels per day to just 391,000 barrels in 1942.[8] The excess was either shut in or stored in local tank farms.

Here, too, production stopped, workers were laid off, local economies suffered. As early as the beginning of March 1942, Texas oilmen began to talk about reducing production by at least 10 per cent to ease the growing demand for above-ground storage.[9] As early as the end of the first week of March 1942, US east coast stocks were 10.74 million barrels lower than they had been the previous year. British imports of both crude and petroleum products fell from 12.3 million tons in 1941 to 9.9 million in 1942, while total imports of refined gasoline, for both motor and aviation use, dropped from 4,768 tons per week in 1941 to 4,115 per week in 1942,[10] even as demand rose. Put simply, the Germans were sinking tankers far faster than Allied shipyards could replace them.[11] On March 12, 1942, Prime Minister Winston S. Churchill wrote President Franklin D. Roosevelt: "I am most deeply concerned at the immense sinkings of tankers west of the 40th meridian and in the Caribbean Sea."[12] General George C. Marshall, Chief of Staff of the US Army and Roosevelt's chief military advisor, was somewhat less prone to dramatic statements than Churchill, but even he told Admiral Ernest J. King in June 1942, "The losses by submarines off our Atlantic seaboard and in the Caribbean now threaten our entire war effort."[13]

* * *

For civilians in the United States, Great Britain, and Canada, the most direct link between the slaughter of the tankers and their daily lives was the elaborate system of rationing and controls over energy consumption that would ultimately limit their ability to drive their cars or heat their homes and workplaces. But civilian demand was not a significant part of the problem. It was the newly and greatly expanded demands for war production – oil for everything from chemicals to explosives, plastics, and rubber; to pave new runways; to provide rainproof ponchos for the infantry; hydraulic fluid for brakes for motor vehicles and aircraft; and, most importantly, aviation gasoline for bombers and fighters. The ultimate solution to these shortages was to destroy the Nazi regime and the submarine fleet it had created; everyone knew that would take some time. In the interim, there were three solutions: in the short term, rationing; in the intermediate term (for the United States), transporting oil without tankers; and in the long run, building more tankers and sinking more submarines. Britain had imposed rigid controls over consumption, distribution, transport, and storage of oil and petroleum products almost as soon as the war began in the fall of 1939. In July 1942, even further cuts were made and no gasoline was allowed for civilian uses except for 40,000 individuals living in remote areas of the British Isles who were allowed to drive up to 120 miles per month.[14]

In the late winter and spring of 1942, Americans too began to feel the pinch. An immediate halt in the manufacturing of private cars and the shift of the auto industry to army vehicle and aircraft production essentially froze potential growth in the number of civilian vehicles on the road. On April 16, 1942, supplies of gasoline to retail dealers was curtailed by one third. A month later a temporary ration of five gallons per week for civilians was imposed on the east coast. On July 1, a permanent limit of 16 gallons a month was imposed on drivers west of the Appalachians; 12 gallons a month for those on the eastern seaboard.[15] Fuel oil to heat private homes was also cut; citizens were told to convert to coal and to keep their thermostats at 65° F during the daytime and 55° F at night.

The British, Americans, and Canadians faced two separate but related problems. For Britain, the tanker shortage meant less oil and petroleum products, period. For Canada and the United States, lost tankers created internal distribution problems of the first magnitude. Sufficient

oil was produced and refined in the Canadian West to take care of local demand, but eastern Canada was heavily dependent on tanker-borne
supplies from the Caribbean. Tanker losses – and shifting tankers from
the eastern Canadian trade to other purposes – put enormous pressure on
Canadian supplies. In the United States, the tanker shortage had its most
serious impact on the east coast because of the roughly 576,000 barrels a
day that flowed into the region; 414,000 barrels from the Gulf of Mexico
and 115,000 from the Caribbean came by tanker and only 46,000 barrels
by rail, truck, barge, or pipeline.[16] The challenge, then, was to replace as
much of the tanker-borne oil flowing to the east coast as possible by other
means.

Railway tank cars were the most obvious replacement. In June 1941,
Harold L. Ickes, Roosevelt's Interior Secretary and recently appointed
Petroleum Coordinator for National Defense, began to press the oil companies to make greater use of tanker cars to supplement sea-going tankers.
But Ickes faced several obstacles. First, it was at least ten times as expensive to move oil by rail as it was by sea. Second, any open move to coordinate the flow of oil by rail between the largest oil companies and railroads
might easily be construed by the Department of Justice as a form of trust
or monopoly, and thus prohibited by America's tough antitrust legislation.
Third, the railroads had little infrastructure to handle large traffic in oil,
either loading and storage facilities or branch lines to enough terminals.
Ickes thus urged the Interstate Commerce Commission to allow railroads
to set rates for petroleum shipment that would produce richer returns
while he won permission from the Department of Justice to encourage
the major oil companies to pool reserves and coordinate shipments with
the Association of American Railroads.[17] At the same time, grain cars and
liquid gas cars were converted to oil carriers.

Ickes' campaign produced quick results. By March 1942, some 13,500
tanker cars a week brought more than 435,000 barrels of oil to the east
coast. Given the greatly increased wartime demand for oil and petroleum
by-products in March 1942 over March 1941, it wasn't nearly enough, but
it was a start. Tanker trucks were also used to supplement rail deliveries –
the War Production Board issued exemptions for the manufacture of tank
trucks while states (Pennsylvania Turnpike) suspended rules that barred
tanker traffic from some trunk highways.[18] Another, though much slower,

alternative was to ship oil and other petroleum products by barge up the Mississippi and via the great rivers of the eastern and central states such as the Ohio and the Tennessee, as well as the extensive barge canals built in the nineteenth century. A barge starting at Corpus Christie, Texas, might make its way as far as Pittsburgh, or via the Missouri River, the Illinois Waterway, and the Great Lakes to Cleveland and even Buffalo.[19] Lake tankers were pressed into service to off-load oil from barges at major terminuses and carry it to lakeside cities such as Chicago. In June 1941, about 95 per cent of the petroleum deliveries to the east coast were made by tanker. In April 1945, only 22 per cent were brought by tanker, while 30 per cent came by rail and 8 per cent by barge. But, by then, the remaining 40 per cent came by pipeline.[20]

* * *

There were about 100,000 miles of oil pipelines moving more than three million barrels of crude oil and petroleum products in the United States in 1940, none larger than eight inches in diameter, and none from the major oil-producing regions in the southwest to the east coast. The trunk lines that did exist connected loading facilities in Texas, Louisiana, Oklahoma, and New Mexico to the Gulf of Mexico and its large tanker loading docks. A number also connected the southwest to California and to midwest refineries.[21] The heavy reliance on sea-borne oil to sustain the east coast posed no significant inconvenience to the oil industry or the public until the start of the war in 1939, and most particularly the surrender of France in June 1940. The subsequent slaughter of the tankers in the North Atlantic and off the US east coast after Pearl Harbor as well as in the Caribbean after February 16, 1942, convinced almost everyone involved in the business of organizing wartime energy supplies that the only real way to replace the tanker traffic was via a new emergency pipeline from Texas to the east coast.

Ickes had warned Roosevelt in July 1940 that "the building of a crude oil pipeline from Texas to the East might not be economically sound [in peacetime]; but in the event of an emergency it might be absolutely necessary."[22] The problem was that the United States was not yet at war and in 1940 no one could foresee if it would join the conflict, or when. In the meantime, the president's chief objective was to begin the long and complex job of building up American military strength. The United States was so weak in the summer and fall of 1940 that it could barely field a single modern, well-trained, armored division.

On May 27, 1941, Roosevelt declared an unlimited state of national emergency; the next day he named Ickes oil tsar. Ickes' war emergency duties were aided by the establishment by executive order of the Petroleum Administration for War (PAW) on December 2, 1941, with him as Petroleum Coordinator for War. As Secretary of the Interior, Ickes had been a strong conservationist and no friend of the oil industry, but he saw immediately that he could not succeed in preparing the United States to fight a modern oil-based war without industry's cooperation. He overcame Big Oil's misgivings by inviting the managers of the nation's largest companies to join him to coordinate war supplies for energy, while he also persuaded the Department of Justice to relax its investigations, and prosecutions, of companies that cooperated with each other, especially in the shipping of oil.

The companies responded with the creation of the Petroleum Industry War Council, a voluntary association of all the major American oil producers, with several important subcommittees such as the Petroleum Economics Committee and the Transportation Committee. Their intention was to help the war effort in any way possible but primarily to – in effect – create one large oil consortium for the duration. As one dramatic example, companies even pooled gasoline, despite the near religious fervor with which one company's brand had been extolled over another's during peacetime. Ickes and the oil industry accomplished a great deal before the United States was dragged into the war, but it was not nearly enough to offset the disastrous loss of tankers that began in February 1942.

When the impact of the tanker losses was first felt, Ickes and the oil companies bore down hard on solving the growing shortage of east coast oil by using existing pipelines, railroads, and barges; by pooling, exchanging, and sharing facilities and equipment; and by speeding up construction of tankers, of which more than 800 were built by the United States alone by 1945.[23] But Ickes remained convinced that a war emergency pipeline from Texas to the east coast was the only long-term answer. As early as September 1941, he had formally requested the Supply Priorities and

Allocations Board (SPAB) – responsible for allocating all strategic materials and supplies in the immediate prewar period, soon to be replaced by the War Production Board – to provide enough steel to build a 24-inch pipeline under the auspices of the newly incorporated National Defense Pipe Lines Company. The SPAB had rejected the request.

Ickes pushed on. He had the authority to quickly get rights-of-way for construction from the Cole Pipeline Bill, signed into law by FDR on July 30, 1941, which bestowed on the president the power to designate any proposed pipeline as necessary to the nation's defense and to confer on its builders the right of eminent domain – that is, the right to expropriate property at fair market value, if necessary, over the objections of the property's owners.[24] The pipeline industry was heavily involved in Ickes' bid, with eleven companies offering to finance and build the new line. Such a project would require hundreds of thousands of tons of new steel for pipe, pumps, valves, storage tanks, and loading facilities. The SPAB rejected the project again. As far as it was concerned, the steel was needed for more important things – ships, tanks, aircraft, guns, helmets, even bayonets. Oil transportation would have to make do for the moment. Two days after Pearl Harbor, the National Defense Pipe Lines Company was dissolved.

The SPAB held fast to its opposition to the pipeline through the rest of the fall of 1941 and into the first half of 1942 – by which time it had been transformed into the War Production Board (WPB). On February 24, 1942, the latter rejected a third proposal for a Texas–East Coast pipeline. As *The Oil Weekly* declared in its edition of March 9, 1942:

> In turning down Ickes' application [of the previous week], WPB accepted the SPAB [Supply Priorities and Applications Board] ruling of last November that the value of the line as a defense project was not great enough to justify the high priority ratings that would be necessary, and pointed out that materials shortages since that time have increased rather than decreased.[25]

On March 23, engineers and management representatives of 67 oil and pipeline companies, all members of the Petroleum Industry War Council, gathered for three days at the Mayo Hotel in Tulsa, Oklahoma, to hammer out a domestic pipeline strategy for the war emergency period. The Petroleum Industry War Council's Temporary Joint Pipeline Sub-Committee was made up of some of the industry's foremost experts in pipeline construction and management, storage and shipment, and traffic control. They examined virtually every mile of existing pipe, every pumping station, every tank farm, and drew up a plan to effectively re-jig the existing national network for moving oil so as to increase the amount flowing to the northeast and to other areas where vital war activities such as ship construction were going on.

The resulting Tulsa Plan contained numerous recommendations for the reconfiguration and extension of existing pipelines, the increased use of alternate means of fuel delivery, specific measures to be taken by companies to increase the efficiency of supply, the use of old pipe, pumps, and meters on new lines, the reversal of pumping direction on some lines, and the tearing up of old pipe to extend existing lines. Even if those measures were to be adopted, however, the Transportation Sub-Committee concluded that the east coast would still be left far short of its minimum daily oil requirements. "There does not seem to be any solution but to build two big pipelines from the Texas area thru to the Atlantic Coast if tankers are not going to be available."[26] The Tulsa Plan was warmly received by the industry and accepted by Ickes on May 11, 1942. *The Oil Weekly* was optimistic that the next time Ickes, armed with the plan, went to the WPB for steel, his reception would be somewhat warmer: "It is expected the Office of Petroleum Coordinator will make another effort and there are indications that WPB may now adopt a more liberal attitude toward the project."[27]

Ickes did go back to the War Production Board on May 25. This time his argument was strongly supported by both the army and the navy. It was also dramatically illustrated by the brutal reality that some 100 tankers had been sunk off the east coast, and in the Caribbean, since the beginning of 1942. On June 10, the WPB approved a 24-inch crude oil pipeline – dubbed Big Inch – but only from Longview, Texas, to Norris City, Illinois. There, the WPB proposed that the pipeline feed into a tank-car loading facility for rail transportation further east. Four months later, the Petroleum Administration for War urged the WPB to approve allocation of steel and other materials for the line's extension to the east coast. Permission was granted on October 26 for a single 24-inch line from Norris, Illinois, to Phoenixville, Pennsylvania, where two 20-inch lines would be built to New York City and Philadelphia.

At the same time, PAW told the Board that a request for a second line would shortly be made. That line, the Little Big Inch, was to be a 20-inch pipeline for carrying refined petroleum products from Beaumont, Texas, to Linden, New Jersey. PAW made the application on January 18, 1943. It was initially granted permission for the line to be built to Norris, Illinois, but then, on April 2, received the WPB's blessing for its extension to Linden, New Jersey.[28] Thus, whereas in the fall of 1941 the United States had no pipelines to carry crude or crude products from Texas to the east coast, now there were to be two large lines with a potential capacity of half a million barrels of product delivered every day. Concurrently, other national emergency pipelines and extensions of previous pipelines were built – such as the Plantation Line from the Gulf Coast through the Old South to Richmond – which considerably alleviated the oil shortage in the southeast.

But how to finance, build, and manage the Big Inch and Little Big Inch, and do so quickly enough to offset the growing tanker shortage?[29] That was the next and greatest challenge by far.

* * *

The $35 million cost for the first stage of the Big Inch line was advanced by the government-owned Depression-Era Reconstruction Finance Corporation. For the duration of the war, the line was to be owned by the Defense Plant Corporation and managed by War Emergency Pipelines, Inc., a consortium of 11 companies. WEP was also the prime contractor for the construction of the line, using dozens of private companies to do the surveying, trenching, pipe-laying, backfilling, and other essential work, including bridging and underground boring. W. Alton Jones, President of Cities Service Oil Company, was named president of WEP; Burton E. Hull as vice president and general manager.

A bluff, weather-beaten Texan, Hull had graduated from Texas A & M University in 1904 with a degree in engineering. Short and stocky, he was plain-spoken, energetic, and a born leader. He was considered one of the best pipeline engineers in the business and was determined to bring the line into operation on time and on budget. On June 23, 1942, three days before the official contract was even signed, he put 15 surveying parties into the field to stake out a 531-mile right-of-way between Longview, Texas, and Norris City, Illinois. Williams Brothers, one of the country's most experienced pipeline layers, was selected as prime contractor. The National Tube Company, a subsidiary of US Steel, shipped the first trainload of 24-inch seamless pipe, 3/8-inch thick, on July 18; on August 3, construction got under way near Little Rock, Arkansas.

Hull's assistant, Major A. N. Horne, arrived in Little Rock on July 1 with nothing but a briefcase; Hull joined him two days later. The two men worked out of a steamy hotel room, sending telegrams and letters, making phone calls and hiring a traffic manager to work with them as they assembled the management team they would need to get construction launched. A few days later, they went to an auction to buy some used office furniture. On July 11, Hull summoned to Little Rock representatives from every pipeline contractor in the nation with the men and ability to handle heavy pipe; together they formulated a plan to get the job done by the beginning of January 1943.

The Big Inch line was to run 530.36 miles along a right-of-way 75 feet wide with pumping stations approximately every 50 miles. Eight principal pipe-laying crews, each between 300 and 400 men, worked on different sections of the pipe at the same time. Other crews – 18 of them – tackled the 33 river and stream crossings that were necessary, including one under the Mississippi River. It took 11 weeks to blast a trench in the river bottom and prepare it for the pipe. The standard method of construction on land was to dig a trench four feet deep and three feet wide, and lay the pipe into the trench with a side-boom tractor, usually from truck trailers. The inside of each section of pipe was swabbed by pulling a (necessarily small) man through the pipe on a cleaning pad with rags on his hands. The ends were then prepared for welding, after which the pipe was coated, wrapped and covered, laid back in the trench, and backfilled. Most of the necessary pipe bending was done on the spot. Rivers were crossed using 4,800-pound river clamps to hold the pipe to the bottoms.[30]

The pipe was laid through forests and swamps, over the Allegheny Mountains, across rivers, lakes, creeks, and tidal marshes, beneath streets and railroad rights-of-way, and through backyards. It traversed 95 counties in ten states. At the same time, work began on large tank farms in Long-view, Texas, and Norris, Illinois, and at a large facility at the latter to load tank cars either directly from the pipe or from the storage tanks. It was an incredible feat of wartime construction – some five months after the work on the initial section of the pipeline had begun, the first flow of oil from Longview to Norris (about 60,000 barrels a day) started through the pipe. The line had been completed so fast that sufficient storage tanks were not yet ready at Longview or Norris. No matter, oil was immediately transferred to tank cars and sent east. At the same time, construction on the eastern connection to New York and Philadelphia, started in November 1942, pro-ceeded at a rapid pace. Most of this line was welded, not seamless, pipe supplied by Youngstown Sheet and Tube.

On July 19, 1943, the final weld was made on this eastern extension of the Big Inch at Phoenixville, Pennsylvania. By that time, 100,000 barrels a day were already being pumped into the pipe from fields near Longview, Texas, and from other fields in the state connected to Longview by newly built pipes. It moved along at 40 miles a day; when the pipeline was filled from end to end, it held five million barrels of oil. The trick was to run the pumps and operate the valves in such a way as to reach its full capacity of 300,000 barrels a day, which was done by the fall. At the same time, Little Big Inch construction moved rapidly ahead. This pipe was to carry gasoline, heating oil, and other products using rubber balls slightly larger than the diameter of the pipe, and inserted into the pipe, to separate the products being sent through. It could also carry crude, if necessary. The combined carrying capacity of the two lines was 500,000 barrels a day.

In the words of one major survey of the American oil industry in World War II, "The completion of Big Inch [in July 1943] marked the beginning of the end of the supply problem on the East Coast."[31] The com-pletion of Little Big Inch to the east coast in March 1944 "virtually solved the transportation deficiency to the east coast,"[32] according to the official history of the PAW. Although these two pipelines (and many others) that had been rushed to completion or extended added capacity that had not existed before 1941, older means of shipping also improved greatly. By 1944, pipelines delivered 662,559 barrels a day to the east coast (38.7 per cent of the total), rail cars 646,113 barrels or 37.7 per cent of the east coast supply, and barges and lake tankers (via Lakes Michigan, Erie, and

Pennsylvania section of the war emergency 24-inch pipeline to carry oil from Texas fields to eastern refineries, completed in July 1943. Willis Garner about to tack weld a section of pipe. Source: Library of Congress, Prints & Photographs Division, FSA/OWI Collection [reproduction number LC-USW3-015067-D (b&w film neg.)].

Ontario) another 127,641 barrels or 7.5 per cent. Thus, the total amount of oil flowing to the east coast from overland routes was 1.4 million barrels a day by 1944, which was more than the total amount (tankers included) in 1942 or 1943.

The railroads did an outstanding job in collecting, modifying, and pooling tank cars and increasing tank-car deliveries to the east coast from 35,000 barrels a day in 1941 to 841,905 (or 61.3 per cent of the total) by 1943, but railway cars were in very high demand throughout North America to deliver all manner of war goods. Thus, as soon as the pipelines began to take up the slack with the completion of Big Inch and Little Big Inch, thousands of tank cars were diverted elsewhere to deliver other liquid goods or to be converted to carry dry goods.

The other factor that helped to ease the oil supply situation in the East was a crash program to build tankers. Sea-borne oil delivered to the east coast reached its high point in 1941, accounting for 1.4 million barrels a day or 92.5 per cent of the total. But that was prior to the war. After Pearl Harbor, tankers were sunk in large numbers, but tankers (including much new construction) were also grabbed up by the Royal Navy and the US Navy as fleet tankers. The US Navy had developed the art of refueling at sea in the interwar period in order to be able to reach the vast expanses of the Pacific Ocean. The Royal Navy was much slower to take up that challenge but was well into the practice by 1943. Hence, both used tankers for two purposes – to move supplies from port to port, but also to sail with their task forces and fuel at sea. Navies claimed first priority on new tanker construction.

The world has long been aware of the incredible feats of ship construction in the United States, Great Britain, and Canada that, by early 1943, was launching many more new bottoms than U-boats could sink. US "Liberty" ships were mass-produced, sometimes in a matter of days. But so, too, were tankers. One of the best new modern designs was the T-2 tanker, adopted from two prototypes – *Mobilfuel* and *Mobilube* – whose construction had started before the war. They set the pattern for 481 other T-2s built by 1945, many of which were taken over by the US Navy. All were over 500 feet long, displaced at least 10,000 tons, and were powered by steam-turbine engines with maximum speeds exceeding 16 knots. Put simply, they were larger, faster, and more powerful than most tankers afloat in 1940. And that meant that one of the unforeseen side effects of Dönitz's war against Allied tankers was the rebuilding of a major part of the prewar tanker fleet to deliver more product, more quickly. By 1945, tanker deliveries to the east coast had increased from an average of 159,563 barrels per day in 1943 to 450,665 barrels a day.[33]

By Christmas 1944, the holiday lights were on again in most of the United States and the United Kingdom. Both had more than made up for the huge losses in tankers suffered in 1942 and 1943 on the eastern seaboard of the United States, on the trans-Atlantic routes, and on the runs from Venezuela to the Netherlands Antilles and on to Canada, the US east coast, and the United Kingdom. The U-boats would continue to haunt the Caribbean in late 1942 and 1943 – indeed, they would in small numbers revisit for most of the war – but the outcome of the struggle had already been determined by the rapid construction of just over 1,000 miles of pipeline. The virtual closing of the Caribbean oil supply in 1942 had been a brilliant feat of the German submarine service, but the industrial capacity of the Allies, especially the United States, was more than a match for the U-boats. When led, organized, and driven by men such as Harold Ickes and Burt Hull, the thousands of welders, pipe fitters, pipe layers, engineers, earth-moving equipment operators, truck drivers, surveyors, riveters, pile drivers, boring machine operators, drillers, and just plain laborers neutralized the largest submarine force the world has ever seen.

THE ALLIES STRIKE BACK

By the end of 1942, the Caribbean was interlaced with 27 different convoy routes. The intricate convoy network required far more escort vessels than were available. The three main navies fighting the Battle of the Atlantic – the Royal Navy, the Royal Canadian Navy, and the United States Navy – were hard pressed in 1942 to provide escorts in the Caribbean because all three suffered an acute shortage of destroyers and other escort vessels. Almost all new US Navy destroyers were being pushed into the Pacific as fast as they were commissioned. The Royal Navy had been so hard pressed for escorts in the late summer of 1940 that it had been forced to conclude the destroyers-for-bases deal in order to obtain 50 World War I–era American destroyers. The British possessed some 200 escort vessels of all types in the spring of 1942 – including the Flower-class corvettes, lightly armed, slow, coastal defense vessels pressed into service for mid-ocean duty. But the ships were spread out over half the globe: ten (all corvettes) had been transferred directly to the US Navy; 61 were in the South Atlantic, Mediterranean, or the Pacific; 37 were in British home waters; 78 were on the North Atlantic run; and six were on convoy duty to Russia. Only Escort Group B5 was available for full-time duty in the Caribbean, supplemented from time to time by a half dozen other vessels.[1] The Royal Canadian Navy was only able to detach four corvettes and the occasional destroyer to escort oil convoys to and from Canada's east coast. Although the Caribbean was a vital war theater, the North Atlantic remained the most important area of operations, and it came first for all three navies.

* * *

The US Navy was singularly unready even to fight a war in the Atlantic, let alone the Caribbean. There were about 100 destroyers in the Atlantic Fleet at the outbreak of war, but only Task Force 4, operating out of Argentia, Newfoundland, made its destroyers available for convoy escort duty, and only a few of those were detached to subordinate commands such as the Caribbean Sea Frontier. As Vice Admiral Royal E. Ingersoll, Commander in Chief, Atlantic Fleet, later recalled:

> The Sea Frontiers (i.e., Caribbean Sea Frontier) had their own coastal forces but they were small ships and were not really good anti-submarine vessels. The subchasers, of which the navy had a lot, weren't very good. They were one of Mr. Roosevelt's fads.... The submarine chaser was no craft to combat submarines on the high seas.[2]

And yet, after Pearl Harbor, the Americans had little choice but to use their motley fleet in the Caribbean; they had little else. The Naval Operating Base at Port of Spain, Trinidad, was almost denuded of defenses in December 1941, with two 500-ton converted yachts, two patrol craft, four Catalina flying boats of VP-31, and one utility transport, all guarded by a mixed force of 172 Marines and "bluejackets," as the navy called its noncommissioned personnel.[3] Rear Admiral Hoover's odd collection of ships was directed mainly from three headquarters: San Juan, Puerto Rico; Guantánamo, Cuba; and Trinidad. The only rapid reinforcement sent after the attack on Aruba was one additional squadron of PBYs.

Hoover's Caribbean fleet included Roosevelt's cherished sub-chasers. Most displaced 450 tons and had a top speed of 21 knots. They were armed with two three-inch guns, five 20-mm rapid fire anti-aircraft/anti-motor-torpedo-boat guns, two depth-charge throwers, and two depth-charge racks astern. As well, Hoover had at his disposal so-called "Q-ships." They had first been used in ASW during World War I, and followed a long tradition of navies disguising warships at sea as merchant vessels in order to draw in and then destroy unsuspecting enemy warships. In January 1942 the US Navy ordered several "Q-ships" fitted with four-inch guns, .50-caliber machine guns and 20-mm anti-aircraft/anti-torpedo boat guns. The first of these ships – USS *Atik* – was sunk by *U-123* on the night of March 26, 1942. No one was recovered. No wreckage was found. A German radio report announced that "a Q-boat ... was among 13 vessels sunk off the American Atlantic coast ... by a submarine only after a 'bitter battle', fought partly on the surface with artillery and partly beneath the water with bombs and torpedoes."[4]

Two of the American "Q-ships" were deployed to the Caribbean. USS *Asterion* operated out of the American naval base on Trinidad and made a few convoy runs westward of Aruba before returning to New York at the end of December. It encountered no German submarines. USS *Big Horn* worked out of Trinidad and Curaçao. On October 16, 1942, it was steaming in the wake of convoy T-19, eastward from Trinidad, when two freighters were torpedoed. Its gunners trained on a periscope but could not open fire without damaging some of the ships in convoy. Not long after, another chance to shoot arose, but a sub-chaser crossed in between *Big Horn*'s guns and the target. On November 10, *Big Horn* was sailing in convoy TAG-20 when U-boats attacked the gunboat USS *Erie*, just 1,000 yards off its starboard bow. But again, the "Q-ship" had no chance to avenge the attack. *Big Horn*'s failure to actually engage a submarine during two convoy attacks and the sinking of the *Atik* on its maiden voyage are apt testimony to the uselessness of the "Q-ship" program.[5]

At President Roosevelt's initiative, the United States mounted a major ship construction program in the late 1930s. Beginning in Fiscal Year 1938, 73 destroyers of the Porter, Somers, Benson/Gleaves, and Bristol classes were authorized, bringing the total of modern destroyers up to 100 by the time of Pearl Harbor. There were minor differences in these different classes of ships, but in the main they were fast and well-armed. Most were immediately sent to the Pacific after December 7, and many of the rest were dispatched to the United Kingdom.[6] One of these was the USS *Lansdowne*. Displacing 1,630 tons, it carried four modern five-inch dual purpose guns, a variety of heavy automatic weapons, depth charge racks and throwers, and a multiple torpedo launcher. On July 13, 1942, *Lansdowne* was designated flagship of Destroyer Division 24 and deployed off the east coast of Panama to help with the recent U-boat onslaught.

Also operating in those waters was *U-153*, a Type IXC boat under the command of Wilfried Reichmann. The Korvettenkapitän was a member of the Class of 1924, and at age 36, senior in years. But he was inexperienced – while bringing *U-153* from the Baltic Sea to the Bay of Biscay in November 1941, he had collided with another brand new boat, *U-583*; the latter was lost with 45 men on board.[7] Reichmann left Lorient on his first war patrol on June 6, 1942. On June 25, he sunk his first ship, the 5,268-ton British steamer *Anglo-Canadian* (ironically, this ship had survived a Japanese air attack on April 16, 1942, in the Bay of Bengal), southeast of Bermuda. His next victim, sunk two days later, was the 6,058-ton American vessel *Potlatch*, with a cargo of trucks and tanks; 49 crew members abandoned ship and spent 32 days in a lifeboat drifting from one uninhabited island to another before landing in the Bahamas. *Potlatch* was east-northeast of Guadeloupe Island when it sank – *U-153* was moving across the Caribbean toward the coast of Colombia. Sure enough, two days later, another American ship, the 4,833-ton *Ruth* fell to Reichmann's torpedoes northwest of Great Inagua Island.[8] By then, *U-153* was hunting close to the coast of Panama. Reichmann may have been unaware of the large number of US planes and aircraft operating out of Cristóbal. On July 5 and 6, he was twice caught on the surface by B-18 bombers of 59th Squadron north of Bahía de Portete, Colombia. In the second of these attacks, the bomber carried out a beam attack and dropped four depth charges before *U-153* slipped beneath the surface.

USS *Mimosa* was a small, 560-ton net tender armed with a single three-inch gun, and with a top speed of 12.5 knots. It was about 60 miles off Almirante, Panama, in the evening of July 11, having laid antisubmarine nets off Puerto Castilla, Honduras. *Mimosa* should have made easy pickings for *U-153*, which fired no fewer than five torpedoes at this relatively insignificant target, but all missed. Reichmann's very poor shooting may have resulted from a gross underestimation of *Mimosa*'s draft and size, or from desperation. No one will ever know. *Mimosa* immediately radioed word of the attack, bringing a PBY out at 4 a.m. It detected *U-153* on the surface, dropped flares and then four depth charges. The sub went deep, but not deep enough. Over the next 24 hours Patrol Craft 458 – which had followed the PBY to the scene – and several other aircraft kept vigil over the general area where the U-boat was last seen. At 10:13 a.m. on July 13, PC-458 spotted an oil slick and dropped its remaining complement of six depth charges. The aircraft followed suit but, other than the oil slick, there was no sign that the U-boat had been destroyed.

That same day, USS *Lansdowne* arrived with a convoy at Cristóbal. Its skipper, Lieutenant-Commander William R. Smedberg III, was ordered out of harbor at flank speed to join the hunt. *Lansdowne* reached the scene at 6:30 p.m. and slowed to begin a sonar search. Fifteen minutes later: contact! Smedberg maneuvered his ship into attack position and dropped 11 depth charges. At first, there was only the usual roiling of the sea. Then came the unmistakable sound of an underwater explosion, followed by a great spreading slick of fuel oil. *Lansdowne* claimed a kill; postwar records confirmed the death of *U-153*.[9] After *U-157* (killed by the Coast Guard cutter *Thetis*), *U-153* was the second Caribbean submarine destroyed by the US Navy or Coast Guard.

The US Navy might have done better sinking submarines in the Caribbean in 1942 – given that it was hardly present in the North Atlantic, and given that it had a surfeit of World War I era "flush-deck" destroyers. Although narrow-beamed and top-heavy, they were perfectly suited for Caribbean operations. Yet three were lost or heavily damaged in 1942 – *Blakeley*, *Sturtevant*, and *Barney*.[10] One US Navy officer concluded in a Naval War College paper written in 1996:

> The United States Navy had no effective system of promulgating "Lessons Learned" to units not previously involved in antisubmarine warfare. Therefore, the lack of organization and experience among newly assigned ships and squadrons hastily deployed to the Caribbean, significantly improved the survivability of the U-boats they engaged.[11]

* * *

The Royal Navy's Escort Group B5 began operations in Caribbean waters in mid-May 1942. It initially consisted of the destroyer HMS *Havelock* as escort leader and four Flower-class corvettes. *Havelock* had been under construction for Brazil in a British shipyard when war broke out and was requisitioned for service with the Royal Navy. In early July, B5 covered the WAT/TAW convoys from Key West (later convoys sailed from Guantánamo as GAT/TAG) to Aruba and to Trinidad and back. It was joined for about eight weeks by HMS *Churchill*, a four-stack Town-class transferred to the Royal Navy in September 1940. B5 covered a lot of sea miles in the Caribbean but sank no U-boats.

Churchill picked up 37 survivors from the American tanker *Franklin K. Lane*, torpedoed by *U-502* on June 9, 1942, northeast of Cape Blanco, Venezuela, and then sank the wreck by gunfire. It rescued another 50 survivors from the *Delmundo*, also an American ship, sunk by *U-600* south of Cape Maisí, Cuba, on August 13, 1942.[12] But several days later, B5 lost two merchant ships – the tanker *British Consul* and the cargo vessel *Empire Cloud* – from a 14-ship TAW convoy outbound from Trinidad, to *U-564*.[13]

Canada, too, became enmeshed in the war in the Caribbean when Chief of the Naval Staff Admiral Percy Nelles ordered two British destroyers under his command – HMS *Burnham* and *Caldwell* – to begin escorting Canadian tankers and ore carriers up from Caribbean ports in late April 1942. Canada had already experienced shortages in oil stocks due to diversions of tankers away from the Canadian trade and the loss of tankers at sea or into repair yards. By the end of April 1942, naval fuel stocks at Halifax and St. John's, Newfoundland, had dwindled to 45,000 tons. On May 22, Ottawa instituted a full-fledged tanker convoy route between Halifax and Port of Spain, Trinidad (HT and TH). The Canadians were so short of escorts that they could only spare four corvettes for the task, two for outbound convoys and two for inbound, thus severely limiting the number of tankers on the runs via either the Mona or Windward Passages.[14] In early July, an additional corvette was assigned in each direction. No tankers were ever lost on these runs.

Late in August 1942, the Commander of Allied Forces Aruba-Curaçao, a US naval officer, ordered the Canadian corvettes HMCS *Halifax*, *Oakville*, and *Snowberry*, as well as the Dutch minelayer *Jan Van Brakel*, to link up with convoy TAW-15, outbound from Trinidad via Aruba to Key West. In the morning of August 27, the convoy – now 29 ships and escorted as well by the US destroyer *Lea* and three small patrol craft of the "Donald Duck Navy" – was spotted by *U-94*, commanded by Oberleutnant Otto Ites.

Ites was only 24 years old but already a decorated veteran with four successful war patrols and a total of 15 ships sunk of 76,882 tons. He first spotted the convoy, arrayed in six columns, just after 6 p.m. as the sunset's

Otto Ites. Another Knight's Cross winner, Kapitaenleutnant Ites commanded *U-94* on four Atlantic and one one Caribbean patrol. He sunk 15 Allied ships of 76,882 tons, before being sunk by depth charges from an aircraft and severe ramming by the Canadian corvette HMCS *Oakville* on 28 August 1942. Ites was among the survivors, and after captivity in the US until May 1946, he became one of the very few U-boat commanders to accept service in the new West German Navy (Bundesmarine). Source: Deutsches U-Boot-Museum, Cuxhaven-Altenbruch, Germany.

radiant beams of orange and red spread over the dark waters. Ites tracked TAW-15 for six hours, while radioing its position; Friedrich Steinhoff's *U-511*, also in the Caribbean, picked up Ites' signals and began to head for the convoy. The sky was clear with the moon shining brightly above. Ites brought *U-94* to the surface and crept into position about three miles astern of the convoy. He penetrated the escort screen through a gap between *Oakville* and *Snowberry*. Suddenly, the dreaded cry from the bridge watch: "Alarm! Aircraft at 235o!" It was a US Navy PBY Catalina, about a quarter mile off the port beam. Its pilot was Lieutenant Gordon R. Fiss, and it carried four MK 29, 650-pound depth charges. Fiss later reported that "the submarine was visually sighted in the moon path ... fully surfaced."[15] The Catalina was flying at 500 feet and Fiss knew immediately that he had a good chance of making his bomb run while *U-94* was still on the surface. He flew low over the U-boat and released his depth charges. "A quick glance astern a few seconds later revealed the conning tower becoming obliterated by the bomb upheaval. Members of the crew in the waste hatch stated the stern of the submarine was raised clear of the water." To mark the spot, Fiss dropped a flare into the dark waters.

Ites was lucky. Fiss had set the depth charges at 50 feet; at 25, they would have finished *U-94* off there and then. Still, the boat took a terrible pounding and the pressure wave from the explosions drove it up to the surface. *Oakville* was closest when the Catalina attacked.[16] Sub-lieutenant Hal Lawrence later recalled the moment when he first spotted *U-94*:

> Four plumes of water from the aircraft's depth-bombs were subsiding into a misty haze and showing small, ethereal rainbows in the moonlight.... The aircraft circled in a continuous, tight bank, making S's in Morse code with its signal light.... A flare drifted down, its ghostly radiance matching the moon.[17]

Oakville's skipper, Lieutenant-Commander Clarence A. King, heeled the corvette over hard and headed for the swirl where the submarine had disappeared. "Fire a five-charge pattern," he barked out. As the submarine broke the surface about 100 yards from *Oakville*, King gave new orders: "Stand by to ram!" But Ites managed to maneuver away, and the corvette only struck *U-94* a glancing blow. Then *Oakville*'s four-inch gun, one of

HMCS *Oakville*, the only Royal Canadian Navy vessel to have a confirmed submarine kill within the Caribbean theatre of the Second World War. Image courtesy of the Royal Canadian Navy MC-2725.

its 20-mm Oerlikons, and a .50-caliber machine gun opened up, striking *U-94* on the conning tower and scoring hits on the hull. Again, King turned to ram. But again *Oakville* only scored a glancing blow. King ordered more depth charges, right under the U-boat. Then he swung out for another try at ramming it.

The second and third set of depth charges crippled the boat. Machinery broke loose from its mountings, valves blew, oil hoses ruptured, gauges broke, and urine buckets rolled in the bilge. One of the four-inch shells had blown the boat's 8.8-cm deck gun off its base; another had hit *U-94* "squarely abaft the conning tower." All the while, *Oakville*'s machine guns raked the crippled U-boat with intensive fire; Ites took two bullets in the leg then yelled: "Abandon ship!" *Oakville* turned into the submarine again; this time it rode right over *U-94*. Some of *Oakville*'s sailors threw empty Coke bottles and yelled "partisan baseball invectives" at the Germans, barely 20 feet away. *U-94* wallowed in the moderate sea, without artillery or power.

Lieutenant Hal Lawrence and Stoker Petty Officer A. J. Powell. The unorthodox boarding of *U-94* by Lawrence and Powell became legendary within the history of the Royal Canadian Navy. Image courtesy of the Royal Canadian Navy, H-04137.

King ordered "Away boarding party ... Come on Lawrence! Get cracking!" Lawrence, Petty Officer Art Powell, and about 12 other hands made ready to board the submarine as King maneuvered alongside. It was ten feet down from the corvette deck to *U-94*. As the two vessels grated alongside, Lawrence and the others jumped to the submarine's deck. He later wrote: "I was always a romantic youth, and from age ten onward, stories of the Spanish Main were a large part of my literary diet." As he hit the U-boat's deck, he thought, "Mother of God! I really *am* boarding an enemy ship on the Spanish main." Suddenly the belt of his tropical shorts broke; the shorts fell to his feet. He stumbled and kicked them off and with pistol in hand he lurched up the deck towards the conning tower. He and Powell were alone. Someone from *Oakville* opened fire with a machine gun. Bullets snapped through the air and pinged on the submarine's superstructure; Lawrence and Powell jumped into the water. Then when the shooting stopped they were swept back onto the submarine with the next wave. They moved cautiously toward the conning tower and shot two crew members advancing on them from the hatch. Then Powell shepherded the escaping Germans aft while Lawrence went below to search; he found nothing of value. Sea water had reached the boat's batteries, producing chlorine gas. Lawrence and Powell abandoned the vessel, which sank shortly after. *U-94* was the only submarine destroyed by the Royal Canadian Navy in the Caribbean.[18]

Less than a week later, the Royal Navy scored its first and only kill in the Caribbean theater, just outside the Windward Islands. On December 18, 1941, the British battleship HMS *Queen Elizabeth* was damaged by an Italian human torpedo attack in Alexandria harbor, Egypt. The Royal Navy ordered the battleship to Norfolk, Virginia, for repairs via the Cape of Good Hope and the Azores, escorted by the destroyers HMS *Vimy*, *Pathfinder*, and *Quentin*. On September 3, American destroyers took over escort duties, and the three British destroyers headed toward Port of Spain, Trinidad.

The group was about two hundred miles from its destination at 6:05 p.m. when *Pathfinder*'s radar obtained a contact about 1,200 yards off the port bow. The officer of the watch, Lieutenant C. R. Halins, made ready to attack, but *Pathfinder*'s skipper, Commander E. A. Gibbs, instead ordered the ship to stop so that it could carry out a full sound sweep. The decision was almost fatal. As Gibbs later remembered: "Whilst investigating, [high-speed screw sound] from a torpedo was heard … and a torpedo broke surface ahead and, running on the surface, circled widely to port and narrowly missed *Quentin*."[19] *Pathfinder*'s ASDIC immediately picked up a solid contact at 600 yards.

The "contact" that had fired the torpedo was *U-162*, commanded by none other than Fregattenkapitän Jürgen Wattenberg, one of the Caribbean "aces." He was on his third war patrol and had just torpedoed 30,481 tons of Allied shipping in 11 days, raising his overall total to a whopping 85,662 tons. The watch had spotted one of the destroyers coming straight on from ten miles away, and Wattenberg had taken *U-162* down to periscope depth and prepared a nasty surprise for what he thought was a lone destroyer. For some reason, he was unaware that he had picked a fight with three well-armed British destroyers. The single bow torpedo that he had fired had broached and run in a circle, "going up and down like a dolphin," passing beside *Quentin*, which was turning for its life. At that moment, the submarine's hydrophone operator screamed out that there was not one, but rather three "tin cans" on a line bearing almost abeam, one mile apart.[20]

No doubt relieved that the torpedo had missed, Commander Gibbs ordered a Mark VII ten-charge pattern, some set at 150 and others at 300 feet. The hammer blows of the 300-lb. warheads shook *U-162* violently, disabling its hydrophones and damaging one of its dive tanks. After *Pathfinder*'s attack, *Quentin* also ran in and dropped a six-charge pattern.[21] The deadly bracket of explosions and the accompanying pressure wave again rocked *U-162*, this time damaging its diving planes and rudders as well as causing a leak in the engine room. Wattenberg had enough. He headed southward, running silently, and took the boat deep: "A+120," an incredible 200 meters. The British destroyers lost contact.

There is no real certainty of what happened aboard *U-162* during these punishing attacks, but Wattenberg now knew he was facing three destroyers and that his boat had suffered major damage. For three hours, he eluded his attackers, heading off in the opposite direction whenever one approached, and slipping behind them when they seemed on the verge of attacking. Once he had put distance between himself and his attackers, around 11 p.m., Wattenberg surfaced. The night was pitch black. A few rain squalls were coming in from the direction of Barbados and Tobago. Good weather for an escape. *U-162* began the run of its life.

But Wattenberg's adversary was equally wily: Commander Gibbs detailed *Vimy*, the only one of the three destroyers with the new Type 271 centimetric radar, to stay put, while he took *Pathfinder* and *Quentin* to sweep eastward. It was a bold move, for by now *U-162* was heading back to the Caribbean on a course of 315 degrees. But luck was with Gibbs: within ten minutes, *Vimy* obtained a contact at 2,800 yards. Its skipper, Lieutenant-Commander Henry G. de Chair, ordered "full ahead both" and opened fire with the forward four-inch gun. But de Chair could not see his target in the glare of the forward gun and ordered cease fire. Suddenly, he saw the submarine stern-on and prepared to ram.

The V-Class Destroyer HMS *Vimy* at anchor. Source: Ken Macpherson Photographic Archives, Library and Archives at The Military Museums, Libraries and Cultural Resources, University of Calgary.

While Wattenberg and de Chair were locked in mortal combat, Gibbs at 11:27 p.m. saw two red signal rockets to the westward, followed by gun flashes and white star shell bursts. Wattenberg had tried one last desperate ruse – two red lights was the current recognition signal for British submarines. Gibbs was not deceived. He ordered *Quentin* and *Pathfinder* to make for *Vimy* at 30 knots. "It was plain to me that either 'Vimy' was sinking the submarine or the submarine was sinking 'Vimy' – I was not at all sure which."[22]

Jürgen Wattenberg knew that he was finished. For the second time in his life – after having been party to the decision to scuttle the "pocket" battleship *Graf Spee* in Montevideo Harbor in 1939 – he ordered a crew to abandon ship. *Vimy* bore in. When the two vessels collided, *U-162*'s hydroplane sliced into *Vimy*'s hull above the waterline and damaged its port propeller. "We were left wallowing alongside the U-boat," de Chair later wrote, "whose crew were on deck wearing lifebelts." No quarter was asked and none was given. De Chair ordered star shell from one of the stern guns and a charge fired at 50 feet over the submarine; all the while, he raked the U-boat with machine-gun fire. The depth-charge explosion was the last blow: *U-162* went down fast. Wattenberg and all his crew save one were rescued by the British destroyers.

* * *

It was obvious from the beginning of the war in the Caribbean that aircraft would play a major role in the defense of shipping. At best, aircraft could destroy U-boats; at worst, they could force them to submerge, robbing them of their speed, range, and mobility. The first major problem with aircraft was that virtually all those deployed in the Caribbean by the British and the Americans were short-range. Very Long Range (VLR) aircraft, particularly the B-24 Liberator, were in great demand to cover North Atlantic convoys, but very few were available in 1942 even for that most important mission. By the fall of that year, the lack of long-range aircraft in the Caribbean was essentially resolved as the Allies put into use a ring of airfields and seaplane bases from Cozumel, Mexico, to Waller Field, Trinidad, and from San Juan, Puerto Rico, to France Field, Panama. All manner of aircraft were deployed. The British used the venerable Swordfish, the Lockheed Hudson bomber, and the Avro Anson trainer. The US Army Air Forces deployed predominantly B-18 "Bolos" and A-20 "Havocs," but also the occasional B-24 in 1943. Fighters such as the P-39 Aerocobra and P-40 Warhawk flew reconnaissance. The US Navy used both the PBY Catalina and the PBM Martin Mariner; the latter accounted for the greatest destruction of U-boats in and near the Caribbean over the course of the campaign.

The second major problem was getting sufficient aircraft in place to make certain that no submarine could run on the surface, day or night, and that no periscope could pop up by day, without being spotted from the air. At first, Army Air Forces officers were reluctant to employ their longer-range aircraft for antisubmarine patrol. To a certain extent, the navy was too. Both clung to the already obsolete concept that the greatest danger in the Caribbean was either enemy carrier-borne aircraft or aircraft which would use secret fields in Central or South America. The army was also unwilling to place its aircraft under navy command. Thus, in April 1942, the total number of US aircraft available for several thousand miles on both sides of the Panama Canal were 28 heavy bombers (mostly B-17s based in Panama), of which 15 were equipped with Anti-Surface Vessel (ASV) radar, 30 medium bombers, 16 light bombers, and 34 Navy PBYs, in addition to fighter planes.[23] The lack of ASV radar was especially acute when the campaign began, but by the summer of 1942 virtually all army and navy aircraft were so equipped.[24]

The network of small airfields and airstrips along the island chain and in Central and South America was anchored on the complex of air bases in the South Florida area, the Canal Zone, Puerto Rico, and Trinidad. Trinidad alone hosted four main bases – Chaguaramas, Piarco, Waller Field, and Edinburgh Field. Chaguaramas Naval Air Station was home to more than 14 squadrons of Catalinas and Mariners. There were usually up to 60 at a time using the base for routine patrolling of the sea lanes and convoy escort. Waller Field was the US Army Air Forces' primary base in the eastern Caribbean. Intended to be the main center for combat flying, Waller was increasingly used as a major link in the trans-Atlantic air route between the United States and the Middle East via Africa. As a result, most combat aircraft in the eastern Caribbean were transferred to Edinburgh Field in central Trinidad, which initially had been designed merely as a satellite runway complex for Waller Field.

While the Royal Navy played a very small role in the air campaign against the U-boats in the Caribbean, the Americans learned how to sink U-boats by studying British methods. As the official history of the US Army Air Forces in World War II states,

> ... both Army and Navy antisubmarine forces were able to draw largely on the experience of the British for their initial stock of technical data, and they made extensive use of their opportunity. Of particular aid to the AAF units was the help given by two liaison officers sent to the United States in February [1942].[25]

With the help of the British, effective tactical doctrine was worked out over several months. The Americans learned that there was no point bombing submarines more than 15 seconds after a crash dive. The British strongly recommended guns mounted in forward turrets rather than fixed firing guns, giving gunners greater ability to sweep the target as they approached. It was found that depth charges and depth bombs worked best at minimum depth, so that when they bracketed a surfaced U-boat, they would explode shortly after they dropped into the water, severely damaging the submarine's underside. It was also stressed that the entire load should be dropped at the same time.[26]

As the number of aircraft increased, so too did attacks on surfaced U-boats. July 1942 was especially busy. On July 5, a plane flying off Aruba sighted a submarine three miles ahead, dove to 400 feet and dropped a full stick of depth bombs. The next day, an aircraft hit a surfaced submarine off the coast of Venezuela, but the bomb rolled into the sea without exploding. In the waters off Trinidad, on July 11, a night attack was carried out, but the bombs were dropped without visible effect. The next night, another attack was made off Cristóbal, but again with no impact. On July 16, a ferry plane spotted a surfaced U-boat and dove on it; the depth bombs exploded all around the boat just as it disappeared below the surface. No damage was reported. On July 19, a submarine was sighted between Cuba and Jamaica and four depth bombs were dropped. Once again, all were duds. On July 20, three aircraft attacked a surfaced submarine off Georgetown harbor, Jamaica, but no damage was done. That same day, another attack was carried out on a U-boat between Aruba and Curaçao, but with no effect. And a night attack on a submarine on July 29 produced no damage.[27]

In all these cases, pilot or bombardier technique (or error), faulty equipment, poor tactics, or sheer bad luck saved a U-boat to fight another day, but the frequency and the intensity of the attacks was a worrisome sign to Dönitz's captains that opportunities would grow far slimmer in the coming months. Radar-equipped aircraft made it increasingly unsafe to surface day or night, keeping the crews sealed in their hot and humid hulls and making the captains super-cautious whenever planning attacks. Doubtless, the American flyers would only get better. But the air attacks were not completely fruitless: on August 22, 1942, *U-654* was destroyed north of Colón, Panama, by a B-18 from Bomb Squadron 45; and on October 2, another B-18 sank *U-512* off Cayenne, French Guiana.

When "Teddy" Suhren, already holder of the Knight's Cross with Oak Leaves, brought *U-564* into West Indian waters in the summer of 1942, he and his crew suffered quite a shock. They were proceeding on the surface when the startled cry that the crew dreaded to hear burst from the bridge: "*Flieger!*" A large enemy aircraft was closing rapidly from out of the sun, flattening out only 20 meters above the waves and heading rapidly into a low-level attack. Suhren ordered "Emergency Dive!"

> With mere metres of water over her bridge, three well-placed bombs bracketed the U-boat, severely shaking the hull and causing fresh chaos aboard. A thin jet of flame shot from the closed hatch to number five torpedo tube.... However, there was no water leakage.[28]

The boat sank to nearly 200 meters – dangerously close to crushing depth – before Suhren was able to regain control. *U-564* eventually reached the surface, only to be destroyed the following year under a new skipper.

With the increase in air coverage, the number of submarine attacks fell off rapidly. As the official history of the US Army in the western hemisphere in World War II put it: "Losses throughout the Caribbean area were the lowest in six months.... The cyclical pattern of the U-boat assault had already manifested itself. That the October lull would be followed by renewed activity was expected, but it was impossible to foretell precisely how high the new peak would reach."[29]

13

A HARD WAR:
HARTENSTEIN AND *U-156*

The loss of Otto Ites in *U-94* and Jürgen Wattenberg in *U-162* made clear to Admiral Karl Dönitz that Allied antisubmarine warfare in the Caribbean basin had improved dramatically. The enemy was now able routinely to detect the "gray sharks" and then to direct airplanes and flying boats with new "direction finders" (radar) to destroy them. The seven Type IXC boats that sailed to the Americas in July 1942 had torpedoed 23 ships of 130,000 tons – about three ships of 18,500 tons per boat. The five Type IXC boats that sortied in the Caribbean in August 1942 did even better, sinking 30 ships of 143,000 tons – six ships of 28,600 tons each. But the overall balance sheet was negative: in 1942, American, British, and Canadian shipyards produced 7.1 million tons of new merchant shipping (including 92 large tankers), about one million tons more than the U-boats destroyed.[1] Obviously, the "gray sharks" were not making a major dent in the Allies' merchant-ship pool of 30 million tons. The constantly updated charts at U-Boat Headquarters made clear to Dönitz that the Allies were beginning to win his "tonnage war."

There were other setbacks. On October 21, 15 of 90 B-17 and B-24 bombers from General Ira Eaker's American Eighth Air Force Bomber Command in Britain dropped 60,000 lbs. of bombs on Lorient, killing 160 Germans and 180 conscripted Belgian or Dutch laborers. They did little damage to the massive steel-reinforced concrete U-boat pens at Kéroman. In nine follow-up raids to January 3, 1943, 357 of 870 bombers reached Lorient and dropped their loads.[2] The raids drove home the point that the Allies had seized the initiative. After several follow-up bombing runs by the Royal Air Force in January and February 1943, most of the civilians not employed in the German war effort fled Lorient. The port became part devastated ghost town and part German war base, a purely military target.

Also critically, and of course unknown to Dönitz, Bletchley Park's code-breakers (ULTRA), using the Short Weather Keys recently taken off *U-559* after it had been attacked by three Royal Navy destroyers in the Mediterranean, managed by December 13, 1942, to get enough "cribs" finally to break the German four-rotor "Triton" Enigma keys.[3] By the end of the month, they were able to achieve solutions in about 12 hours.

But there was much bleaker news from the larger war front. In November, the Allies dramatically altered the strategic situation in North Africa. By November 4, General Bernard Montgomery's Eighth Army had stopped Erwin Rommel's Afrika Korps dead in its tracks at El Alamein, ending forever German hopes of taking the Suez Canal and gaining access to the vast oil reserves of the Middle East. Four days later, Anglo-American ground forces landed at Morocco and Algeria (Operation TORCH), meeting opposition only from some Vichy French forces stationed in Algeria. The Afrika Korps was now sandwiched between two Allied armies. Benito Mussolini's cherished "Italian lake," the Mediterranean, was again firmly in Allied hands.

The greater disaster, of course, came in the East. On January 31, 1943, the newly promoted Field Marshal Friedrich Paulus surrendered the Sixth Army at Stalingrad: a Soviet counteroffensive by Aleksandr Vasilevsky and Georgi Zhukov resulted in the death of 147,000 and the capture of 91,000 German and Romanian soldiers. It was a strategic defeat of the first magnitude. Operation Blue's initial successes of the summer in the South had turned to disaster. Visions of German control of the Eurasian heartland as far eastward as the Caspian Sea receded from view, as did those of control of the Transcaucasian oilfields. Germany had lost the initiative in the war in the East as well.

For Dönitz, these setbacks seemed to demand a drastic reassessment of the Battle of the Atlantic. Instead, his war diary (KTB) revealed a strange composite of operations manual and pep-rally script. On August 11, 1942, he yet again insisted that the U-boat war was purely an operational art form, and not a strategic design. It was all a simple matter of "sinkings, regardless where and whether [the ships were] loaded or in ballast."[4] One of Grand Admiral Erich Raeder's staff officers could not help but sarcastically note "truly remarkable" in the margin of the document where Dönitz had juxtaposed "strategic pressure" and tactical "sinkings"; he wondered whether "Commander U-Boats has been misled in these matters or simply does not want to understand them"![5]

One month later Dönitz acknowledged that Operation Neuland had run its course.[6] "After the disappearance of single-ship traffic, the area [Caribbean] no longer bears fruit. Strong aerial surveillance makes an attack approach against a convoy difficult, if not impossible." The few successes that his skippers recently had scored were due "mainly to chance." Moreover, "these successes have been paid for by relatively high losses … presumably due to air attacks." German radar detectors such as the FuMB-1 produced by the French firms Metox and Grandin had proven ineffective against Allied "direction finders," and the tropical heat hardly made "lengthy sorties promising." From now on, Commander U-Boats would dispatch boats to the Caribbean only in small groups of two or three.

In November 1942, Dönitz again consigned his purely operational thoughts about the Battle of the Atlantic to the war diary: "The tonnage war is the primary task of the U-boats, perhaps the decisive contribution by the U-boats to the outcome of the war," he wrote. "It must be conducted where the greatest successes can be gained at the least cost." All available boats needed "to be concentrated for this primary task," even "at the cost of thereby creating gaps and weaknesses in other areas."[7] The frequency of such entries into his official war diary allows the comment that they seem almost to be mental reminders as well as operational justifications for the simple "tonnage war" that he was waging against the Allies.

By spring 1943, the pep-rally rhetoric totally dominated Dönitz's war diary. Early in May 1943, he "demanded" of his captains that they "continue resolutely to take up the struggle with the enemy," that they counter "his cunning and technical innovations" with their own "ingenuity, ability and iron will."[8] Later that month, he called on them to sink and sink, fight and fight, and "force the enemy to undergo a permanent bloodletting, one by which even the strongest body must slowly and inevitably bleed to death."[9] Dönitz had chosen his words carefully: the term "bleed white" had been used by General Erich von Falkenhayn, Chief of the General

Staff, in his Christmas Memorandum of 1915 to justify the war of attrition planned at Verdun the following spring.

Dönitz's professional life soon was radically changed by events over which he had no control. Adolf Hitler had increasingly become disenchanted with the role of the surface fleet in the war – beginning with the scuttling of the "pocket" battleship *Graf Spee* in Montevideo in December 1939, through the sinking of the brand-new battleship *Bismarck* in the Atlantic in May 1941, to the interminable delays in working up its sister ship *Tirpitz*. The last straw in that seemingly endless list of failures and disappointments had come on December 31, 1942, when the heavy cruisers *Hipper* and *Lützow* failed to dispatch Convoy JW.51B in the Arctic, with *Hipper* limping back to port heavily damaged.

Over lunch on December 30, the Führer had bitingly informed Vice Admiral Theodor Krancke, Raeder's representative at Military Headquarters, that the German Navy was but a carbon copy of the British Royal Navy, "only more miserable, the U-boats excepted." The fleet "lacked the willingness to engage; [the] ships lay around in fjords like so much scrap metal; are totally useless." By New Year's Day, Hitler had worked himself up into a full lather. Furiously pacing up and down in his bunker at the Wolf's Lair near Rastenburg, East Prussia, he spat out the full measure of his venom at Krancke. "Ships totally worthless; because of the lying about inactive and the lack of willingness to engage, are only a hearth of revolution; this means the death of the High Sea Fleet." The reference to the mutiny in the fleet that had precipitated the revolution of 1918 was intended to cut to the quick. More, German surface ships fired only on unarmed freighters; unlike the Royal Navy, they did not "fight to the bitter end."[10] To rub salt into the wound, Hitler ordered his comments to be taken down in writing, with Field Marshal Wilhelm Keitel, head of the Supreme Command of the Wehrmacht (OKW), as witness.

Still, Hitler was not done. On January 6, 1943, he subjected the Commander in Chief Kriegsmarine to a 90-minute tirade about the ineffectiveness first of the Prussian and later of the German surface fleet. It had remained "without effect" during the Franco-Prussian War (1870–71); it had remained "without effect" during World War I; and its role in the revolution of 1918 and the scuttling at Scapa Flow in 1919 constituted "no page of honor" in its history. Now, as then, the light forces, and especially the U-boats, were carrying the main burden of the war. The Führer demanded that the "big ships" be mothballed and their guns deployed as land-based artillery.[11]

Raeder was crushed. He immediately requested that he be allowed to retire, effective January 30, 1943 – the tenth anniversary of his service to the Führer. As a possible successor, he suggested two men, Rolf Carls and Karl Dönitz. Given Hitler's rant against the surface fleet (with which Carls had served both as its Chief and as Naval Commander North), the choice was simple: on January 31, Dönitz was promoted to the rank of grand admiral and appointed Commander in Chief Navy. As well, he continued in his role as Commander U-Boats, elevating Generaladmiral Hans-Georg von Friedeburg to chief of staff, with the formal title of Commanding Admiral, U-Boats.

With regard to the Battle of the Atlantic, Dönitz's immediate mission was to destroy as much of the American supply line to the fighting fronts in North Africa as possible. In March 1943, the greatest convoy battles of World War II took place in the turbulent waters of the North Atlantic: while the U-boats were able to destroy 84 merchant ships of 505,000 tons, they lost 14 boats and 650 men of their own. For the first time in the war, aircraft sank more "gray sharks" than did surface craft in a given month.

Dönitz apprised Hitler of the seriousness of the situation.[12] "Losses are high. The U-boats' struggle is hard." Still, there was no alternative other than to continue the fight, to destroy 100,000 to 200,000 tons more shipping per month than the Allies could build. To "bleed the enemy white" – a term that he used again – the Reich would have to escalate U-boat production from 27 boats per month in the first half of 1944 to 30 by the end of 1945. Losses among the roughly 425 to 438 boats on patrol continued at a relatively steady rate: 19 in February, 15 in March, and 14 in April 1943. But in May, when Dönitz learned that the Allies had destroyed 38 U-boats – which he termed "a frightful total" – he came to the "logical conclusion" of withdrawing the boats from the North Atlantic. From now on, they would be deployed in the "area south-west of the Azores."[13] This made sense because the Allies, highly concerned about the loss of tankers in the Caribbean basin, had directed them to take their oil to Cape Town, South Africa, where naval units operating in the Indian Ocean could draw fuel.

Dönitz was fully aware that he was losing the "tonnage war." On January 6, 1943, President Franklin D. Roosevelt personally had delivered his annual Budget Message to the Congress. It was a blockbuster. He demanded that the Republic raise its production of aircraft from 45,000 in 1942 to 75,000 in 1943, and of tanks from 45,000 to 75,000 during that same period. His figures for new merchant-ship construction were still more staggering: from 1.1 million tons in 1941 to six million in 1942 and to ten million by 1943. The American press calculated that this would translate into "a plane every four minutes in 1943; a tank every seven minutes; two seagoing ships a day."[14]

While Hitler and his propaganda minister, Joseph Goebbels, decried Roosevelt's demands as constituting "hysterically inflated figures," "absolute nonsense," and Jewish-inspired "truly criminal politics,"[15] Dönitz appreciated that the United States was becoming the major player in the Battle of the Atlantic. In 1942, the Republic built more than 15,000 aircraft and, in 1943, almost 25,000, as well as 108 warships in 1942 and 369 in 1943. With regard to merchant shipbuilding, US yards produced 5.5 million tons in 1942 and a staggering 11.4 million in 1943.[16] By war's end, the United States had launched 5,800 vessels (including 600 tankers), while the U-boats had destroyed 733 American merchantmen. To stem this flood of American-produced shipping, Dönitz rushed his boats out into the South Atlantic at a frenetic pace.

* * *

Werner Hartenstein in *U-156* and Albrecht Achilles in *U-161* belonged to a group of eight boats that Dönitz had dispatched to the South Atlantic in the fall of 1942. Dönitz could rely on Hartenstein and Achilles, both veterans of the first wave of New Land boats, to find fresh opportunities south of the Azores. In August, *U-156* joined three other Type IXC boats as Group *Eisbär* (Polar Bear), supported by the "milk cow" *U-459* and tasked with carrying out a surprise attack on Cape Town. In September, *U-161* became part of Group *Streitaxt* (Battle-Axe) and was redirected to Grid Quadrant ER 50, where it was to be resupplied from the tanker *U-461* before heading for the waters off Brazil.

U-156 sailed out of Lorient on August 20. Its passage of the Bay of Biscay was a litany of having to dive to evade enemy aircraft. Running submerged depleted the batteries, and Hartenstein was forced repeatedly to bring *U-156* up on the surface to recharge them and to ventilate the boat.[17] After torpedoing the 5,941-ton British ore freighter *Clan Macwhirter* about 600 nautical miles west of Casablanca late on August 26, Hartenstein crossed the equator and on September 12 stood 550 nautical miles south of Cape Palmas, Liberia. At 7:07 p.m., he fired two bow torpedoes at what he took to be "an old passenger freighter." He had torpedoed the 19,695-ton Cunard White Star Line troop transport *Laconia*. In what U-boat historian Michael Hadley has called "a scenario unique in nautical lore,"[18] Hartenstein asked Dönitz to arrange a "diplomatic neutralization of the scene of the sinking," that is, a sort of unofficial cease-fire, and under a Red-Cross flag to rescue as many of the survivors as he could. All the while, an American B-24 Liberator piloted by Lieutenant James D. Harden for two days made numerous bombing runs at the U-boat, incredibly failing to destroy it. Dönitz was livid. On September 17, he issued his Triton-Null Order (soon to be known as the *Laconia* Order):

> All attempts to rescue the crews of sunken ships will cease forthwith. This prohibition applies equally to the picking up of men in the water and putting them aboard a lifeboat, to the righting of capsized lifeboats and to the supply of food and water. Such activities are a contradiction of the primary object of war, namely, the destruction of enemy ships and their crews.[19]

Just for good measure, he added (but omitted from his *Memoirs*): "Be harsh, having in mind that the enemy takes no regard of women and children in his attacks on German cities." And he awarded Hartenstein the Knight's Cross.

Out in the South Atlantic, *U-156* continued its war patrol. On September 19, Hartenstein torpedoed the 4,745-ton British steam freighter *Quebec City*, out of Cape Town and bound for Freetown with a cargo of 6,600 tons of cotton and wool. But the rest of the patrol proved to be uneventful. On November 16, Hartenstein returned to Lorient. *U-156* in 88 days at sea had covered 11,887 nautical miles, 373 of them submerged.

* * *

From the start of the U-boat war in the Caribbean, the Allies knew that airpower was going to be crucial to stop the slaughter of tankers and other merchant ships, and that the United States would be called on to supply most of the aircraft for the fight. At the beginning of the air campaign against the U-boats, both aircraft and effective antisubmarine weapons were scarce. American aircrews, both Army Air Forces (AAF) and Navy, were inexperienced. But even as these problems were slowly resolved in the summer and fall of 1942, one major headache remained – there was still no unity of command between the navy and the army. One side in this war – the Germans – fought under one commander pursuing a single overall objective and organized by a single headquarters, with one view of what was necessary to pursue victory in this campaign. The other side – primarily the United States – did not move to a single purpose, nor even to two single purposes, but to several different commands at different times.

As early as May 1942, General H. A. "Hap" Arnold, Chief of Staff of the United States Army Air Forces, proposed to Admiral Ernest J. King that the two services establish an American version of the Royal Air Force's Coastal Command. The latter was one of several commands operating under Air Chief Marshal Sir Charles Portal, Commander in Chief RAF (other commands being, for example, Bomber Command and Fighter Command). But Coastal Command worked very closely with the Royal Navy and was effectively under its operational control. Though sometimes difficult, the culture of inter-service cooperation had developed in the early years of the war to the point where air force and navy were able to operate effectively in the escort of convoys and attacks against U-boats in the Bay of Biscay and the North Atlantic.[20]

Such was not the case in the United States, where airpower culture was still strongly influenced by the airpower theorists of the 1920s and 1930s, who believed that the overall objective of an air force was offensive – primarily to bomb the enemy's productive capabilities. Thus, as much as the AAF was, in effect, dragged into the Caribbean campaign (and that off the US east coast as well) in early 1942, it was still uncomfortable in the role and unhappy at operating under naval control. To complicate matters, elements of three air forces – First, Third, and Sixth – were called into the antisubmarine campaign on an emergency basis between February and October 1942. It was a chaotic situation, made even more complex in July 1942 when the AAF set up the Antilles Air Task Force at Borinquen Field, Puerto Rico, under command of Sixth Air Force.

Arnold's suggested US "Coastal Command" never materialized; instead, the AAF established the Antisubmarine Command in mid-October 1942, using resources of I Bomber Command, First Air Force, operating out of several air fields on the US east coast and Florida. It assigned I Bomber Command's 26th Wing to Miami to cover the Gulf of Mexico and the Caribbean Sea, while 25th Wing was headquartered at New York to provide air cover for the east coast. As a history of the Antisubmarine Command put it: "The most serious [operational] problem [encountered by the 26th Wing] came ... not from the climate or the native population, but from the command situation into which the AAF Antisubmarine Command squadrons were plunged."[21] Antisubmarine Command squadrons operating in the Caribbean were under the control of the Caribbean Sea Frontier, a naval command, and the Caribbean Defense Command, but "many lesser headquarters existed between the highest echelon and the single AAF Antisubmarine Command squadron serving at Trinidad."[22] There was no end of command, control, and logistics problems afflicting these and the other AAF squadrons operating in the area. The most substantial benefit of these mostly ad hoc arrangements was that several dozen B-24s finally began to flow into the Caribbean air campaign.

The B-24 Liberator was a four-engine, long-range, strategic bomber developed in the 1930s to strike targets deep in enemy territory. It was heavily armed with .50-caliber machine guns for self-defense and carried a bomb load of 5,000 pounds, with a range of 2,200 miles. Stripped of most of its machine guns and with additional gas tanks installed in the bomb bay, it became the best antisubmarine aircraft of World War II. The problem was that the AAF wanted every Liberator it could get its hands on for strategic bombing in Europe and the Pacific. Both Coastal Command and the US Navy had to beg for any scraps they could get. The flow of B-24s to the RAF and the USN eased greatly after the Casablanca Conference between President Franklin D. Roosevelt and Prime Minister Winston S. Churchill in December 1942 and January 1943. Most of the

B-24s obtained by the RAF were sent to Ireland and Iceland, while Army Air Forces' B-24s began operating near Trinidad, in the south Atlantic, and off the coast of Brazil, flying out of Natal and Ascension Island. It was one of these aircraft that had given Hartenstein such a difficult time.

The beginning of the final resolution of American command problems in the Atlantic, Caribbean, and Gulf of Mexico was the establishment by Admiral King of the new US Tenth Fleet on May 1, 1943. King made the decision following the Casablanca Conference – where the president and the prime minister emphasized the importance of defeating the U-boat threat preliminary to D-Day – and the Atlantic Convoy Conference held in Washington in March 1943. The latter was a gathering of US, British, and Canadian naval chiefs aimed at deciding how to best carry out the aims of the Casablanca Conference. On April 6, King had named Rear Admiral Francis S. Low chief of staff for antisubmarine warfare, who subsequently recommended that the whole US Navy ASW campaign be placed under the command of Admiral Royal E. Ingersoll, Commander in Chief Atlantic Fleet (CINCLANT in naval parlance).

While this did not solve the AAF/Navy impasse, or the AAF's own internal command problems, it did consolidate and rationalize USN antisubmarine activities. In fact, King went a step further and placed Tenth Fleet as a "fleet without ships" under his own command to co-ordinate ASW operations, using resources from existing formations such as CINCLANT. Low actually ran Tenth Fleet and exercised control over all US Atlantic Sea Frontiers. Tenth Fleet used centralized intelligence gathered from signals intelligence and other Allied sources and coordin-ated ship movements with the Royal Canadian Navy and the Royal Navy, while directing the US Navy's shore establishments, aircraft, and antisub-marine assets at sea – especially the new hunter-killer groups formed around escort carriers, or "baby flat tops," carrying Wildcat fighters and Avenger torpedo bombers.[23]

The creation of Tenth Fleet – and the obvious implication that the USN would continue to control its own antisubmarine aircraft, includ-ing its B-24s – made the AAF Antisubmarine Command redundant. At first, a temporary solution was to place an Army Air Forces general in command of Tenth Fleet, but the AAF continued to insist that its aircraft hunt submarines. The USN, on the other hand, demanded convoy escort

– that is, aircraft over the convoy and not chasing submarines around the western Atlantic or Caribbean. In May 1943, King asked Arnold to send a squadron of B-24s to Newfoundland to close the North Atlantic air gap. The planes went north, but with instructions to hunt submarines, not escort convoys.

It was frustrating for all. Finally, Arnold offered to get out of the antisubmarine campaign entirely. The AAF would turn its B-24s, already configured for ASW operations, over to the Navy in return for an equal number of as-yet unmodified B-24s out of the Navy's allocation. The USN thus took over all airborne antisubmarine operations on September 1, 1943, and for the first time in the war, airborne ASW operations from Panama to the Outer Antilles (not to mention the Atlantic Ocean) bent to the will of a single commander – Admiral Francis S. Low of Tenth Fleet. This change of command, along with more and better long-range aircraft, improved ASW weapons, amended sub-hunting tactics, an increase in escort vessels, and the hunter-killer groups, would deal a further heavy blow to the "gray sharks" in the Caribbean.

HIGH NOON IN THE CARIBBEAN AND THE SOUTH ATLANTIC

U-156 remained in one of the "dry" pens of the Kéroman bunkers for almost two months. The attack by the B-24 Liberator had caused a good deal of damage. A diesel compressor, cooling-water flange, wireless set, sounding gear, and hydrophones were replaced. The sky periscope was repaired and its stuffing boxes renewed. Nineteen damaged batteries were replaced. The battery compartment was newly welded as several seams had cracked, allowing up to 150 liters of seawater per hour to flood the compartment. The entire pressure hull was inspected for cracks.

As ever, Admiral Karl Dönitz had some new technology to be installed. *U-156* was outfitted with the so-called Bridge Conversion I: twin two-cm FLAK[1] machine guns were mounted on a platform half the height of the bridge and abaft the conning tower. It was quickly dubbed the "winter garden" by the crews. The boat also received the new "Metox" VHF-heterodyne receiver, together with the "Biscay Cross," a primitive wooden-frame aerial designed as a direction finder.

U-156 underwent the usual rotation of officers and ratings. About one-third of the crew was retained, one-third was transferred to other boats, and one-third went on leave. Commander Werner Hartenstein, Second Watch Officer Max Fischer and Chief Engineer Ernst Schulze remained with the boat. Executive Officer Gert Mannesmann was sent to officer's command school and replaced by Lieutenant Leopold Schuhmacher. And since Dönitz had selected Hartenstein to command yet another grueling transatlantic war patrol, he assigned *U-156* a Third Watch Officer, Lieutenant Silvester Peters.

Hartenstein took *U-156* out of Lorient on January 16, 1943, on its fifth war patrol. Destination: the Caribbean. It was one of just four Type IXC boats dispatched to the Americas in January. Hartenstein was confident of success: he was still the top Caribbean "ace," with 11 ships sunk and two damaged in a single patrol. His orders were to proceed to Trinidad via the Cape Verde Islands, where, Dönitz radioed him on January 26, he might run across stragglers from a recently sighted Allied convoy. For days, the freshly baked Knight's Cross holder raked the waters of the mid-Atlantic in Grid Quadrants DH, DT, and EJ. To no avail. Finally, on February 12, U-Boat Command gave Hartenstein orders to shape a course for Quadrants EO 50 and EO 20 and cleared him to scour the waters off Brazil, which had entered the war in August 1942. He was to lie close in to shore and to use the coming new moon period to advantage – as he had off Aruba in February 1942. Paris assured him that hostile surface escorts remained weak and "only slight[ly] trained," while "convoys have strong air protection."[2]

Finding no traffic in EO, Hartenstein sailed off northwest toward Trinidad. After a month of being cooped up inside the hull in sweltering heat and humidity, he allowed the crew to come up in small groups to recharge their lungs with fresh air – or to foul them with smokes. Some took to fishing, others to porpoise watching. A few went into the water off the stern, despite the "Great Lion's" admonition that such aquatic recreation be banned on U-boats. For, on September 11, 1942, one of his veteran skippers, Rolf Mützelburg, had made a head-first dive from the conning tower of *U-203* – just as a lazy swell rolled the boat over and the Kapitänleutnant had struck the boat's saddle tank with head and shoulders. He died the next day.

On February 16, Dönitz dashed off an Enigma message to both *U-156* and *U-510*. Johann Mohr in *U-124* had just returned from Trinidad and Tobago after an 81-day war patrol, during which he claimed to have torpedoed eight ships of almost 46,000 tons. On the basis of Mohr's after-action report, Dönitz assured both boats that they could expect "little air reconnaissance" in their new operations area, that there had been "no aircraft radar" reports "so far," and that enemy air did not fly at night. He admonished both to remain on the surface, "also during the day," since Allied escorts were "completely untrained." He closed this bizarre radio message with one of his customary exhortations: "In any case, go at them even in shallow water and utilize every chance to shoot."[3] The crew of *U-156* took time out for a modest celebration on February 27. While the *Smutje* managed a cake on the small galley stove, the officers broke out the "medicinal brandy" and doled it out in small portions. It was the Old Man's 35th birthday. He had celebrated the last one off Aruba, and so it was only fitting that this one, too, took place in the Caribbean.

Whenever *U-156* cruised on the surface, Hartenstein had the men erect the "Biscay Cross" on the bridge and check the "Metox" set for the tell-tale "pinging" of enemy "direction finders" (radar). They heard nothing – despite the fact that several times the watch alerted the skipper to the noise of heavy four-engine bombers. In fact, Trinidad had become an antisubmarine warfare stronghold. US Navy and Royal Navy surface craft routinely put out from Chaguaramas to escort convoys up to Guantánamo Bay, Cuba, and beyond, as did PBY Catalina flying boats. Fleets of American B-18 and British Lockheed Hudson bombers at all hours of the day and night took off from Edinburgh Field on the lookout for "gray sharks." The days of Albrecht Achilles' early and easy successes in the Gulf of Paria were but a distant memory.

Near dusk on March 2, an American B-18 "Bolo" bomber of 9th Reconnaissance Squadron, flying out of Trinidad, spotted *U-156* racing on the surface after a small convoy, TB-4, just north of the Grand Boca.[4] Hartenstein saw the hostile at the last moment. "Emergency Dive!" But the B-18 swooped in low and fast, its twin 930-hp Wright engines droning over the swirl of the slowly disappearing U-boat, and dropped four aerial depth charges. *U-156* endured a savage bombardment, probably sustaining some damage. It had been a close call, one without warning from the "Metox" device.

The B-18 pilot radioed in his position and later that night another B-18, this one from 80th Bomber Squadron at Edinburgh Field, re-established radar contact with *U-156*. Unable to spot the dark U-boat in the pitch black night, the pilot switched on his landing lights – and immediately drew fire from Hartenstein. Again, the B-18 radioed in the new location of *U-156* but broke off the attack due to accurate and heavy fire from Lieutenant Fischer's anti-aircraft gunners and lack of visibility. Hartenstein executed yet another crash dive.

It was all too bewildering for the veteran commander. He had not experienced such strong integrated ASW defenses during his previous two cruises in the Caribbean. As per instructions from Dönitz, he stayed on the surface because there existed "no aircraft radar" and enemy planes did not fly at night! Hartenstein brought *U-156* back up to the surface, certain that the aircraft had returned to its base. Wrong. As soon as the submarine broke the surface, the B-18, which had patiently circled above, swooped down for another attack. Four more bombs crashed about the boat. Again, they did only secondary damage – likely smashing glass and gauges and bursting pressure hoses – mainly because the pilot had set them for 25 feet, and *U-156* was still on the surface. Unknown to Hartenstein, one of the bombs had caused a tear in a fuel tank and the boat was now leaving a thin oil slick in its wake.

Hartenstein took *U-156* east to recharge its batteries in what remained of the night and to allow Schulze and the technical crew to carry out repairs. Above, a host of American aircraft trailed his every move. Throughout March 3, Airship K17 followed the oil slick until it got a read with its Magnetic Anomality Gear, a secret new detection device. The pilot dropped three depth charges, which did no damage to the submerged boat, but alerted its skipper to the fact that he was leaving some tell-tale sign behind. Later that day, PBY Catalinas and surface patrol boats cruised in the area, but made no contact.

On March 4, Convoy TE-1, comprised of only four merchantmen but guarded by four destroyers, left the Bocas. It stumbled upon *U-156*, and the USS *Nelson* dropped nine depth charges in the general area of the U-boat's oil slick, probably causing further damage to the boat. The oil slick grew larger. Later that day, Hartenstein crash dived to evade yet another B-18 "Bolo." He stayed submerged for the next 30 hours, by now suspecting that the enemy was tracking his every move due to the boat's oil trail.

Hartenstein put U-Boat Command in the picture concerning the radically changed nature of enemy ASW actions, sending off an Enigma radio message either late on the evening of March 5 or in the early hours of March 6:

Strongest possible air cover. New radar. Metox useless. Very accurate attacks [at night] without searchlights. Impossible to operate against 'Testigos' convoy. Turned away. Still have [all] eels.[5]

It was to be Hartenstein's last Enigma message. He had, of course, made his first encounter with the Allies' new airborne centimetric radar as well as with "Huff-Duff" (High-Frequency Direction-Finder). Edinburgh Field picked up his transmission, as did Seawell Airport in Barbados. Through triangulation of the Enigma source, they obtained a new fix on the whereabouts of *U-156*.

For two days, Hartenstein took *U-156* still further east, most of the time submerged. Above, relays of American PBY flying boats and B-18 bombers searched for the intruder. By the morning of March 8, he had put some 300 miles between Trinidad and his new position. Sure that he had eluded his tormentors, Hartenstein surfaced to ventilate the boat and to recharge the batteries.

Just after breakfast on March 8, John "Duke" Dryden, promoted lieutenant (jg) only a week earlier, flew his PBY Catalina P-1 of US Navy Squadron VP-53 out of Chaguaramas. He pointed east in search of the reported U-boat, quickly climbing to 4,500 feet. Nothing in sight. After five hours, Dryden decided to head home. He turned the controls over to one of the crew, Captain J. M. Cleary, and retreated to the navigation compartment to check his course plot. At 1 p.m., Cleary spotted the fully surfaced gray shadow of a U-boat bearing 265° relative to the PBY at a distance of eight miles.[6] Because of good visibility (15 to 18 miles) the flying boat had switched off its radar.

Dryden was back in the pilot's seat in less than a minute. Cleary returned to his nose turret. Taking advantage of the PBY's white camouflage paint, Dryden turned right and ducked behind one cumulus cloud and then another. At 1,500 feet, range one-quarter mile, he left the cloud cover. At 1,200 feet, he pushed the flying boat into a 45-degree dive at 140 knots. The sun was "directly behind the plane and almost overhead." By 1:15 p.m., the PBY was 75 to 100 feet above the water at a target angle of 150 degrees. Second Pilot S. C. Beal pulled the two manual release switches, dropping four 325-pound Mark 44 Torpex aerial bombs set to explode at 25 feet. Dryden then sharply banked the craft away from the U-boat to evade anti-aircraft fire.

Hartenstein and his lookouts were caught completely off guard. They became aware of the PBY only from the roar of its twin 1,200-hp Pratt & Whitney engines. By then, it was too late. At 400 yards, the plane's nose and port guns raked the U-boat's deck and open conning tower hatch with 100 rounds of .30-caliber and 15 rounds of .50-caliber ammunition, while its tunnel hatch fired 30 .30-caliber shells on the deck forward of the U-boat's conning tower. Two sailors, evidently sunbathing on the deck, were killed instantly. Hartenstein had not even had time to take the tarpaulin off his forward deck gun. Then four bombs crashed into the sea. The port-side crew of the flying boat saw two splash into the water "10–15 feet to starboard and just abaft the conning tower"; the other two exploded further away. On the Catalina, port blister gunner J. F. Connelly saw the submarine lift, break in two in the middle, the center sections going underwater first, the bow and stern rising into the air and then going under.

Simultaneously, a high-order explosion occurred causing debris, smoke, and water to cascade 30 to 40 feet into the air in great profusion.

Werner Hartenstein had paid a terrible price for his momentary lack of vigilance. One can only imagine the chaos that must have reigned in *U-156*. Many of the crew undoubtedly were killed instantly by the powerful blasts and pressure wave from the Mark 44 bombs; others perhaps by the subsequent explosion, most likely of the boat's torpedo warheads. Still others would have drowned as the seawater rushed into the hull, filling compartments and making escape impossible. And some might have cowered in a water-tight compartment, waiting for the "terror-filled drop into the depths of the Atlantic."[7]

Dryden brought the PBY back to the spot of the attack. A terrible sight greeted him. A "large patch of foam, 150–200 feet across" floated on the water – as did a "silver green oil slick," which then turned a dark brown. Wreckage in the form of deck planks and large cylinders (most likely torpedo storage tubes) floated on the surface – as did 11 survivors. Dryden took half a dozen pictures to confirm the "kill." He then dropped two rubber life rafts into the water – one failed to inflate because a line attached to the operating lever slipped off as the raft left the PBY – as well as an emergency ration kit attached to two "Mae West" life jackets.[8] He ignored frantic hand signals from the survivors to land and to pick them up, for the sea was extremely rough and he was running low on fuel.

Of the 11 survivors, two clung desperately to one of the silver-colored cylinders and four to another black cylinder. The latter group disappeared beneath the waves "almost immediately" and the former within "10–15 minutes." That left five survivors. They managed to scramble into the inflated raft. Four were clad only in shorts. The fifth, possibly an officer, had a shirt on as well. "He was heavier and apparently older than the others, who all appeared to be in their late teens." Was it Hartenstein? Surviving pictures of the Old Man and his crew with Admiral Dönitz clearly show that Hartenstein was at least 15 years older and much bulkier than his young crew. Only Chief Engineer Schulze at 31 was close to Hartenstein's age, and he was much thinner. The likelihood of his having gotten out of the engine room in the stern of the boat after Dryden's deadly attack is remote. One of the survivors "was seen to shake his fist" at the Catalina.

The fate of *U-156*'s five survivors remains a mystery. Immediately after his "kill," Dryden had radioed the whereabouts of the German sailors to all shore and sea units within range, including the merchantman *Aldecoa España* and the tanker *Gobeo*, both Spanish. As well, the destroyer USS *Barney* was dispatched to the site. None found any survivors. Most likely, the raft had drifted out to sea and its occupants died of exposure to the broiling sun, or of shark attacks. In all, three officers and 49 ratings were lost.

U-156 was the first "kill" for the Chaguaramas flying boats. "Duke" Dryden was promoted to lieutenant and awarded the Distinguished Flying Cross; the rest of the crew were awarded the Air Medal for their sterling actions. Formal after-action evaluations accorded all involved a grade of "A."

After repeated unanswered calls between March 8 and 24, 1943, Admiral Dönitz declared *U-156* "potentially lost" on April 18 and "formally lost" on November 16, 1943. *U-156* disappeared at Latitude 12° 38′ North, Longitude 54° 40′ West, northeast of Trinidad and east of Barbados. It came to rest 3,500 meters beneath the sea.

Dönitz fully appreciated that he had lost another "ace," a veteran Knight's Cross commander with 100,000 tons sunk to his credit. In a way,

Hartenstein had come full circle: from the brilliant attack on San Nicolas, Aruba, in February 1942, to the sinking off Trinidad and Barbados in March 1943. One of just two remaining memorabilia of the Dönitz era at the Villa Kerillon at Kernével is a small tiled coffee table that the admiral obviously had ordered to be made. Its center tile is the "Plauen" conning tower crest of *U-156*.[9]

* * *

Hartenstein's demise was only one of a growing epidemic of destruction of U-boats by Allied aircraft. The problem was particularly acute in the Bay of Biscay, where both out- and in-bound submarines were especially vulnerable to attack, but also on all the other sea frontiers where the Battle of the Atlantic was being fought. Dönitz's answer was to beef up the submarines' anti-aircraft defenses – as had been done with Hartenstein's boat – and eventually, in the early spring, to order his commanders to stay on the surface and fight it out with attacking aircraft if they thought there was any chance at all of surviving. Dönitz may have decided this after he learned of the encounter of the outbound *U-333*, commanded by Oberleutnant Werner Schwaff, with an RAF Wellington in the Bay of Biscay on March 4, 1943. Schwaff's boat was on the surface at 9:31 p.m. local time when the Wellington switched on its Leigh Light and caught *U-333* fully in its glare. Schwaff's crew opened fire just as the bomber dove to the attack and dropped two depth charges; the Wellington caught fire and crashed with the loss of all six of its crew.[10]

The new defensive tactics called for even more modifications on the submarine fleet. Special "U-Flak" boats were built, carrying heavy anti-aircraft armament on their conning towers: two quad (four-barrel) 20-mm guns and a 37-mm flak gun, for a total of five. They were specifically designed for service in the Bay of Biscay,[11] but virtually all other U-boats (such as Hartenstein's) had their anti-aircraft capability increased in one fashion or another. The more heavily armed boats made their way into the Caribbean or South Atlantic. Allied pilots suddenly began reporting these encounters – *High Noon*-type shootouts – and noted that the U-boats usually opened fire with their new 20-mm guns at 600 yards, well outside the drop zone for aerial depth charges, and were deadly effective at

300 yards. During these attacks, the U-boats continuously turned toward the incoming aircraft.

Allied flyers also reported that the decision to stay on the surface seemed to rest with individual U-boat commanders since both tactics – diving and staying up – were being used. There seemed to be no standardization of the new anti-aircraft armament.

> Reports from all areas where the U-boats were increasingly active showed ... one 3.46″ gun forward of the conning tower, 0.79″ gun and four MG's on the bridge; twin Italian 12.7 mm (0.46") mounted with the 0.79″ gun on either side of the bridge, all in addition or, in some instances, in place of 20 mm cannon. Experimentation by Axis submarines in the use of dual purpose deck guns was reported.[12]

But no matter what the actual armament, the common tactic was "to throw up as heavy a barrage as possible."[13]

The antisubmarine planes were "somewhat defenseless" against these new arms and tactics. Even the mighty B-24 Liberator had been modified to drop antisubmarine weapons, not to kill submarines with gunfire. Virtually all American planes in the region had to be beefed up as quickly as possible. As a first step, many of the .30-caliber machine guns on the B-18s and other medium, twin-engine aircraft were replaced by .50-caliber machine guns. These guns and their ammunition added extra weight to the aircraft but packed a much more powerful punch. Bell P-39 Airacobra single-seat fighters based on Aruba and Curaçao sported a large 37-mm cannon firing through the propeller hub and were especially effective. The B-24s were given .50-caliber guns in nose turrets of various kinds, but the up-arming on these aircraft was unsatisfactory, largely because of poor visibility directly ahead. As a result, their noses were lowered slightly, giving the forward gunner a much better view. Armor plate and a bullet-proof glass shield were also added.

One important result of these alterations was a modification in the plane's center of gravity, giving it a nose-down alignment in flight – which resulted in better vision for the pilots. In addition, specially modified B-25s mounting 75-mm guns were deployed. New tactics were also introduced.

The slower and more vulnerable B-18s began flying in pairs, one equipped with normal demolition bombs and flares, the other with aerial depth charges. New tactics for dropping flares or more long-burning floating lights were also developed.[14]

The Battle of the Atlantic climaxed in May 1943, when the Allies destroyed 41 submarines. German codes yielded a bounty of intelligence; high-frequency direction-finders, both long- and short-range, told Allied radio operators where submarines were transmitting from; the air gap had been closed with an abundance of USN, RAF, and RCAF Liberators and other long-range aircraft; and American hunter-killer groups as well as British support groups were now available to find submarines on the surface and destroy them. The Kriegsmarine could not afford such heavy losses, and Dönitz ordered most of the U-boats out of the central Atlantic. The submarines were to be updated with new radar detectors, new types of torpedoes, new armament, and eventually *Schnorkels* to allow them to run on their diesel engines just under the surface. In the meantime, the U-boats were ordered to the South Atlantic and a small number sent back to, or kept in, the Caribbean Sea.

In May and June 1943, the submarines sank only three vessels in the Caribbean – a British cargo ship of 4,748 tons and two small tankers, one Cuban of 1,983 tons, and one American of 2,249 tons. Off the Brazilian coast, however, after a seven-day chase by air and sea units, *U-128* had been spotted by two US Navy Mariners based at Aratu on May 17. The two aircraft dove to the attack. The submarine managed to slip beneath the sea – only to surface a short time after, no doubt with heavy damage. The two Mariners were then joined by the destroyers USS *Jouett* and *Moffett*, and *U-128* was hit repeatedly by their gunfire. After four direct hits, the U-boat crew abandoned ship as the submarine rolled over and sank. About 50 survivors were picked up.[15]

Suddenly, in July, the U-boats returned to the Caribbean. On July 1, a small Brazilian cargo ship of 1,125 tons was torpedoed northeast of the Windward Islands, beginning a toll of destruction that lasted through the month. Fourteen ships went down in or near the Caribbean in the next four weeks, ranging in size from the schooner *Harvard* (114 tons) destroyed on July 14, to the *BP Newton*, a 10,324-ton Norwegian flagged tanker destroyed on July 8. *BP Newton* was one of only two tankers sunk in July – the other was the 3,177-ton Dutch *Rosalia* – in a month that cost the Allies a total of 66,383 tons in the Caribbean area. Over the last two weeks of July, 11 aircraft in the Antilles Department engaged in running gun battles with surfaced U-boats, but, in the end, it was the submarines that paid the heaviest toll by far. It took time for the Americans to adapt, but once they did, the destruction of U-boats in the Caribbean began to match that for May in the North Atlantic. July 1943 saw 21 attacks on U-boats by aircraft and 9 by surface craft with the following results:[16]

July 9, 1943: U-590 was on its first war patrol near the Amazon estuary when it was caught on the surface by an American PBY. It was sunk with all hands.[17]

July 15, 1943: U-759 was attacked by a US Navy Mariner east of Jamaica in the Caribbean Sea. It was sunk with all hands.[18]

July 19, 1943: U-513 was attacked off the coast of Brazil by a US Navy Mariner stationed at Rio de Janeiro. The U-boat at first put up a curtain of heavy anti-aircraft fire, but Lieutenant (jg) Roy S. Whitcomb swung the big aircraft over the submarine and dropped six Mark 44 depth charges, then banked quickly away to avoid the boat's anti-aircraft fire. The tail gunner yelled "we got him, we got him" and when Whitcomb flew back in the direction of *U-513*, the crew spotted floating debris and about 15 survivors in the water.[19]

July 21, 1943: US Navy Catalina 94-P-7 took off at 2:10 a.m. local time from Belém to rendezvous with convoy TJ-1 about 300 miles off the Brazilian coast. After arriving in the vicinity of the convoy three and a half hours later, bow gunner F. J. DeNauw spotted the surfaced *U-662* three miles distant, just off the PBY's starboard bow. Pilot Lieutenant (jg) R. H. Rowland was flying at 1,200 feet and nudged the aircraft to the left, heading for the submarine in a shallow dive. The plane's bow gun would not fire and the U-boat put up a persistent and heavy barrage, making no effort to dive. The PBY was hit in several spots and the radioman wounded, but Rowland pressed home his attack. He swung the aircraft to the right a bit to give the right blister gunner a chance to fire. He next eased the aircraft to the left; then to the right again; and when he was about 75 feet above the surface of the sea, flew over *U-662*. He attempted to drop four Mark 44 aerial depth charges set at 25 feet. One of the charges "hung up" under the plane's wing, but the others did the job.[20] Three of the U-boat crew survived.

July 26, 1943: A US Navy Mariner attacked and destroyed *U-359* in the Caribbean south of Santo Domingo by aerial depth charges. All hands were lost.[21]

July 28, 1943: *U-159* suffered the same fate. A Navy Mariner piloted by Lieutenant (jg) D. C. Pinholster spotted the submarine south of Haiti. It was proceeding on the surface at about 15 knots and opened fire as Pinholster turned the aircraft on an intercepting course. The Mariner's bow gun jammed, but Pinholster's plane bore on, taking hits and suffering two crew wounded. The right blister gun poured fire into *U-159* while Pinholster dropped four Mark 44 aerial depth charges on it, then orbited one-and-a-half times so that the right blister gun and the tail gun could continue firing. The bombs exploded and *U-159* seemed to lose head-way. Still, with one gun out of ammunition, one gun not firing, and two wounded crew members, Pinholster turned for base. Suddenly, one of the waist gunners saw a large explosion that engulfed the U-boat's whole conning tower. The next morning, a large oil slick was spotted from the air at the position of the attack. That was the last visible sign of *U-159*.[22]

August 3, 1943: *U-572* was sunk by depth charges dropped by a US Navy Mariner northeast of Trinidad. All hands were lost.[23]

Thus, in total, Dönitz lost eight submarines in the Caribbean and the South Atlantic between the end of May and the beginning of August 1943. The losses were bad enough; even worse was the sharply diminished opportunity of his U-boats to sink anything of importance. There were just too many Allied aircraft, the radar was too good, and the co-ordination between shore establishments, surface ships, and aircraft too effective. When Dönitz tried to shift his efforts from the Caribbean to the South Atlantic, the result was the same: American and even Brazilian aircraft continued to appear from nowhere to hammer his submarines to the bottom. Just as the Kriegsmarine had lost the Battle of the Atlantic in May 1943, it lost the battle of the Caribbean and the South Atlantic in the high summer of 1943.

15

GUNDOWN: *U-615* AND *U-161*

Ralph Kapitzky was an unlikely hero. He commanded only one boat, did not execute a single bravado war patrol, and did not come close to winning a Knight's Cross. Yet his war patrol to the Caribbean in the summer of 1943 was the stuff of legends, of modern-day buccaneers on the Spanish Main.[1] A member of the Crew of 1935, Kapitzky did brief tours of duty on the old battleship *Schlesien* and the light cruiser *Karlsruhe* before being seconded to the Luftwaffe. He flew a Heinkel-111 during the invasion of Poland and survived being shot down; posted to Caen, France, he flew 100 missions in Stuka and Ju 88 bombers during the Battle of Britain.[2]

The Kriegsmarine recalled Kapitzky in December 1940. After U-boat training, he was posted Executive Officer on *U-93*. On June 1, 1942, just before his 26th birthday, Kapitzky was promoted Kapitänleutnant and given a brand new Type VIIC boat, *U-615*. His first war patrol in September off Newfoundland was a litany of ice as well as hostile destroyers and airplanes.[3] Then, on October 11, he torpedoed the 4,221-ton Panamanian freighter *El Lago*, in ballast from Reykjavik, Iceland, to New York. Twelve days later, en route to La Pallice, he dispatched the 12,656-ton British passenger/cargo ship MV *Empire Star*.[4] It was a good first patrol. U-Boat Command agreed. "The operations against convoys were severely impaired by poor weather and visibility. The sinking of the two lone freighters, including the refrigerator-ship 'Empire Star,' is gratifying."[5]

Kapitzky took *U-615* out of La Pallice on November 25 and again shaped a course for the North Atlantic. It was a miserable war patrol.[6] Violent gales whipped the gray waters into mountains of foam and spray; heavy escorts protected the few convoys that Kapitzky sighted; and enemy aircraft repeatedly drove the boat under. Returning to base on January 9,

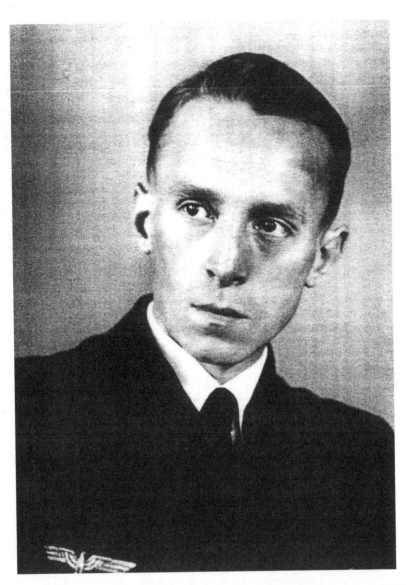

Ralph Kapitzky. Kapitaenleutnant Kapitzky undertook only four patrols in *U-615*, sinking but four ships of 27,231 tons. His fame rests on an epic battle that he fought for days in August 1943 in the Caribbean against a crushing Allied superiority of aircraft and warships. He was last seen clinging to the bridge with legs shot off, but still directing fire against the enemy. Source: Deutsches U-Boot-Museum, Cuxhaven-Altenbruch, Germany.

1943, without a "kill," Kapitzky knew that the formal evaluation would not be kind. He was right. "Hardly a satisfying operation."[7] Captain Eberhard Godt, Dönitz's chief of operations, charged Kapitzky with "overestimating the escorts," with "prematurely diving" after sighting the destroyers, and with "having remained submerged for far too long." The lesson was clear: "The commander must not allow himself to be distracted too much from his primary goal – namely, to get close to the ships – by way of sustained dives and turning away [from hostile escorts]."

Kapitzky returned to Newfoundland for his third war patrol in March 1943. Again, it was sheer misery: Force 6 to 7 gales. One violent storm after another. The ocean became a mad fury of towering waves and cascades of seawater. Then heavy snow showers. Sheets of ice clung to men and boat alike. Periscopes froze. Guns became inoperable. For four days, Kapitzky was swept up in the biggest convoy action of the war. But U-615 had no success. Time and again, Kapitzky broke off his attack run at the approach of destroyers. Time and again, he permitted B-24s to drive him under the sea. Humiliated, he informed Dönitz: "All eels [still on board]." Finally, on April 11, he torpedoed the 7,177-ton American Liberty ship *Edward B. Dudley*, bound from New York to Liverpool via Halifax with a cargo of 4,000 tons of munitions, food, and cotton. Shrapnel from the explosion showered the bridge of U-615 and struck the skipper in the right arm, shoulder, and ear.

Captain Godt raked Kapitzky over the coals in his evaluation of the war patrol. The commander still had not learned how to drive home an attack. He still was too ready to evade attacks by enemy air and surface escorts. "Do not attempt to dive when aircraft is already too close. Repel 1st attack surfaced." While in the biggest convoy battle of the war, Kapitzky again had been too content to take on a reconnaissance role. "The chances offered to attack were not exploited. Never postpone an attack without cogent cause." Godt ended his evaluation on an ominous note: "The commander must take advantage of his experiences to date. A healthy portion of self-confidence and optimism is not only justified but also necessary."[8] It was a clear warning.

* * *

U–615 spent nearly two months in the repair bunkers at La Pallice. As part of "Bridge Conversion II," it received two twin-barreled 2-cm guns on the upper platform abaft the conning tower as well as a four-barreled 2-cm *Vierling* anti-aircraft system on a new lower platform, the so-called "winter garden." The 8.8-cm gun on the foredeck was removed and replaced with a semiautomatic 3.7-cm cannon.

Dönitz sent 12 Type VIIC and one Type IXC boat to the Caribbean in June 1943. Two U-tankers, *U–487* and *U–488*, were to refuel them west of the Azores. *U–615* was to act as a scout south of the Windward Passage and to engage Allied vessels only if "circumstances were entirely favorable."[9]

During the first week of June, the crew of *U–615* reassembled at La Pallice for the boat's fourth war patrol.[10] After the mandatory examination for venereal disease – captain exempted – there was a raucous sendoff party. Then the men packed their private belongings into trunks, placed their last will and testament on top of them, and offered both up to Flotilla Command for safe-keeping. On June 12, the new Second Watch Officer, Klaus von Egan-Krieger, joined the boat. As Executive Officer Herbert Schlipper backed *U–615* out of the U-boat pen, a mysterious box was handed over.[11] The men would later learn that Dönitz had chosen their boat to test out a new "ship finder" called *Nachtfernrohr*.[12] It was to be operated by one of the lookouts on the bridge, from where a cable would be run down the conning-tower hatch to the control room and the radio shack. Both "pings" via the headphones and "pips" on a screen would warn of approaching hostiles. Kapitzky alone had the keys to the box.

As *U–615* steamed out into the Rade de la Pallice, the captain broke out the boat's store of shorts, light shirts, canvas slippers, and pith helmets. The men were delighted: this would not be another patrol in the cold, gray wastes of the North Atlantic. To test the effectiveness of the new quadruple anti-aircraft guns, U-Boat Command had bundled five boats together for mutual protection against Allied "bees." It was a wise decision. For ten hours throughout June 14, the flotilla was savagely attacked and depth charged in the Bay of Biscay by Sunderland flying boats as well as Whitley and Wellington bombers. Gunner Helmut Langer managed to shoot down a "four-engine bomber," which could only have been a B-24 Liberator. But in the process, that aircraft's machine guns shot Gunner Heinz Wilke through the stomach; he bled to death on the upper deck and was buried at sea three days later. The boat sustained only minor damage.

After taking on 20 cbm of fuel oil from *U-535*, U-Boat Command ordered Kapitzky to shape a course for Curaçao. *U-615* no longer was to act as scout for the other boats but to operate off the Dutch island during the "next favorable moon" as convoys were expected to assemble there. The orders were open-ended: "Further cruising according to your own judgment."[13] It was a nice birthday (June 28) present for the 27-year-old Kaleu.

On July 13, *U-615* entered the Caribbean via the dangerous coral-reef Anegada Passage in the British Virgin Islands and then headed south. As he approached Curaçao, Kapitzky must have remembered Werner Hartenstein and *U-156*'s surprise attack on the lake tankers off San Nicolas in February 1942. Could Kapitzky repeat that "happy time"?

He could not. Allied ASW had been beefed up and fully integrated since those days. For two weeks, *U-615* was repeatedly forced to execute emergency dives due to constant surface and aerial observation. Kapitzky remained submerged by day, coming up at night to ventilate the boat and to recharge the batteries. He could detect only small coastal vessels. Radar tracked his every move. His hydrophones were useless close in to shore.

On July 28, Kapitzky got lucky. The lookouts sighted the unescorted 3,177-ton Dutch lake tanker *Rosalia* ten miles south of Willemstad. Kapitzky fired two bow torpedoes into the hapless victim; it burst into flames and sank. Then his luck ran out: an Enigma message to U-Boat Headquarters – "Sank a 6,000 ton tanker" – was picked up by the Allies. There ensued, in the words of U-boat historian Clay Blair, "one of the most relentless U-boat hunts of the war."[14]

It began one day after the destruction of the *Rosalia* when an American B-18 out of Aruba found and attacked *U-615*. It continued on July 31 when a Mariner flying boat out of Trinidad made contact and dropped depth charges as well as bombs in the direction of *U-615*. And it continued on August 1 when a B-24 Liberator out of Curaçao found and attacked the boat. None of the pilots spotted "visible evidence of damage."[15]

On August 2, Kapitzky came across Convoy GAT-77 east of the Dutch islands and set out to attack it. At a range of 1,100 yards, the American patrol craft PC-1196 sighted the U-boat's periscope. It launched five

U-615. One of 568 commissioned Type VIIC 870-ton boats, *U-615* was commissioned in March 1942 and commanded by Kapitaenleutnant Ralph Kapitzky; on her fourth patrol *U-615* fought the longest ongoing battle with aircraft southeast of Curacao, before being depth charged by US Mariner and Ventura aircraft. Source: Deutsches U-Boot-Museum, Cuxhaven-Altenbruch, Germany.

depth charges and made four "mousetrap" runs over the swirl of *U-615* as it executed an emergency dive. The patrol craft detected diesel oil on the surface but could make no certain damage assessment. But "damage" had been done: *U-615* was now in the crosshairs of every American warship and aircraft in the Caribbean. Perhaps acting on Dönitz's recall order, Kapitzky headed due east for Galleon's Passage between Trinidad Tobago and the open ocean.

For four days, *U-615* eluded its pursuers, mainly by running submerged. The inside of the boat became a veritable hell of heat and humidity, sweat and stench. Kapitzky took *U-615* up for short spells under the cover of darkness, but he could not escape the prying eyes of enemy radar: on the afternoon of August 5, the destroyer USS *Biddle* obtained an ASDIC contact and depth-charged the raider. Kapitzky released a *Bold* sonar decoy to slip away. Yet again, the enemy had a specific "fix" on his position, northwest of Trinidad.

Twin-engine Mariner flying boats scoured the area, ignoring the steady rain and approaching darkness. At midnight[16] on August 5, Lieutenant J. M. Erskine in P-6 of Patrol Squadron VP-204 obtained a radar contact 40 miles northwest of Blanquilla Island, Venezuela. It was *U-615*, running on the surface on an easterly course at a leisurely six knots, likely to conserve fuel and to reduce its wake. Erskine fired off two flares to illuminate the target and then swooped down for the "kill." At an altitude of 1,600 feet, he dropped two bombs. They exploded with a bright red flash – but *U-615* continued to run on the surface. Apparently, its captain figured that the attack was over.

He was wrong. Erskine banked the Mariner and came back at the submarine. As he flew over its conning tower, he pulled the manual release switches – only to discover to his horror that three of the four depth charges hung up in the bomb bays. The fourth fell harmlessly 150 feet off target. Kapitzky ordered "Emergency Dive!" By the time Erskine could bring the Mariner back for a third attack, the raider was gone.

Another Mariner out of Chaguaramas, P-6, had also arrived on the scene. At 2 a.m. on August 6, it obtained a radar contact. Both Mariners depth-charged the area of the contact – and were aghast to discover that they had bombed not a submarine but a small inter-island schooner. Kapitzky had used it as a radar shadow, and he now resumed his easterly course on the surface.

Kapitzky's clever escape infuriated Allied shore commands. "Huff-Duff" stations on Trinidad, Antigua, and Dutch Guiana triangulated his position. A Harpoon ASW bomber from VB-130 and two B-18 bombers from Edinburgh Field joined the Mariners in the hunt. The Americans knew precisely where *U-615* was and had established its general course – directly toward the largest US antisubmarine base in the world! They also knew through Enigma decrypts that most of the other Caribbean U-boats were racing for home. Only *U-615* and *U-634* remained in the once "Golden West." It was just a matter of time.

And time it would take. Instead of immediately concentrating on Kapitzky, the Americans pulled many of their best submarine hunters (including the five tracking Kapitzky) out of the search for *U-615* and assigned them to guard four large convoys then in Trinidadian waters. It was a major tactical blunder as it left only a single Mariner, Lieutenant A. R. Matuski's P-4 of VP-205, to take care of *U-615*. For much of the morning of Friday, August 6, Matuski flew a barrier search over the U-boat's last reported position. Kapitzky tracked the Mariner through his sky telescope and timed the American's approaches. A former Luftwaffe flyer, he made a rough calculation that he would have ten minutes between Matuski's "loops" to surface and charge his batteries for the run past Trinidad. He brought *U-615* to the surface. "Both Engines! Full Ahead!" *U-615* knifed through the water at 17 knots. The batteries were coming back to life and fresh ocean air was sucked into the boat. Around 1:30 p.m., Kapitzky ordered a routine "Dive!" as his stopwatch told him that Matuski was due back soon. A last 360-degree sweep by the bridge watch showed nothing in the sky.

Kapitzky never knew what hit him. Four depth charges exploded all around the boat at roughly 50 meters depth. *U-615* began to violently whip up and down – now by the bow, then by the stern. All the while, it continued its rapid descent. The terrified crew in the control room saw the depth indicator needle dip past 240 meters, twice the builder's maximum limit. The pressure hull groaned and creaked. Chief Engineer Skora was finally able to trim the boat by blowing the ballast tanks. Machinist Mate Reinhold Abel later recalled: "Damages: water break-in in the engine room – lights out and loss of the depth regulator – pressure hull bulkheads bent in 1.5 m[eters] near the air intake manifold."[17] In layman's language, *U-615* with its cracked pressure hull and flooded engine room could now operate, if at all, only on the surface. Further investigation showed that both electric motors and the port diesel engine were out of commission, that numerous high-pressure air lines had blown, and that the lubricating oil tank had ruptured and its contents spilled into the bilge. Kapitzky decided to surface.

Lieutenant Matuski could not believe his good fortune: almost directly below him, a heavily damaged German U-boat had shot up out of the sea bow first in a gigantic swirl of foam and air. He immediately notified Chaguaramas and then, like any good pilot, powered up his two 1,700-hp Wright engines and swooped in for what could only be a certain quick "kill."

Kapitzky ordered the gun crews up on deck to man all ten anti-aircraft guns as well as the 3.7-cm semiautomatic cannon. The boat spewed out a deadly fire of more than 5,000 rounds per minute at the incoming Mariner. They struck home with lethal force. "P-4 damaged – damaged – Fire" was the last message Matuski sent off just before the Mariner and its crew of 11 crashed into the sea and exploded. A broken wingtip float, an uninflated dinghy, and a waterlogged cardboard box were all that was left of Matuski and P-4.[18]

Kapitzky took stock of his situation. The bilge pumps in the engine room could not keep water from rising in the stern. Both diesels and both electric motors were down. Hostile air forces undoubtedly were already on their way for a final attack. Nightfall was still six or seven hours off. The closest land was 250 kilometers away. He had to make the most critical decision of his life – and fast. Undoubtedly, Captain Godt's scathing after-action reports raced through his mind. After the second patrol, Godt had chastised Kapitzky for "prematurely diving" at the approach of hostile ASW forces; after the third, for diving "when an aircraft is already too close" and for not repelling "1st attack surfaced." Finally, Godt had challenged the Kaleu to develop "a healthy portion of self-confidence and optimism."

Ralph Kapitzky decided to show U-Boat Command his mettle. While Skora and the technical crew labored to restart one of the diesels and to work the bilge pumps in the stern, Kapitzky, Schlipper, Egan-Krieger, and Chief Petty Officer Hans-Peter Dittmer supervised the transfer of the remaining stocks of 2-cm and 3.7-cm ammunition up on deck. Just in time: at 3:30 p.m., the watch reported an aircraft approaching at 11,000 meters.

"Battle Stations!" The attacker was another Mariner, P-11 of Patrol Squadron VP-205 out of Chaguaramas. At the controls sat Lieutenant (jg) L. D. Crockett, an experienced pilot and one thirsting for revenge ever since his copilot had been killed by gunners from *U-406* just three weeks earlier.[19] Crockett circled the crippled U-boat below him, radioed his position back to base, and then began his attack run. The P-11's anxious gunners opened fire with the twin Browning .50-caliber machine guns in the nose turret a mile from target. Kapitzky held his fire until the Mariner was 300 meters away. Gunners Langner and Dittmer were sharp as ever: their first bursts holed the aircraft and one 2-cm shell ripped through the starboard wing root, starting a gasoline-fed fire. Crockett released two MK-17 aerial bombs. They landed harmlessly off the U-boat's port quarter.

With the Mariner pouring out a steady plume of smoke and fire and in danger of exploding at any moment, Crockett pressed home a second attack. Navy Machinist A. S. Creider climbed into the Mariner's wing root and with a spare shirt tried to smother the flames. For a second time, the two antagonists blazed away at each other. As he passed over the submarine's conning tower, Crockett released four MK-44 depth charges. He then banked away from the deadly wall of anti-aircraft fire and saw four gigantic columns of water arise all around the U-boat. He had landed a deadly punch. Numerous new cracks opened in the boat's hull and the sea began to rush in. The men inside the boat were working in water up to their knees. U-615's stern settled ever deeper into the sea, while its bow rose concomitantly. The boat was in danger of sliding into the depths by the stern. Kapitzky and Skora urged on the men at the pumps and ordered others to join them. The boat was turning in circles as the last attack had jammed one of the rudders hard-a-starboard. U-615 was a sitting duck.

Less than an hour after Crockett's second attack, a Ventura PV-2 Harpoon bomber, B-5 of VB-130, arrived on the scene and joined the Mariner in a combined attack. They approached the stricken submarine flying just 50 feet over the water. Kapitzky instructed his gunners to ignore the shattered Mariner and to concentrate on the Harpoon and its five machine guns. Roaring in at 280 knots, Lieutenant T. M. Holmes' B-5 flew through not only Kapitzky's tracer bullets, but also Crockett's .50-caliber shells. It then bracketed U-615 with four 325-lb. bombs. It should have been the end – but instead of ripping the U-boat apart, the simultaneous explosions of the depth charges drove U-615 under the sea, taking its tethered bridge personnel and those inside the craft with it and washing its gunners into the sea. After what must have been a terrifying 15 seconds, U-615 came back up. Some of the gunners swam back to the boat and manned their weapons. As Crockett came in for a third attack, which he took to be a certain "kill," he was met instead by yet another withering hail of machine-gun fire and had to veer off sharply.

How long could this go on? Despite the best efforts of the men at the pumps, *U-615* was sinking by the stern. A single electric motor had been made operable. Most gauges and instruments had long been smashed. Damage control as such was non-existent. Incapable of diving or of steaming, *U-615* had been reduced to a beleaguered (and sinking) gun platform.

At 6 p.m., Mariner P-8 of VP-204, Lieutenant (jg) John W. Dresbach at the controls, arrived at the scene and joined P-11 and B-5 in a concerted effort finally to sink *U-615*. Dresbach came in low from the stern at 190 knots. A burst of fire from Kapitzky's quadruple anti-aircraft guns smashed through the Mariner's nose, killing its pilot instantly and knocking out the plane's radar and automatic pilot. The four depth charges that Dresbach had released just before dying exploded harmlessly in the water. Inside the cockpit of P-8, copilot Oran Christian grabbed the control yoke with one hand and Dresbach with the other, until the crew could claw the dead pilot out of his seat. In anger, Christian wiped the blood off the cockpit windshield and barreled in for a second attack. He released two depth charges at 1,500 feet; they exploded some 300 feet off the submarine's port side. The wind fairly whistled through the gaping holes that Kapitzky's gunners had made in P-8. The two attacks lifted *U-615*'s stern out of the water, smashed its recently jerry-rigged rudder, and shredded its aft diving planes. More holes in the pressure hull. More water in the boat. *U-615*'s stern sank below the sea again. Shore installations by now had all tuned in to the reports coming from Crockett for none could believe that the German submarine was still afloat. US Navy Command ordered three warships out of Grenada and the brand-new destroyer USS *Walker* out of the Gulf of Paria to rid the Caribbean of Kapitzky and *U-615*.

The last attack had again been costly for *U-615*.[20] Chief Petty Officer Dittmer, a veteran of 13 previous war patrols, had been shot through the head by one of the Harpoon's shells and blown clean overboard. Gunner Langner, who had brought down the four-engine bomber in the Bay of Biscay and had just helped destroy Matuski's Mariner, had taken a heavy-caliber bullet to the knee; he would later bleed to death. Some of the Mariner's other shells had torn into Kapitzky's thigh. He lay slumped in a corner of the bridge, bleeding heavily, the shattered leg crazily drooped across his chest. He was given morphine and propped up against the periscope standard. His last orders were to transfer command to Schlipper and to be remembered to his parents.

At about 6:30 p.m., yet another Mariner hoved into sight: P-2 from VP-205, piloted by Lieutenant-Commander R. S. Hull. Yet again, Crockett directed an attack on the U-boat by all four aircraft. It was another bitter disappointment: the Mariner's bomb doors opened prematurely ("failure of the release mechanism") and its stick of depth charges exploded harmlessly 600 feet astern *U-615*. Furious, Hull took his machine up to 1,500 feet and then roared in for a visual bombing run – both bombs splashed harmlessly into the water some 500 feet from the sub. *U-615*'s gunners were as deadly as ever, and Hull was forced to take his battered Mariner back to base. At 6:40 p.m., the Harpoon also informed Crockett that it had to return to base because it was running low on fuel. Darkness finally fell on hell.

But all was not calm. At the last twilight, yet another tormentor arrived: Lieutenant Milton Wiederhold's B-18 bomber out of Edinburgh Field. The indefatigable Crockett set up yet another attack run on *U-615* – but, to his dismay, it was gone. Darkness and a tropical rain storm had swallowed up the boat. Navy Airship K68 had also arrived on the scene. At 9:15 p.m., its pilot, Lieutenant (jg) Wallace Wydean, spotted *U-615* on the surface between two rain squalls and guided Wiederhold's B-18 in on its attack run. For the last time, *U-615*'s gunners put up a blistering hail of anti-aircraft fire. The depth charges from the B-18 rocked the U-boat, but they were not close enough to sink it. By the time the American bomber returned on a second run, rain squalls again had enveloped *U-615*. Wydean had been so engrossed in the action that he forgot to check his fuel situation; when ordered home, he was too far away and had to crash-land K68 on Blanquilla Island. Heavy winds tore the beached blimp to shreds the next day. It was *U-615*'s last victim.

U-615 had been depth-charged 14 times by seven different aircraft. It barely remained afloat. Its ammunition had been shot off. Its engines were down. Its rudders and aft dive planes were shattered. Some of its bulkheads had been caved in and its pressure hull compromised. Up above, a dozen Mariners were still searching for it. Schlipper assembled the crew on the foredeck in order to press down the bow and thereby raise the stern. He asked Machinist Mate Abel to go below to take charge of damage control. Miraculously, Abel kept the bilge pumps going through the night and even occasionally blew high-pressure air into the diving tanks to prevent the boat from sinking.

* * *

The story of the last night on *U-615* became (and remains) the subject of wild speculation and myth-making, increasing in drama with every telling. In his original (1988) account, historian Gaylord T. M. Kelshall had Kapitzky quietly bleeding to death propped up against the periscope standard.[21] Ten years later, after an interview with Executive Officer Schlipper – who allegedly gave Kelshall a "Kapitzky Diary" (improbably saved off the sinking *U-615*!) – the story of Kapitzky's last hours took a much more dramatic turn.[22] The skipper, mortally wounded and profusely bleeding, managed to greet and to shake hands with every member of the crew as they came up on deck. He even joked with some of them. As the seas grew rougher during the night, Schlipper had Kapitzky and Langner placed in a rubber dinghy. An exceptionally high wave swept the small craft off the deck. Seaman First Class Richard Sura dived into the dark waters to retrieve it, but as he brought the dinghy alongside the U-boat, his body slipped beneath the sea. Others took up the cause and eventually brought the dinghy, as well as Kapitzky and Langner, back safely. The Old Man was still in a joking mood, telling his Executive Officer that he now qualified for the "Silver Wounded Badge."

Sometime around 1 a.m., Kelshall relates, the commander who "had fought the greatest battle of the war against aircraft" died. There then ensued a Wagnerian funeral befitting the opera stage at Bayreuth. Amid the "background sound of snarling hunter's [*sic*] engines overhead, complimented [*sic*] by lightning and rolls of thunder, with the lashing rain soaking everyone," the sailors sewed Kapitzky's corpse into a hammock and weighted down his feet. Then they lustily sang "the traditional naval hymn, the words of which were heard in the fierce wind and rain." As "the body of their much beloved commander slid over the side," the boat's "gunners stood to their weapons." Chief Engineer Skora recalls a more simple order of events. "The commander was committed overboard to the Caribbean Sea during the night 6/7," August 1943.[23]

There is no doubt that the final night on board the barely floating hulk that once was *U-615* must have been frightening. The men had suffered two days of aerial depth charges and strafing runs. Their captain and their best gunner lay dead on the deck. Their much-loved petty officer, Dittmer, had been shot to death and hurled overboard by the machine-gun blast. Nineteen sailors were wounded, some bleeding profusely. The casings were awash with seawater. It was pitch black, with rain, thunder, and lightning flashing all about. They were in shark-infested waters about 250 kilometers from the nearest land. Their prospects were not good.

The first rays of light brought a "smoke smudge" on the horizon. Rescue? Or death? The sailors grabbed life vests and floats, took to the water, and grouped around the rubber dinghy. Schlipper and Skora joined Abel inside the boat and blew the last remaining high-pressure air out of Diving Tank No. 3, allowing seawater to rush in. By the time they returned on deck, they were standing in deep water. *U-615* slipped beneath the sea around 5 a.m. on August 7, 1943. No one bothered to take along the "top-secret" *Nachtfernrohr*, for it had detected not a single attacker. The last thing the survivors of *U-615* saw was its conning tower emblem: a torpedo across which a winged aerial bomb had been superimposed. Fitting!

The "smoke smudge" on the horizon was the USS *Walker* under Commander O. F. Gregor. At 5:25 a.m., *Walker* sighted red flares off the port bow, and 22 minutes later the conning tower of a "submarine apparently submerging" at 16,000 yards. Leery of possible U-boats in the area, *Walker* approached the source "zig zagging radically at high speeds." At 6:07 a.m., Gregor spied survivors in a raft. He began rescue operations at once, and after a brief interruption at 6:55 a.m. caused by a false "contact" report, hauled "3 officers, 40 enlisted men and 1 dead enlisted man" out of the water. He ordered medical attention to three survivors for gunshot wounds, one for shrapnel wounds, and 15 for "superficial lacerations, contusions and abrasions."[24]

Pilots Crockett, Christian, and Dresbach (posthumously) were awarded the Distinguished Flying Cross for their valor in destroying *U-615*. Both *The New York Times* and the *Reader's Digest* eulogized their actions in feature articles in November 1943.

For Karl Dönitz, the destruction of *U-615* was but the most dramatic episode in the "killing" of eight of the 13 boats that he had dispatched to the Caribbean in June 1943, in return for a mere three ships of 17,000 tons sunk. After several futile attempts to reach Kapitzky on August 11, 14, and 18, Captain Godt declared *U-615* "potentially lost" on August 30, 1943, and "formally lost" on May 18, 1944.[25] On August 18, 1943, U-Boat Command had fired off an ominous last Enigma message: "No refueling possible for Kapitzky (615)."[26] It was a fitting epitaph.

* * *

In January 1943, Admiral Karl Dönitz had awarded Albrecht Achilles the Knight's Cross for his spectacular attacks on Port of Spain and Castries. It brought the young "ace" no luck. A harrowing fourth war patrol to Newfoundland and Rhode Island in Force 6 to 7 seas battered the slender craft, and its only "kill" was the 250-ton brig *Angelus* of Montreal, loaded with molasses. A despondent Achilles did not even submit an after-action report for the wretched patrol. U-Boat Command laconically commented, "Patrol by a single craft which, despite long duration, brought no special success. Only minimal traffic encountered, but strong sea and air patrols. The sinking of a brig by artillery is the only consolation. Nothing else to be noted."[27]

As always, Dönitz had technological innovations on hand. The entire conning tower was reinforced with protective shielding. As well, *U-161* became the first boat to receive the new "W. Anz g 1" direction-finder receiver of the Hagenuk Company. Formally introduced into the U-Boat Service as FuMB-9, it became *Wanze*, or "bedbug," in crew parlance. Dönitz was certain that he had found the "cure" for Allied "direction-finding." Due to the frequency of enemy air attacks, he added a medical doctor to the crew of each boat. In the case of *U-161*, this was Oberleutnant Dr. Thilo Weiss. U-Boat Command obviously had taken the Kaleus' after-action reports concerning Allied air attacks seriously.

Achilles took *U-161* out of Lorient on August 8. His orders were to rendezvous with Shinji Uchino's submarine *I-8* west of the Azores. Thereafter, *U-161* was to proceed to the coast of Brazil. After crossing the Bay of Biscay, *U-161* steamed down the coast of Iberia and headed for the prearranged rendezvous with *I-8*.[28] Codenamed *Flieder* (Lilac), the Japanese blockade runner carried an extra crew of 48 sailors; Uchino

The Crew of *U-161* musters topside for the loading/unloading of torpedoes. Source: Ken Macpherson Photographic Archives, Library and Archives at The Military Museums, Libraries and Cultural Resources, University of Calgary.

was to take possession of the new *U-1224*, a gift from Hitler to Emperor Hirohito. After firing the recognition signal, Achilles sent one officer and four radio operators as well as a brand-new *Wanze* over to *I-8* by rubber dinghy. Uchino then made for Brittany to load aircraft engines, torpedoes, and anti-aircraft guns as well as German advisors for the return leg of the journey. But Hirohito would be denied the Führer's present, grandiosely misnamed "Marco Polo II": bound for Japan, the renamed *RO-501* was sunk by depth charges from the destroyer USS *Francis M. Robinson* northwest of the Cape Verde Islands on May 13, 1944.

Achilles headed for Brazil. He met the homeward-bound *U-198* in the South Atlantic on September 5, handed over a *Wanze* receiver, and took on fuel oil. The next day U-Boat Command ordered *U-161* to proceed to Bahia to begin the war patrol. Around 5 p.m.,[29] on September 19, off Martin Vaz Rocks, Achilles found the unescorted 5,472-ton British steam freighter *St. Usk*. It was armed and loaded with 6,500 tons of Brazilian rice, tinned meat, and cotton seed, bound from Rio de Janeiro to Freetown. Achilles fired a single torpedo, hitting the target aft but not sinking it. Master G. H. Moss immediately ordered a zigzag course. Darkness set in. Achilles pursued and just before midnight fired two torpedoes. Despite hearing "powerful muffled thuds," Moss's ship refused to go under. Again, Achilles pursued. At 6:50 a.m. the next morning, he loosed a single torpedo at the plucky vessel; it struck aft of the port side in the No. 5 Hold. The ship's main top-mast came down, the derricks were shattered, and the propeller was blown off. Moss ordered "Abandon Ship!" and the *St. Usk* sank by the stern an hour later.

Achilles made for the lifeboats. After apologizing for sinking the ship, he handed the survivors coffee and water, ordered Dr. Weiss to attend to the wounded, and then presented a small-scale chart on which he drew a course for Bahia, 500 miles away. Acting on Dönitz's standing order, he took its master prisoner but released the rest of the crew of 47 in the lifeboats. Chief Officer E. C. Martyn remembered the submarine's commander, dressed in khaki shorts, tropical jacket with battered epaulettes and "forage" cap, as "a young man, in his early thirties," with "hair of medium colouring, and face sunburned," with small but "well kept hands," and fluent in English, "with no trace of an accent." The crew he judged to be all very young, with the exception of the doctor, who was "grey-haired."[30]

On September 26, off Maceió, Achilles came across another unescorted loner, the 4,998-ton Brazilian packet ship *Itapagé*, bound from Rio de Janeiro to Belém with 600 tons of general cargo.[31] At 3:50 p.m., he fired two torpedoes into its starboard side. The vessel sank within four minutes. Eighteen of its crew and four passengers were killed; 85 survivors made it in lifeboats to São Miguel dos Campos. That same day, *U-161* possibly also found the 300-ton sailing ship *Cisne Branco*, carrying a load of salt, and sank it with a single torpedo. Six of its complement of ten were eventually rescued.[32] Later that evening, Achilles took his boat to the mouth of the São Francisco River, north of Aracajú, in hopes of encountering other unescorted merchantmen.

By now, US Navy ASW forces stationed in Brazil were scouring the coast off Bahia in search of the raider. They had read "all pertinent [radio] traffic" emanating from *U-161* through ULTRA intercepts and began to close the noose. During the night of September 26, "Huff-Duff" operators got a rough fix on the U-boat, and next morning several aircraft flew barrier sweeps over the suspected location. One of those was a Martin PBM-3C Mariner flying boat, P-2 of Patrol Squadron VP-74, piloted by Lieutenant (jg) Harry B. Patterson.[33] The blue-gray Mariner had lifted off the waters at Aratú at 9:29 a.m. It was a clear, cloudless day with unlimited visibility. Patterson climbed to 4,500 feet. At 10:50 a.m., Radioman D. A. Bealer made a contact at 38 miles. It was a submarine bearing eight degrees to port and making flank speed, 18 knots. Why, Patterson must have wondered, this extravagant use of fuel? Was it pursuing fresh prey?

The P-2 quickly closed range. Within five minutes Lieutenant (jg) Charles Fergerson, second pilot, made visual contact at 18 miles, attracted by the fully surfaced submarine's wide white wake. Patterson sounded battle stations, brought the Mariner up to 180 knots, and executed a shallow turn to the left "to take advantage of the sun." The flying boat's oyster-white bottom was perfectly suited to the conditions. Patterson decided to attack the raider's stern. At a range of seven to eight miles, the U-boat's gunners suddenly sent up steady bursts of anti-aircraft fire, all of it short, leaving only "white puffs" in a line across P-2's approach. Achilles had well remembered Dönitz's admonition: "In case of doubt, stay up top and shoot!" But he had overestimated the range of his new guns. *U-161* turned to port to keep the flying boat off its stern. Patterson was confused: in the past, the boats had always presented a beam target.

Second Watch Officer Detlef Knackfuss's gunners were superb, and a stream of fire from the twin and quadruple mounts of 2-cm guns whistled past the Mariner, buffeting it with air turbulence. At 3,000 yards, Bow Gunner L. V. Schocklin opened fire with the .50-caliber Browning machine guns. He squeezed off 1,000 rounds and was certain that he had wounded or killed several gunners on the U-boat's deck. But the "twin fifties" were still out of range. By now, only 75 to 100 feet above the sea, Patterson approached *U-161* over the port stern. Copilot Fergerson dropped a string of six Mark 44 Torpex-filled bombs. Patterson then banked the Mariner into a sharp left turn to escape Knackfuss's lethal fire.

Hell broke loose all around *U-161*. The Mariner's tail and waist gunners saw one depth charge explode off the starboard side "abeam quarter," another off the starboard stern. The U-boat was engulfed in sea spray. The Americans assumed some damage to the submarine. VP-74 Squadron Commander G. C. Merrick later estimated that the six depth bombs "were slightly over to starboard, believed within damage range of submarine's stern."

Achilles opted to stay on the surface and fight it out. He used his 10.5-cm deck gun whenever P-2 was off in the distance and his 2-cm cannons whenever it closed range. After reaching 800 feet, Patterson renewed the attack. The fire from the U-boat was "heavier and more accurate," the shells exploding just off P-2's port side. Then one struck the flying boat forward of the galley door, the "shrapnel and aluminum" severely injuring Radioman Bealer and Ensign Oliver J. Brett, the bombardier. Patterson continued his run. Fergerson dropped the remaining two Mark 44 bombs as the Mariner passed over the target from stern to bow at 150 feet. Both bombs exploded near the submarine. *U-161* slewed almost to a halt and "maneuvered erratically and violently." Patterson observed: "after deck awash," "light grey smoke … in addition to diesel fumes," and "small fire believed begun near [conning tower] base on after deck." The submarine's stern "vibrated." Still, the gun crews were putting out a "continuous fire." Commander Merrick later recorded that the last two charges "were on the starboard quarter," and again "within damage range of the submarine's stern."

Patterson brought the flying boat up to 2,500 feet. He then went back for a final look. At 11:22 p.m., he saw *U-161* "submerge." He dropped a marker over the swirl and headed back to base to get medical treatment for his crew. A US Navy Lockheed Ventura bomber from VB-129 arrived a short time later but saw neither the U-boat nor its survivors.

U-161 disappeared beneath the waves 250 miles south of Recife, Brazil, in two miles of water. No sign of either the boat or its 52-man crew (and Master Moss of the *St. Usk*) was ever seen. Thus, one can only speculate on "Ajax" Achilles' final moments. From all reports by the crew of the P-2 Mariner, the six Mark 44 aerial bombs had started a "small fire" near the conning tower, brought the craft to a virtual halt, forced the stern to "vibrate," caused the craft to emit "light grey smoke," and so damaged the aft that *U-161* "maneuvered erratically and violently." Under these conditions, and knowing that the Mariner undoubtedly had reported his position and called in reinforcements, "the ferret of Port of Spain" apparently had decided to seek safety by going deep. Historian Gaylord Kelshall speculates: "If this is the case then the occupants of U161 must have died under nightmare circumstances, diving deeper and deeper, with the boat out of control, until the sea finally claimed them."[34]

When Achilles failed to reply to urgent Enigma messages on October 8, 9, 10, 12, 14, and 19 – including a dire warning that he was thereby violating Paragraph 354 of the *Captain's Handbook* – U-Boat Command deemed *U-161* to have been "sunk by Brazilian air force."[35] But how? Had the Hagenuk *Wanze* failed to detect enemy aircraft? Had the anti-aircraft guns malfunctioned? Or were even the quadruple cannons inadequate to bring Allied planes out of the skies? On April 5, 1945, Dönitz posthumously promoted Achilles to the rank of Korvettenkapitän.[36]

Within one calendar year the "Great Lion" had lost his four Caribbean "aces": Otto Ites of *U-94* in August and Jürgen Wattenberg of *U-162* in September 1942; Werner Hartenstein of *U-156* in March and Albrecht Achilles of *U-161* in September 1943. There no longer were sufficient boats and veteran skippers to mount attack waves against the font of Allied oil refining. Two-thirds of the U-boat tanker fleet (*U-487*, *U-459*, *U-461*, *U-462*, and *U-489*) needed to support operations in distant waters had been destroyed in less than four weeks, from July 13 to August 4, 1943. Allied air and surface ASW with its unfathomable new technologies – "Huff-Duff," Leigh Lights, ASV radar – had gained the upper hand in the Battle of the Atlantic.

For the time being, there was nothing that Dönitz could offer his commanders, save more exhortations. "Take Advantage of any chance to attack." "Bring honor to your name." "Go after 'em at top speed." "Something must be sunk out of this convoy tonight. At 'em." "You have only tonight left, so put all you have into it."[37] For the late fall of 1943, he placed his hopes in German engineers to come up with antidotes to the deadly Allied ASW technologies. But the days of the "Golden West," when half a dozen U-boats could wreak havoc with Caribbean oil, were a thing of the past. Operation New Land had run its course.

CONCLUSION[1]

After Adolf Hitler's declaration of war against the United States on December 11, 1941, the U-boat war was extended to the eastern seaboard of the United States. Under the auspices of Operation Drumbeat (Paukenschlag), the "gray sharks" found harbors and ships well lit; and in what Samuel E. Morison, the official historian of the United States Navy in World War II, termed a "merry massacre," a mere five Type IX U-boats in January–February 1942 sank 25 ships of 156,939 tons. Vice Admiral Karl Dönitz, Commander U-Boats, sent out several more waves of U-boats over the next six months, eventually bagging 397 Allied ships of more than 2 million tons off the US coast, losing only seven boats in the process. Drumbeat was a spectacular surprise attack. It showed what massive destruction a mere five, slender 1,000-ton U-boats could wreak in the hands of experienced commanders.

On January 15, 1942 Dönitz, obviously emboldened by the first news of Drumbeat successes, decided to extend American operations to the vital nerve center of the Allied bauxite and oil supply: the Caribbean basin. He invited five U-boat captains and two Hamburg-Amerika Line captains to his command post in the Villa Kerillon at Kernével. The U-boat skippers were regular navy men, and had served with the surface fleet before joining the "Volunteer Corps Dönitz." Their assignment under Operation Neuland (New Land), was as straightforward as it was demanding: "Surprise, concentric attack on the traffic in the waters adjacent to the West Indies Islands. The core of the attack thus consists in the surprising and synchronized appearance at the main stations of Aruba a[nd] Curaçao."[2] The group was to commence operations during the new moon period beginning on February 16: Günther Müller-Stöckheim's *U-67* off Curaçao;

Werner Hartenstein's *U-156* and Jürgen von Rosenstiel's *U-502* off Aruba; Albrecht Achilles' *U-161* off Trinidad; and Asmus "Nicolai" Clausen's *U-129* off the coast of British and Dutch Guiana.

Formalized on January 17, 1942, "Operations Order No 51 'West Indien'" identified primary targets to be the lake tankers and bauxite freighters as well as the oil refineries on the islands—most notably the Standard Oil of New Jersey Esso Lago complex at San Nicolas, Aruba, the largest in the world; the Trinidad Leaseholds' refinery at Pointe-à-Pierre, the largest in the British Empire; and the Royal Dutch Shell Shottegat plant at Curaçao. The Hamburg-Amerika Line captains had briefed the vice admiral on the nature of the oil traffic. "The oil is brought to Aruba as well as Curaçao from the Gulf of Maracaibo in shallow-draft tankers of about 1,200 to 1,500 tons with a draft of 2 to 3 m[eters], is refined there and loaded onto large ocean-going tankers." Trinidad offered an especially target-rich environment: apart from housing the oil refinery and tank farms, it was the transshipment site for the bauxite needed for Allied air industries as well as the departure point for seaborne traffic bound for Cape Town. The U-boats undertook the great circle trips across the Atlantic—Aruba 8,000 nautical miles and Trinidad 7,200 nautical miles return—using only one diesel engine to save fuel. This left the captains two to three weeks at most on station in the Caribbean; the so-called "milk-cows" (Milchkühe), U-boat tankers, were not yet operational. The attacks were driven home precisely "five hours before day break" on February 16 to assure surprise.

Moreover, Operation New Land was a departure from Dönitz's customary operational tactics. This time the "gray sharks" were assigned specific targets to attack in a specific region. Their captains were free to interpret their zones of attack liberally and independently. They were not to hit and run, but to remain in theater to drive home their attacks. "Thus, do not break off [operations] too soon!" They were to deploy their torpedoes first, and thereafter their deck guns (if land targets were in the offing). In eager anticipation, the five skippers spoke of a dawning "Golden West." Still, New Land was a bold, even audacious gamble. It would require the utmost of U-boat crews both physically and mentally. And it would require a great deal of luck.

What did Dönitz and his U-boats do successfully? The element of surprise certainly was with the German raiders. The simultaneous explosions of tankers off Aruba, Curaçao, and Trinidad shattered the tranquility of Paradise. Radio transmitters from Galveston to Caracas blared out warnings of the new danger to shipping. There was a widespread exodus from coastal cities into the cacti-studded countryside on Aruba. Chinese tanker crews at Curaçao went on strike; 15 were shot by the local Dutch militia, 37 others simply "disappeared."[3] Islanders who were young children retained vivid memories of the panic and uncertainty of February 1942 all their lives.

For a short period, the Allied oil supply was put in jeopardy. The roughly 95 per cent of product for the east coast of the United States—59 million gallons per day—that came from the Caribbean and the Gulf of Mexico by tanker at the end of 1941 shrank by 25 per cent as a result of the U-boat onslaught. There was oil aplenty available at the Gulf Coast ports, but fewer and fewer tankers to ship it in. The slaughter of tankers at sea that began at the very start of the war, but which was dramatically increased with Operation Drumbeat in January 1942 and New Land in February 1942, caught the United States by surprise. At that point, no pipelines connected the oil-producing regions of Texas, Louisiana, and Oklahoma with the US east coast as far west as the Appalachian Mountains. Thus, all of New York, New Jersey, New England, and most of Pennsylvania and Virginia were supplied by sea by tanker. The railways had limited facilities for carrying oil to this vital region, while road transport was completely inadequate. The government made deep cuts in the supply of gasoline and fuel oil in the eastern part of the United States, while it and the oil companies sought a solution to the growing shortage. Eventually, a massive effort was made to push two pipelines—the "Big Inch" and the "Little Big Inch"—from the East Texas oil fields to Norris City, Illinois, and Seymour, Indiana, and then on to Philadelphia and New York.[4] The pipelines, together with a major organization of the railway tanker car system, ended the short-run shortage. When large-scale tanker construction eventually added to this inflow of east coast oil by 1944, the shortages of 1942 disappeared forever.

With regard to Britain, Caribbean oil shipments declined from 67 per cent of total imports in 1941 to just 23 per cent by 1943.[5] At the end of that year, oil stocks had shrunk to six months' supply and shipments of refined gasoline by 20 per cent. Royal Navy stocks fell to the "danger level" and Royal Air Force squadrons faced a severe shortfall of vital high-octane aviation fuel. Anecdotally, King George VI strove to overcome the oil shortage by extinguishing central heat at Buckingham Palace and Windsor Castle, by allowing only single light bulbs to burn in bathrooms and bedrooms, and by having red or black lines painted on the inside of bathtubs to restrict hot-water use to five inches of tub.[6]

In launching the assault on Caribbean oil, Dönitz had found one of the few true strategic chokepoints in the Allied war effort. But he did not fully realize it at the time. The admiral knew that oil was a vital commodity in the war, and after he had gathered information from the Hamburg-Amerika Line captains, he knew that Caribbean oil was especially vulnerable. However, he did not know that a great part of US industrial production depended on tankers to carry oil from the Gulf of Mexico to the US east coast. He was ignorant of this because, unlike Britain and the United States, Germany had not prepared adequately for prolonged economic war against its enemies and had not planned a campaign to attack American or British strategic chokepoints. It is true that much of the Allies' planning for economic war was based on false assumptions or poor information, but at least they understood what sort of a war they were in and prepared to fight it.

With regard to bauxite, Britain in 1939 imported all of its raw supply – some 302,000 tons. By 1942, as the U-boats ravaged the waters of British and Dutch Guiana, that figure fell to a dangerous level of just 48,000 tons.[7] Aircraft production was maintained only by drastically increasing the import of finished aluminum – almost exclusively from the United States – from 58,000 to 132,000 tons between 1939 and 1942. Similarly, the U-boats in the Caribbean made a severe dent in the annual shipments of one million tons of bauxite to ALCOA and ALCAN in the United States and Canada in 1942–43.

US Army Chief of Staff George C. Marshall in May 1942 sent Admiral Ernest J. King his assessment of the situation: the New Land boats had destroyed 22 per cent of the Allied bauxite fleet, one out of every four Army ships sent to reinforce the Caribbean theater, and 3.5 per cent of Allied tanker tonnage per month.[8] "Our entire war effort," he warned the

Commander in Chief US Fleet, was now "threatened." Over the first six months, the German raiders dispatched 965,000 tons of Allied shipping in the Caribbean, of which an alarming 57 per cent were tankers. King at times suspended sailings into the area. The US Navy calculated that the sinking of three average ships was equivalent to the damage inflicted by 3,000 successful Luftwaffe bombing sorties.

To give some reference points to these statistics, it took 10,000 gallons of 100-octane aviation fuel per minute to mount a large bombing raid over Germany, 60,000 gallons a day of regular gasoline to keep a single armored division fighting, and the fuel to fill the tanks of one battleship could heat an average family home for 500 years.[9] Prime Minister Winston S. Churchill certainly was aware of the criticality of oil. "The terrible war machine," he had informed the nation in a BBC radio broadcast on June 22, 1941, the day Germany invaded the Soviet Union, "must be fed not only with flesh but with oil."[10]

With regard to the modern-day pirates of the Caribbean, the German skippers who undertook the first assault all enjoyed immense success and were well rewarded. Four received the coveted Knight's Cross: Clausen in March 1942, Hartenstein in September 1942, Müller-Stöckheim in November 1942, and Achilles in January 1943. The fifth, Rosenstiel, was killed by aircraft action in the Bay of Biscay in July 1942 after having sunk or damaged 104,000 tons and thus certainly would have won his *Ritterkreuz* as well.[11] All five became Caribbean "aces": Achilles with 27,997 tons destroyed, Clausen with 25,610, Hartenstein with 44,806, Müller-Stöckheim with 27,795, and Rosenstiel with 46,044.[12]

At first, the descent into the Caribbean basin was a wild success. Within 18 months of its launching, however, it was a significant failure. Why? The Germans' initial successes were scored largely against an enemy that was caught flat-footed, divided politically and militarily, which lacked imagination as to the real threat in the Caribbean, and which had neither the equipment nor the training and capability to weather the onslaught. The United States acted early in recognizing that the security of the Panama Canal was a strategic necessity to it and ultimately to the Allies. Thus, the Americans began in the fall of 1940 to build a ring of concentric defenses on both the Atlantic and Pacific sides of the canal. The "destroyers-for-bases" deal of September 1940 was followed in quick succession by the selection of bases on Britain's Caribbean possessions, and by the initial construction of bases for both the US Army Air Forces and the US Navy. What did not happen until many months after the start of Operation New Land was the forging of a united antisubmarine command, first among the Allies themselves – US, British, and Dutch – then among the Caribbean and Latin American nations that formed the political outer ring of defenses (Cuba, Columbia, Venezuela, and, outside the Caribbean Basin, Brazil), and finally among the US armed forces.

While the Allies stumbled toward unity of command under US leadership, men, artillery, naval vessels, and aircraft finally began to flow south. In time, a ring of airfields and seaplane bases encircled the Caribbean basin – from Cozumel, Mexico, to Waller Field, Trinidad, and from San Nicolas, Aruba to San Juan, Puerto Rico. Initially, the men were untrained, the vessels unsuitable and the aircraft slow, short-ranged, and poorly equipped. But that, too, changed. More and better aircraft – B-18 "Bolos," A-20 "Havocs," P-40 Warhawks, PBY Catalinas, and PBM Martin Mariners – equipped with radar, long-range capability, and effective submarine-killing weapons – flooded rapidly growing Caribbean air bases. When these aircraft were hooked up to inter-island human intelligence, long-range and land-based radar, and a command and control system sited on Panama, Cuba, Puerto Rico, and Trinidad, the Caribbean rapidly became an American lake. By July 1943 – 17 months after Neuland started – no U-boat running on the surface in the Caribbean or in the South Atlantic was safe from attack.

Ultimately, the tactical defeat of the U-boats, in the Caribbean as well as in the Atlantic, came through no single device or effort, but rather through a combination of Allied antisubmarine warfare technology. This began with ULTRA decryption of German Enigma radio signals to U-boats by British scientists at Bletchley Park. Through High-Frequency Direction-Finding, or "Huff-Duff," the Allies were able to triangulate the U-boats' replies to U-Boat Command to within a mile of their source. Thereupon, destroyer escorts were able to pinpoint the U-boats' locations by way of new centimetric radars (Types 271, 286), and Allied aircraft by way of new air-to-surface (ASV) radars. Once located, the U-boats were illuminated by powerful new Leigh Lights attached to the underbellies or wings of a host of Allied aircraft, most notably the American-built

Consolidated B-24 Liberator heavy bomber, the Martin Mariner, the B-18 "Bolos," and the Catalina PBY flying boat.

Germany, by contrast, never developed the necessary research and development loop to counter Allied ASW technology. Neither a host of primitive radar detectors such as the *Funkmessbeobachtungsgerät* (FuMB) nor decoys such as the *Aphrodite* balloons and *Bold* refractors proved effective. Radar remained a mystery. By May 1943, all Dönitz could do was to demand that his skippers overcome what he called Allied "cunning and technical innovations" with their "ingenuity, ability and iron will."[13] Using language that was reminiscent of that used by General Erich von Falkenhayn to justify the battle of Verdun in 1916, the "Great Lion" demanded that his skippers "force the enemy to undergo a permanent bloodletting, one by which even the strongest body must slowly and inevitably bleed to death."[14]

In the Caribbean, the Allied force multiplier was airpower. With the first wave of New Land boats, the United States pressed its rights under the "destroyers-for-bases" deal to militarize the Bahamas, Jamaica, St. Lucia, Antigua, British Guiana, and Trinidad. Eventually, two-thirds of all United States ASW aircraft were based on these British holdings. Jungles were bulldozed and airfields constructed almost overnight. Harbors and inlets were dredged and flying-boat bases established. Of the roughly 90 U-boats that sortied in the Caribbean, US Navy patrol craft destroyed 30, US Army Air Forces bombers four, and the Royal Air Force three.[15] As well, these Allied ASW measures combined in July 1942 to decimate Dönitz's "milk cow" fleet (*U-487, U-459, U-461, U-462,* and *U-489*) off the Azores and Spain, leaving but two U-boat tankers to service the fleet.[16] Even a cursory reading of the war diaries of the Caribbean boats reveals a litany of repeated and prolonged crash dives owing to being spotted from the air.

The Germans also suffered from a host of command and operations problems. From the start and throughout Operation Neuland, there had raged a bitter dispute behind the scenes concerning targeting. While Dönitz was, as ever, fixated on simple "tonnage warfare" against tankers and bauxite carriers, Grand Admiral Erich Raeder, Commander in Chief Kriegsmarine, had demanded that shore installations such as refineries and tank farms be given priority. He had a point. The giant Aruba refineries alone produced 500,000 barrels of gasoline and diesel fuel per day, including 5,000 barrels of critical 100-octane aviation fuel. But Dönitz hoped that Germany could sink ships faster than Britain and the United States could build them. In the end, this simple and naive "strategy" failed.

Raeder, as early as February 11, 1942, had admonished Commander U-Boats that his vessels must "deploy their artillery with incendiary shells against oil tank farms"; five days later, he repeated his demand that Dönitz "inaugurate actions of boats Aruba–Curaçao by shelling tank farms."[17] To no avail. Each time the boats sortied, Dönitz promised compliance with Raeder's injunctions. And each time he ignored them. On one of Dönitz's many rebuttals, in which he again juxtaposed "strategic pressure" and tactical "sinkings," Raeder's staff queried whether "Commander U-Boats has been misled in these matters or simply does not want to understand them!"[18]

Second, there was confusion between the German admirals as to how to proceed operationally. Commander U-Boats' position was that the available craft were to be sent out in rotating waves in order to maintain the element of surprise and to exert maximum pressure on Allied cargo carriers and their escorts. Raeder, on the other hand, demanded that no more waves be dispatched; instead, he wanted what he termed "continuous occupation" of the Caribbean basin by the U-boats.[19] When Raeder on April 2 fired off an acid one-sentence telegram to Kernével, demanding that Commander U-Boats comply with his orders, Dönitz resorted to his customary practice: he did not reply. Instead, he documented his position on the primary mission in the war diary.[20]

Third, there was no clear tactical objective behind Operation Neuland. What to target: the small tankers exiting the Bay of Maracaibo, the ocean-going tankers departing Aruba, Curaçao, and Trinidad, or the refineries on shore? And should whatever course chosen be undertaken by single boats in specific areas or by concentrated "packs" in one area at a time? In the end, it remained a pure "tonnage war."

Fourth, if the Raeder-Dönitz antagonism was not enough, there was the constant meddling of the Commander in Chief, German Armed Forces. Time and again, Adolf Hitler, on the basis of his famous "intuition," ordered U-boats diverted from the Caribbean and reassigned to "threatened" areas. Norway, he constantly lectured Dönitz, remained the "zone of destiny"[21] in the Battle of the Atlantic. In February 1942, at the very launch of Operation New Land, he ordered 20 U-boats to patrol Norwegian waters against an expected Allied invasion. Similarly, in June 1942, as yet another wave of boats headed for the Caribbean, the Führer ordered Dönitz to recall them and to redeploy them around the Azores as well as the Cape Verde and Madeira islands to rebuff what he was sure was an imminent Allied assault on North Africa.[22] And when that operation came in November 1942, Hitler predictably ordered all available U-boats (eight) to execute a "completely victorious operation" against the Allied invasion flotillas.[23]

Fifth, as Karl Hasslinger concluded in a US Naval War College study in 1996, Dönitz failed to apply "overwhelming force" at the "decisive point."[24] He shifted his forces in order to react to enemy moves, rather than to concentrate them on a single target or product. Concentration off Aruba, Curaçao, and Trinidad yielded to concentration off Brazil, which, in turn, yielded to concentration off Africa – and finally to a return to "wolf pack" tactics against almost invulnerable Allied convoys in the North Atlantic. Dönitz's firm conviction that the average sinking per U-boat per day (all documented on neat statistical tables nailed to the walls of his headquarters) gave a true sense of success was flawed, for it ignored the overall number of enemy ships (constantly on the rise) as well as the Reich's limited labor and material with which to replace destroyed U-boats. That Hitler's war in the East consumed ever greater amounts of labor, material, and fuel certainly was beyond Dönitz's power to remedy.

Sixth, the vessels available in 1942 simply were not up to the task demanded of them. The workhorse of the German submarine fleet, the Type VIIC, was a small 800-ton craft with an optimum range of but 8,500 nautical miles at ten knots. The larger 1,541-ton Type IXC boats had an optimum range of 14,035 nautical miles at ten knots. Neither could compare with the 2,400-ton US Navy Gato- or Balao-class submarines, which were air-conditioned. In fact, conditions on board the U-boats in tropical waters were abysmal. Already on his first patrol, Achilles in *U-161* reported temperatures of 40 degrees Celsius and humidity near 100 per cent in the boat. After nine successive crash dives, the men "reached the limits of their physical and psychological capacities."[25] The following month, Hartenstein in *U-156* reported similar conditions after 13 hours underwater. "Humidity is much more troublesome than heat … severe diminution of the crews' efficiency" was the result.[26] His chief engineer, Wilhelm Polchau, in July 1942, took pains to inform Dönitz that water at 30 degrees Celsius failed to cool the engines, that inside temperatures *averaged* 34 degrees, and that at 47 degrees it had been impossible to charge the batteries.[27] A month later, Achilles informed Commander U-Boats for a second time that heat (39 degrees) and humidity (near 100 per cent) in the tropics had taken their toll on the crew: skin sores, boils, digestive disorders, and exhaustion.[28]

Operation New Land showed Dönitz at his operational best but strategic worst. He skillfully employed "operational maneuver as a force multiplier."[29] He thereby forced the Allies to patrol mammoth areas of water while he struck what he considered to be the weak points in their defense. In the process, he sought to ensure "the greatest successes … at the least cost." While the Caribbean campaign dealt the Allies a stunning initial blow both materially and psychologically, Commander U-Boats never managed to sustain that effort. Lacking a base in the Caribbean, he hoped that the "milk cows" would be the answer to the problem of resupply. They were not. At no time did he station small surface tankers there in remote bays as "one-off" supply ships. Nor could he convince Hitler or Raeder to commit overwhelming force to that critical area.

In the end, Karl Dönitz remained wedded to his life-long belief that "tonnage" alone mattered. He adamantly refused – with the brief exception of the Caribbean campaign – to differentiate targets according to their cargo. He never conceded that sinking a ship carrying cotton to Britain was less effective than destroying a tanker bringing 100-octane aviation fuel to the Royal Air Force. Nor did it matter where enemy ships were torpedoed. He put it perhaps best in his war diary on April 14, 1942: "The enemies' shipping forms a single totality. Therefore, in this regard, it is immaterial where a ship is sunk; in the last analysis, it has to be replaced by a new construction."[30] More, it was "incomparably more important" simply to sink ships, whether loaded or in ballast, than "to reduce sinkings by making them in [only] a prescribed area." By then, "new constructions" referred primarily to United States, rather than British, shipyards. "Thus I attack the evil at its root when I assault [US] imports, especially the oil."

Could Operation New Land have succeeded? It was at best a one-time undertaking, one that could neither be repeated nor sustained over time. The February 1942 attack was predicated upon total perfection: intelligence, surprise, coordination, boats, engines, torpedoes, and fuel supply. There was no room for Carl von Clausewitz's notions concerning the "fog of uncertainty," or those of friction and chance. The botched attempt to shell the Esso Lago Refinery by Werner Hartenstein's *U-156* on February 16, 1942, due to the gunnery officer's failure to remove the tampion from the boat's deck gun revealed the "chancy" nature of the operation. Clausewitz's advice that friction could be overcome mainly by "mass" worked to the advantage of the Allies, not Germany. Dönitz's Caribbean "aces" eventually succumbed to Allied "mass" in the air as well as on the sea.

In the end, it was a fool's game. In the critical period between September 1942 and May 1945, as historian Clay Blair has shown, the U-boats sank only 272 of 43,526 Allied merchant ships on the Atlantic run; put differently, 99.4 per cent of Allied ships made it to port safely. Of the 859 "gray sharks" that sortied on war patrols, historian Alex Niestlé has calculated that 648 (75 per cent) were lost – and of these, a shocking 215 (33 per cent) on their first patrol. Allied air accounted for 234 U-boats (36 per cent of all boats lost).[31] The human toll, whatever the final tally, remains even more shocking: a walk through the U-Boat Memorial at Möltenort confronts one with the names of 30,000 submariners killed during World War II etched on bronze tablets.[32] Of two eminent students of the U-Boat war, the American Clay Blair speaks of the U-Boat war by 1943 as being a "suicidal enterprise"; the German Michael Salewski as being "dragged on like a ghostly, senseless, and murderous charade."[33] Operation Neuland at least was conceived within the realm of operational logic and probability. It failed for the same reasons as all of Hitler's campaigns. Step by step, from the invasion of Poland to the attack on the Soviet Union and the declaration of war against the United States, Germany dragged the world into the most far-reaching conflict in history without serious thought as to the long-term requirements, let alone the real possibilities, of waging such a war successfully. Dönitz caught the Allies napping in the Caribbean. But he was incapable of understanding just how vulnerable the route from Lake Maracaibo to the Dutch Antilles was, how totally dependent the United States was on tanker traffic from the Gulf of Mexico to the US east coast, or how completely reliant the Royal Air Force was on Caribbean aviation gas. He himself declared that his submarines were not strategic instruments; it was fortunate for the Allied cause and for the campaign in the Caribbean that he believed that.

GLOSSARY

100 Octane A high-performance fuel commonly used in aviation.

A-20 The Douglas A-20 Havoc light bomber aircraft.

AAF The United States Army Air Forces.

Abaft Relative term meaning in the direction of the stern, or back, of a vessel.

Abeam Bearing at a right angle to a vessel's hull.

Ace U-boat commander who has sunk over 100,000 tons in Allied shipping.

Admiralty The Office of the Lord High Admiral, responsible for all military and administrative functions of the British Royal Navy.

Aft Relative term meaning towards the stern, or back, of a vessel.

Amidships The horizontal or vertical middle of a vessel.

Anglo-American Caribbean Commission A six-member committee created in March 1942 to deal primarily with social issues within the Caribbean.

Aphrodite A radar decoy used to misrepresent the location of a U-boat; a balloon tied to a raft with aluminum foil strips along the rope connecting the two.

Army Air Corps The predecessor of the United States Air Force; active from 1926 to 1942.

ASDIC A method for locating submerged objects through the use of echolocation; an acronym for the Anti-Submarine Detection Investigation Committee.

ASV Anti-Surface Vessel.

ASW Antisubmarine Warfare.

Ato *See* G7a torpedo.

Atlantic Charter An August 1941 statement outlining Allied goals following the conclusion of the Second World War, including no territorial changes without population consent, freedom of the seas, and the disarmament of the Axis powers.

B-18 The Douglas B-18 Bolo medium bomber; originally based on the design of the Douglas DC-2.

B-24 The Consolidated B-24 Liberator heavy bomber.

Ballast Extra weight added to a vessel to improve stabilization, commonly in the form of soft iron; submersible vessels use water as ballast by flooding storage "ballast" tanks.

Barrel (unit) Measurement of volume used for petroleum products; equal to 42 US gallons or 159 litres.

Bauxite A type of ore that contains aluminum.

BdU (*Befehlshaber der Unterseeboote*) Commander-in-Chief of German U-boats Karl Dönitz; also used to informally refer to the headquarters of the U-boat service.

Big Inch A petroleum pipeline from Texas to New Jersey created in 1942–43 as an emergency war measure. Along with "Little Big Inch," the pipeline was viewed as a secure means to transport vital petroleum to the Eastern Seaboard of the United States when German U-boats threatened the traditional maritime means of transportation.

Bilge The floor, or bottom, of a vessel's inside hull where waste liquids collect.

Biscay Cross A type of external antenna consisting of two pieces of wood shaped in a cross that were connected to a VHF receiver. The antenna had to be mounted to the top of the conning tower each time the U-boat surfaced.

Blitzmädchen The German female naval service of the Second World War.

Bluejackets A term used to describe sailors of the United States Navy, based upon the colour of the uniform they wore.

Boatswain The senior non-commissioned officer in charge of all deck work on board a vessel.

Bold Short for *Kobold* (Goblin); a calcium hydride–based, canister-type sonar decoy used by U-boats during the Second World War to produce a false sonar target.

Bow The front portion of a vessel's hull.

Bridge The upper deck location of a vessel where all navigation, deck, and command responsibilities are made by the Captain or Officer of the Watch.

BUNGALOW Plan The US War Department's plan for war with Vichy France.

Canal Zone A strip of territory along the entire length of the Panama Canal under the jurisdiction of the United States.

Caribbean Defense Command United States military command responsible for the defence of the Caribbean basin and for South America.

Caribbean Sea Frontier United States Navy coastal defence area responsible for the protection of Allied shipping in the Caribbean Sea and the Atlantic coast of South America.

Cash and Carry The neutrality policy of the United States to supply resources and material to belligerent states under the stipulation that they provide their own means of transportation, assume all the risks of transport, and immediately pay for all goods in cash.

Catalina *See* PBY Catalina Flying Boat.

Centimetre (cm) Measurement of length; equal to 0.3937 inch or one-hundredth of a metre.

CINCLANT Commander-in-Chief United States Atlantic Command.

Conning Tower The structure of a U-boat between the pressurized hull and exterior bridge.

Creole A rather broadly defined ethnic group typically characterized as people born in Spanish colonies with European, especially French, ancestry.

Cubic Metre (cbm) Measurement of volume; equal to 264 US gallons or 1,000 litres.

Davit Cast iron crane with hoisting gear for raising and lowering objects over the side of a vessel.

Davy Jones' Locker Sailor's idiom for the bottom of the sea and the final resting place of sailors, ships, equipment, or any other article that is lost at sea.

Defense Plant Corporation Company chartered by the United States Congress in August 1940 that had the responsibility of expanding new and existing equipment and facilities for wartime production.

Depth Charge Anti-submarine weapon; uses a hydrostatic fuse to detonate an explosive charge at a preset depth.

DGZ German General Time. UTC/GMT +1.

Die Wochenschau Weekly German newsreel; active from 1940 to 1945.

Dragon's Mouth A series of straits that separates the Paria Peninsula of Venezuela from Trinidad and Tobago.

Eel Idiom for torpedo.

Enigma (*Schlüssel M*) Electro-mechanical rotor machine used by the *Wehrmacht* to cipher and decipher secret messages. The Enigma had an appearance similar to a typewriter. One rotor on the Enigma carried 26 contacts, the number of letters in the German alphabet; several rotors were used in series to further complicate the encryption of a message.

Erster Wach-Offizier (or Executive Officer) Second-in-Command of a vessel, who only reports to the commanding officer.

Eto *See* G7e torpedo.

First Lord of the Admiralty President of the Board of Admiralty within the British Royal Navy; typically held by a civilian and member of the British Cabinet.

Flak Anti-aircraft fire.

Fliegerbombe (*Fliegers* or *Fliebo*) Aerially dropped depth charge.

Fo'c'sle The upper deck section of a vessel that is forward of the main superstructure.

Fore Relative term meaning towards the bow, or front, of a vessel.

Fregattenkapitän Commissioned rank of the *Kriegsmarine* equal to that of a Captain junior grade.

Frettchen German for ferret.

Führer Leader; the title or epithet belonging to Adolf Hitler.

FuMB-9 (*Wanze*) Variation of the FuMB Metox radar detector used by the German *Kriegsmarine* to detect Allied aircraft.

Funkmessbeobachtungsgerät (FuMB) Type of radar detector used by the *Kriegsmarine*.

G7a torpedo An air-driven torpedo of the *Kriegsmarine*.

G7e torpedo An electric powered torpedo of the *Kriegsmarine*.

Gallon (gal) Measurement of volume; equal to 3.79 litres.

Garbage Tour A U-boat tour that must navigate vast distances, such as across the Atlantic Ocean, before arriving at its area of operations.

GAT/TAG An Allied convoy code given to a convoys travelling Guantánamo to Aruba and Trinidad (GAT), or conversely, Trinidad to Aruba and Guantánamo (TAG).

Golden West A term U-boat crews used for the Caribbean.

Good Neighbor Policy United States policy on non-interference within the domestic affairs of Latin America.

Gray Sharks Idiom for German U-boats.

Great Lion Admiral Karl Dönitz; as referred to by his U-boat crews.

Hawser Thick cable or rope used for mooring a vessel.

High-Frequency Direction Finding (HF/DF or "Huff Duff") Type of radio direction equipment used by the Allies in the Second World War to provide a bearing on German U-boats; multiple sets could be used to triangulate the position of a U-boat.

Hot Runner Dangerous situation caused when the motor of a torpedo is running, but the torpedo has failed to exit the tube.

Hydrophone Type of microphone used for hearing underwater sounds.

I.G. Farben German chemical company.

Jetsam Part of a vessel, its equipment, or cargo that has been purposely jettisoned, typically during an emergency.

Kapitänleutnant (*Kaleu*) German *Kriegsmarine* rank equal to Lieutenant Commander; often the rank held by the Commanding Officer of a U-boat.

Kilogram (kg) Measurement of mass; equal to 2.2 pounds.

Kilometer (km) Measurement of length; equal to 0.62 miles.

Knight's Cross German wartime decoration awarded for leadership and bravery in combat that came in several different grades.

Knot Measurement of speed; equal to 1.151 miles per hour or 1.852 kilometres per hour.

Kobold *See* **Bold.**

Korvettenkapitän *Kriegsmarine* rank equal to Commander.

Kriegsmarine Maritime service of German *Wehrmacht*; active from 1935 to 1945.

Kriegstagebuch (KTB) German war diary to provide testament of the decisions, actions, and other activities relating to command.

Leigh Light Extremely powerful 22-million-candlepower searchlight used by Allied aircraft as an anti-submarine device to locate U-boats on the surface at night.

Lend-Lease Act Program through which the United States provided the Allied nations with resources and materials between 1941 and 1945.

Lieutenant (jg) Inferior subdivision within the United States Navy commissioned rank of Lieutenant; equal to a Sub-Lieutenant in the Royal Navy.

List Leaning, or inclining, of a ship to one side; caused by the displacement of cargo or flooding of the hull.

Little Big Inch Petroleum pipeline from Texas to New Jersey created over the 1942–43 period as an emergency war measure. Along with "Big Inch," the pipeline was viewed as a secure means to transport vital petroleum products to the Eastern Seaboard of the United States when German U-boats threatened the traditional maritime means of transportation.

Local Combined Defense Committee Committee of community and military authorities for the coordination of civil defense.

London Blitz Period between September 1940 and May 1941 when the city of London, and the United Kingdom in general, experienced sustained strategic bombing by the German *Luftwaffe* during the Second World War.

Long Tom 155mm field gun used by the United States during the Second World War

Luftwaffe The aerial service of the *Wehrmacht* during the Second World War. The service was active from 1935 to 1945.

Mae West Nickname for the first inflatable life preserver used by Allied servicemen; derived from the physical attributes of actress Mae West.

Magnetic Anomaly Gear Anti-submarine device used to detect variations in the Earth's magnetic field; it was either towed behind a ship or attached to an aircraft.

MAN Diesel German company that produced large diesel engines for marine propulsion.

Meter (m) Measurement of length; equal to 3.28 feet.

Metox High-frequency radar warning receiver for detecting transmissions from patrolling aircraft; used by U-boats to detect Allied aircraft.

Midshipman Most junior commissioned rank in a navy comparable to an officer cadet.

Milchkuh (Milch, or Milk, Cow) Type XIV U-boat; a support U-boat that resupplied, rearmed, and refuelled other U-boats in order to prolong their at-sea operations.

Mortar Type of weapon that fires projectiles at short range with a low velocity and high ballistic trajectory.

Nachtfernrohr Type of night vision telescope used in the *Kriegsmarine*.

Nautical Mile Maritime measurement of length; equal to 6,076 feet or 1,853 metres.

Oberleutnant Kriegsmarine rank equal to a Lieutenant senior grade.

Oil Control Board British government agency consisting of both government and industry representatives; responsible for the wartime rationing and management of petroleum products within the United Kingdom.

"Old Man" Term of endearment used by a ship's company towards the Commanding Officer.

Operation Blue *Wehrmacht* offensive in southern Russia between June and November 1942.

Operation *Neuland* (or New Land) Codename for the February–March 1942 U-boat offensive in the Caribbean.

Operation *Paukenschlag* (Drum Beat) January–August 1942 U-boat offensive against merchant shipping along the East Coast of the United States; known as "the Second Happy Time."

Operation Pot of Gold United States war plan to send over 100,000 soldiers to Brazil by air and sea in response to growing Axis influence within the country; the plan was never carried out.

Operation Torch Allied invasion plan of French North Africa during the Second World War; commenced 8 November 1942.

Panama Declaration Declaration made by American states at the conclusion of the September 1939 Panama Conference; participants reaffirmed their own neutrality in the war; prohibited belligerent submarines from using domestic ports; demanded the cessation of subversive activities by foreign agents; and proclaimed a maritime security zone of 300 miles (480 km) around both coasts of the American continents.

PBM Martin PBM Mariner; a patrol bomber flying boat.

PBY Consolidated PBY Catalina flying boat; a versatile seaplane used in the Second World War for anti-submarine warfare, search and rescue, patrol bombing, and convoy escort.

Periscope Long vertical tube that contains a set of internal lenses and prisms for reflecting and magnifying light to an observer below. Used by U-boats to view the surface while remaining submerged.

Petroleum Administrator for War (PAW) Wartime agency of the United States for the organization and allocation of petroleum products.

Petroleum Coordinator for War Position held by Harold L. Ickes, the United States Secretary of the Interior, which entitled him to enact emergency war measures in the United States petroleum industry.

Petroleum Industry War Council Leaders of the United States petroleum industry who advised on the management, production, and distribution of petroleum resources for the wartime requirements of the United States military and civilian population.

Plantation Line Pipeline in the southeast of the United States spanning 1,261 miles or 2,029 km from Baton Rouge, Louisiana, to Greensboro, North Carolina.

Port Left-hand side of a vessel when facing towards the bow.

Q-Ship Merchant vessel with concealed heavy weaponry; for attacking U-boats unexpectedly while they operated on the surface.

Quadratkarte Type of quadrant chart used by the *Kriegsmarine* for navigation; world's oceans were divided into large squares, which were then further divided into smaller squares. To transmit a location, U-boats would first name the large square (such as "EC") and then the corresponding numbers to the smaller squares inside.

Radiogram Telegraph style of radio message.

RAINBOW 5 Pre-war plan of the United States based upon an alliance with Britain and France.

Screw(s) Propeller(s) of a vessel.

Serpent's Mouth Informal name of Columbus Strait; separates the southwest corner of Trinidad and Tobago from the coast of Venezuela.

Smutje Ship's cook.

Standard Oil United States petroleum company that ultimately grew into the largest oil refiner in the world; dissolved into smaller corporations in 1911 under the Sherman Antitrust Act.

Starboard Right-hand side of a vessel when facing the bow.

Stern The back end of a vessel.

Stevedore Dockworker, or waterfront worker, who assists in the loading and unloading of vessels in port.

Stoker Person who maintains the fire in a vessel's steam engine.

SOS Morse Code international distress signal consisting of three dots, three dashes, and another three dots; often thought to stand for "save our ship" or "save our souls"; however, the letters have no formal significance.

SSS A Morse Code international distress signal consisting of three dots repeated three times; exclusively used to broadcast that a hostile submarine caused the distress.

Supply Priorities and Allocations Board (SPAB) United States precursory agency to the War Production Board.

T2 Tanker Class of oil tankers constructed and produced within the United States during the Second World War.

Tampion Plug used to temporarily close the muzzle of a large gun.

Tarpaulin A tarp; long sheet of material used for its water resistance.

Tin can Idiom used by U-boat crews for a destroyer or other anti-submarine warship.

Ton Measurement of mass; equal to 2,000 pounds or 907 kilograms.

Tonnagekrieg (tonnage war) A German military strategy aimed at destroying as much merchant shipping as possible; based upon a core assumption that merchant ships can be sunk faster than they can be replaced.

Torpedo Directorate German naval office responsible for the design, testing, and virtually anything else relating to the use of torpedoes.

Torpedo Junction Name given by the Allies to the merchant shipping–rich waters between Guiana and Trinidad.

Torrid Zone Waters located between the Tropic of Cancer and the Tropic of Capricorn.

Transocean Press Service Wireless news service; located in Berlin, Germany.

Trinidad Mobile Force United States military force made up of troops from Guantánamo and Trinidad that prepared for the invasion of Vichy France–controlled Martinique; the invasion did not take place.

Triton The four-rotor version of the Enigma machine.

Triton-Null Order (the *Laconia* Order) Order by Admiral Karl Dönitz commanding U-boats to cease the maritime custom of assisting survivors of a sunken vessel.

Tulsa Plan United States wartime petroleum strategy made by experts of the Petroleum Industry War Council. The plan made several recommendations on maximizing the existing petroleum pipeline infrastructure, but further called for the construction of new pipelines between Texas and the east coast to meet escalating demand.

Type VII Most common type of German U-boat in the Second World War; with 703 produced, the Type VII is also the most produced submarine in history.

Type VIIC "C" variant of the Type VII U-boat; 568 were commissioned during the Second World War.

Type IX Class of German U-boat built for prolonged operations away from home-port facilities; carried up to 22 torpedoes.

Type IXC "C" variant of the Type IX U-boat; 54 were constructed.

Twin Fifties Pair of .50-calibre machine guns.

U-boat Submarine operated by Germany in either the First or Second World War.

ULTRA Codename of the British military intelligence organization that deciphered high-level encrypted German signals intelligence during the Second World War.

U.S. Caribbean Defense Command United States military organization that defended Panama and the surrounding Caribbean area.

VLR Abbreviation for Very Long Range.

VP Designation of United States maritime patrol squadrons used for reconnaissance, anti-surface warfare, and anti-submarine warfare.

Volunteer Corps Dönitz Elitist term for U-boat crews that reflected their voluntary nature and the privileges associated with the U-boat service.

W. Anz g 1 *See* **FuMB-9.**

Wanze See **FuMB-9.**

War Production Board Agency of the United States government that oversaw wartime production, including the allocation and rationing of resources.

Wasserbomben (Wabo) *See* **Depth Charge**.

WAT/TAW Allied convoy code for convoys travelling Key West to Aruba and Trinidad (WAT), or conversely, Trinidad to Aruba and Key West (TAW).

Zeiss binoculars Type of prism binoculars, known for folding the optical path and consequently reducing the size of the binoculars.

Zentrale The control room of a U-boat.

NOTES

PROLOGUE

1 Lieutenant in the US Navy; "Kaleu" in German naval parlance.

2 Lieutenant (jg) in the US Navy.

3 German General Time, or DGZ; deduct one hour for Greenwich Mean Time (GMT).

4 War Diary (*Kriegstagebuch*, or KTB), *U-161*, 1. Unternehmung, PG 30,148/1, Bundesarchiv-Militärarchiv (hereafter, BA-MA), Freiburg, Germany.

5 See Jak P. Mallmann Showell, *Hitler's U-Boat Bases* (Annapolis: Naval Institute Press, 2002), 94–106.

6 From Christophe Cérino and Yann Lukas, *Keroman. Base de sous-marins, 1940–2003* (Plomelin: Palantines, 2003), 24–27; and Gordon Williamson and Ian Palmer, *U-Boat Bases and Bunkers 1941–45* (Botley, Oxford: Osprey, 2003), 42–43.

7 See Lawrence Paterson, *Second U-Boat Flotilla* (Barnsley, Yorkshire: L. Cooper, 2003), 74.

8 I am indebted to Rear Admiral Pierre Martinez, Commandant la Marine à Lorient, for taking me on a guided tour of the Villa Kerillon on Saturday morning, July 22, 2006 – HH.

9 The Third Reich's premier civil and military engineering group, roughly akin to the US Army Corps of Engineers.

10 Karl Dönitz, *Memoirs: Ten Years and Twenty Days* (London: Weidenfeld and Nicolson, 1959), 129.

11 See Gaylord T. M. Kelshall, *The U-Boat War in the Caribbean* (Annapolis: Naval Institute Press, 1994), 26–27.

12 Entry for May 21, 1940. I SKL, Teil CVII, Überlegungen des Chefs der SKL und Niederschriften über Vorträge und Besprechungen beim Führer, September 1939–Dezember 1940, PG 32184 Case 230, BA-MA, 146–47.

13 "Operationsbefehl 'Westindien' No 51," January 17, 1942. "Secret. For Commanders Only!" Chefsache, vol. 3, U-Boote, Allgemein, RM 7/2336 BA-MA. Following citations are from this 12-page document, signed "Dönitz." The general contours of Caribbean operations were summarized by Dönitz's son-in-law, Fregattenkapitän Günter Hessler,

after the war: Ministry of Defence (Navy), German Naval History, "The U-Boat War in the Atlantic 1939–1945" (2 vols., London, 1989).

14 See Holger H. Herwig, *Germany's Vision of Empire in Venezuela 1871–1914* (Princeton: Princeton University Press, 1986), 227–31.

15 Thus Ronald H. Spector, *At War at Sea: Sailors and Naval Combat in the Twentieth Century* (New York: Viking, 2001), 224. See also Marc Milner, *Battle of the Atlantic* (Stroud, UK: Vanwell, 2003).

16 See Michael Gannon, *Operation Drumbeat: The Dramatic True Story of Germany's First U-Boat Attacks along the American Coast in World War II* (Annapolis: Naval Institute Press, 2009).

17 Accessed October 4, 2013.

INTRODUCTION

1 British Library of Information, http://www.ibiblio.org.

2 Ed Shaffer, *Canada's Oil and the American Empire* (Edmonton: Hurtig, 1983), 41.

3 Richard Overy, *Why the Allies Won* (New York and London: Norton, 1995), 228–33; Michael Antonucci, "Blood for Oil; The Quest for Fuel in World War II," in *Command*, January–February 1993; also http://www.mikeantonucci.combloodforoil.htm.

4 D. J. Payton-Smith, *Oil: A Study of War-time Policy and Administration* (London: H.M. Stationery Office, 1971), 111–12; See also http://www.warsailors.com/freefleet/shipstats.html.

5 Payton-Smith, *Oil*, 114.

6 Ibid., 13.

7 Henrietta M. Larson, Evelyn H. Knowlton, and Charles Popple, *New Horizons: 1927–1950 (History of Standard Oil Company (New Jersey)* (New York: Harper, 1971), 388–91.

8 Payton-Smith, *Oil*, 155.

9 Martin Gilbert , ed., *The Churchill War Papers*, vol. 3, *The Ever Widening War, 1941* (New York and London: W. W. Norton, 2000), 847.

10 Payton-Smith, *Oil*, 201.

11 Ibid., 201–2; Larson et al., *History of Standard Oil*, 397.

12 John Knape, "British Foreign Policy in the Caribbean Basin 1938–1945: Oil, Nationalism and Relations with the United States," *Journal of Latin American Studies* 19 (November 1987): 279–80; Leonard M. Fanning, *American Oil Operations Abroad* (New York: McGraw-Hill, 1947), 73–83, 256–89.

13 Frederick Haussmann, "Latin American Oil in War and Peace," *Foreign Affairs* 21 (January 1943): 357.

14 Charles Sterling Popple, *Standard Oil Company (New Jersey) in World War II* (New York: Standard Oil Co., 1952), 25; John W. Frey and H. Chandler Ide, eds., *A History of the Petroleum Administration for War 1941–1945* (Washington: US Government Printing Office, 1946), 193.

15 Alfred E. Eckes, Jr., *The United States and the Global Struggle for Minerals* (Austin and London: University of Texas Press, 1979), 90–119.

16 Fitzroy Baptiste, "The Exploitation of Caribbean Bauxite and

Petroleum, 1914–1945," *Social and Economic Studies* 37 (1988): 110–13; Stetson Conn, Rose C. Engelman, and Byron Fairchild, *United States Army in World War II: The Western Hemisphere, Guarding the United States and its Outposts* (Washington: US Government Printing Office, 1964), 337.

17 United States Energy Information Administration.

18 Gerald D. Nash, *United States Oil Policy: 1890–1964* (Westport, CT: Greenwood Press, 1968), 159.

19 Harold F. Williamson et al., *The American Petroleum Industry: The Age of Energy, 1899–1959* (Evanston, Il: Northwestern University Press, 1963), 743–44.

20 Nash, *United States Oil Policy*, 162–63.

21 Conn et al., *The Western Hemisphere: Guarding the United States and its Outposts*, 329.

22 The story of the negotiations is told in detail in Philip Goodhart, *Fifty Ships that Saved the World* (Garden City, NY: Doubleday, 1965), 159–76.

23 Conn et al., *The Western Hemisphere: Guarding the United States and its Outposts*, 355–57.

24 Kelshall, *U-Boat War in the Caribbean*, 8.

1: "RUM AND COCA-COLA": THE YANKEES ARE COMING!

1 Tito P. Achong, *The Mayor's Annual Report: A Review of the Activities of the Port-of-Spain City Council, with Discourses on Social Problems Affecting the Trinidad Community, for the Municipal Year 1942–43* (Boston: US Government Printing Office, 1943), 258–59.

2 Albert Gomes, *Through a Maze of Colour* (Port of Spain: Key Caribbean Publications, 1974), 15.

3 Figures from R. R. Kuczynski, *Demographic Survey of the British Colonial Empire*, vol. III, *West Indian and American Territories* (London, New York, Toronto: A.M. Kelley, 1948–53), 336–39.

4 Following from M. G. Smith, *The Plural Society in the British West Indies* (Berkeley and Los Angeles: University of California Press, 1965), 5, 10–113, 309.

5 Arthur Calder-Marshall, *Glory Dead* (London: M. Joseph Ltd., 1939), 14–15.

6 Gomes, *Through a Maze of Colour*, xi.

7 Robert A. Johnston and James C. Shoultz, Jr., "History of the Trinidad Sector and Base Command," Port of Spain, Caribbean Defense Command, 29.

8 Gomes, *Through a Maze of Colour*, 161–62.

9 Ibid., 28.

10 Following from Fitz A. Baptiste, *The United States and West Indian Unrest, 1918–1939* (Mona, Jamaica: University of the West Indies Press, 1978), 26ff.; P.E.T. O'Connor, *Some Trinidad Yesterdays* (Port of Spain: Imprint Caribbean, 1978), 89–90.

11 O'Connor, *Some Trinidad Yesterdays*, 90–91.

12 16 to 22 cents per hour for skilled agricultural workers, 7 to 23 cents for skilled industrial laborers, 22 cents for stevedores, and 16 to 25 cents for trained transportation workers.

13 Caribbean Commission, U.S. Sector, *The Caribbean Islands and the War: A Record of Progress in*

Facing Stern Realities (Washington: US Government Printing Office, 1943), 70–71. In 1940 it took Trinidadian $4.80 to buy £1 sterling.

14 Calder-Marshall, *Glory Dead*, 271.

15 Ibid., 271–73.

16 Ibid., 57, 246–47.

17 Conn et al., *The Western Hemisphere: Guarding the United States and its Outposts*, 376.

18 Following from Government of Trinidad and Tobago, Office of the Prime Minister, *Historical Documents of Trinidad and Tobago*, Series No. 1, *The Annexation of Chaguaramas* (Port of Spain: Government Printing Office, 1963), 1–24.

19 Conn et al., *The Western Hemisphere: Guarding the United States and its Outposts*, 369.

20 Johnston and Shoultz, "History of the Trinidad Sector and Base Command," 27–28.

21 Ibid., 30.

22 Conn et al., *The Western Hemisphere: Guarding the United States and its Outposts*, 373.

23 *United States Army in World War II. The Technical Services, The Transportation Corps: Operations Overseas* (Washington: US Government Printing Office, 1957), 24.

24 Calder-Marshall, *Glory Dead*, 29.

25 Gomes, *Through a Maze of Colour*, 95–96.

26 Eric Williams, *History of the People of Trinidad and Tobago* (London: Praeger, 1962), 271–72.

27 *The Caribbean Islands and the War*, 79.

28 *The Annexation of Chaguaramas*, 9.

29 Ibid., 12, 14.

30 Conn et al., *The Western Hemisphere: Guarding the United States and its Outposts*, 374–75.

31 Ibid., 403.

32 Kelshall, *U-Boat War in the Caribbean*, 6–7.

33 US Army, Caribbean Defense Command, "Ten Studies on Aspects and Problems of the Caribbean Defense Command During the early and mid-1940s," Office of the Chief of Military History (microfilm), Study 1, "Anti-Submarine Activities in the Caribbean Defense Command, 1941–1946," 9. Hereafter cited as "Anti-Submarine Activities in the Caribbean Defense Command."

34 Ibid., Chapter 3, "The Attack on Aruba," 30.

35 "VI Bomber Command in Defense of the Panama Canal, 1941–45," History of the 25th Bombardment Group, 59th Bombardment Squadron, http://home.satx.rr.com/bombardment/htmdocs/59thbshistorytem.htm

36 Johnston and Shoultz, "History of the Trinidad Sector and Base Command," 165.

37 Ibid., 172.

38 Entry for August 15, 1939. ISKL Iu. U-Boote, Allgemein, RM 7/2319 I, PG 33325a, BA-MA. These figures, compiled at the start of the war, obviously differed from boat to boat and from sortie to sortie.

39 *Erster Wach-Offizier*; Executive officer, or First Watch officer.

40 For a typical Lorient departure, see Wolfgang Hirschfeld, *The Secret Diary of a U-Boat* (London: Phoenix, 2000), 58.

2: ATTACK ON ARUBA

1 Data from Bodo Herzog, *Die deutschen Uboote 1906 bis 1945* (Munich, 1959), 123–28; Eberhard Rössler, *Geschichte des deutschen Ubootbaus* (Munich: J.F. Lehmann, 1975), 192–95.

2 While the army gave calibers in millimeters, the navy used centimeters.

3 From David J. Bercuson and Holger H. Herwig, *Deadly Seas: The Duel between the* St Croix *and the* U305 *in the Battle of the Atlantic* (Toronto: Random House Canada, 1997), 96–98.

4 Ibid., 57–58, 107.

5 Lawrence Paterson, *Second U-Boat Flotilla* (Barnsley: L. Cooper, 2003), 123.

6 War Diary (KTB), *U-161*, 2. Unternehmung, PG 30,148/2, BA-MA.

7 "Erinnerungen des Dr. Götz Roth an die Fahrten mit U 161 (Kapitänleutnant Achilles)," [hereafter "Roth Erinnerungen"], File *U-161*, Deutsches U-Boot-Museum, Cuxhaven-Altenbruch, Germany. Hereafter DU-B-M. Roth was Second Watch Officer on *U-161*.

8 War Diary (KTB), *U-156*, 2. Unternehmung, PG 30,143/2, BA-MA.

9 See Manfred Dörr, *Die Ritterkreuzträger der U-Boot-Waffe* (Osnabrück: Biblio, 1988), 1:111–12.

10 Bernd Gericke, *Die Inhaber des Deutschen Kreuzes in Gold, des Deutschen Kreuzes in Silber der Kriegsmarine und die Inhaber der Ehrentafelspange der Kriegsmarine* (Osnabrück: Biblio, 1993), 97.

11 Erich Topp, *Fackeln über dem Atlantik. Lebensbericht eines U-Boot-Kommandanten* (Herford and Bonn: Mittler, 1990), 91.

12 Gordon Williamson and Darko Pavlovic, *U-Boat Crews 1914–45* (London: Osprey, 1996), 27–28, 58–59.

13 Léonce Peillard, *U-Boats to the Rescue: The* Laconia *Incident* (London: Coronet, 1961), 24.

14 All actions from War Diary (KTB), *U-156*, 2. Unternehmung, PG 30,143/2, BA-MA.

15 Inspection of the lagoon from Rodgers Beach during a research trip to Aruba in February 2006 confirmed the wisdom of Hartenstein's decision.

16 *Pan Aruban*, January 10 and 17, 1942. Archivo Nacionale, Oranjestad, Aruba. In current US dollars, the Fords and Chevrolets would have fetched about $1,300 and the Buicks $2,100.

17 There is wild confusion in the literature concerning the timing of the Neuland attacks. U-boat clocks remained on German General Time (DGZ) and all actions were so logged in the war diaries (KTB). It makes no sense to give the DGZ diary entries, as this would have placed the attack after daybreak. Sunrise came to Aruba at 7 a.m., and hence Dönitz's orders to commence Operation Neuland *precisely* "five hours before sunrise" places the attack at 2 a.m. (0800 in Hartenstein's KTB). Throughout these chapters, Caribbean time is taken to be six hours behind DGZ. David Rooney, Curator of Timekeeping at the Royal Observatory, Greenwich, confirmed in a letter of May 4,

2006, that "the histories of these time zone changes were never well documented."

18 The best account of the San Nicolas action is by William C. Hochstuhl, *German U- Boat 156 Brought War to Aruba February 16, 1942* (Oranjestad: Aruba Scholarship Foundation, 2001); and by the boat's Second Watch Officer, Paul Just, *Vom Seeflieger zum Uboot-Fahrer: Feindflüge und Feindfahrten 1939–1945* (Stuttgart: Motorbush-Verlag, 1979). A highly dramatized version is R. Busch and H. J. Röll, *"Shatten voraus!" Feindfahrten von U-156 unter Werner Hartenstein* as a special edition of *Der Landser Grossband* (Rastatt: Pabel-Moewig, 1996). Contemporary estimates of the general contours of the campaign are in United States, National Security Group, "Intelligence Reports on the War in the Atlantic," 1979.

19 All ship specifications from *Lloyd's Register of British and Foreign Shipping: Universal Register* (London, 1941).

20 *Time Archive 1923 to the Present*, Las Vegas, 23 February 1942.

21 *Miami Herald*, February 17, 1942.

22 *The Nassau Daily Tribune*, February 17, 1942.

23 "The Attack on Aruba." The 37-mm battery at Camp Savaneta could not possibly have sighted the submarine, which was close inshore and probably out of view.

24 Ibid., 38.

25 Ibid.

26 Ibid., February 16, 1942, 24.

3: LONG NIGHT OF THE TANKERS

1 War Diary (KTB), *U-502*, 2. Feindfahrt, PG 30, 539/2, BA-MA. Raeder chastised Dönitz for not having allowed *U-502* to deliver a "surprise attack" already on February 14, when Rosenstiel spied several tankers off Aruba – only to be lectured by BdU that Neuland was to be a concerted surprise attack by all boats precisely five hours before sunrise on February 16, 1942.

2 There is approximately a one-to two-hour difference between the local times for the sinkings as recorded in the U-boat logs and those of the US Army.

3 "The Attack on Aruba," 38–40.

4 War Diary (KTB), *U-502*, 2. Feindfahrt, PG 30, 539/2, BA-MA; "The Attack on Aruba," 40.

5 "The War Years at Lago: 1939–A Summing Up–1945," *Aruba Esso News* special edition. Archivo Nacionale, Oranjestad, Aruba. See also Hochstuhl, *German U-Boat 156*, 31.

6 Peillard, *U-Boats to the Rescue*, 26.

7 Sources differ as to whether the *Arkansas* was at San Nicolas or Oranjestad. Several websites list it as being at the "Eagle Dock" in San Nicolas. The Eagle Dock belonged to the Royal Dutch Shell refinery at Eagle Beach, Oranjestad. Hartenstein never refers to a third tanker struck at San Nicolas in the KTB, but does mention the attack in Oranjestad against the Eagle dock.

8 War Diary (KTB), *U-67*, 4. Unternehmung, PG 30,064/4, BA-MA.

9 Ibid.

10 Hochstuhl, *German U-Boat 156*, 16; also, "The War Years at Lago," 7.

11 Hochstuhl, *German U-Boat 156*, 29.

12 *"U-156 Roundtable Newsletter # 2,"* 30 September 2003. This website is maintained by Don G. Gray of California in memory of that fateful February 16, 1942: http://www.lago-colony.com.

13 The following is from "The War Years at Lago," 4–7, 11.

14 "The Attack on Aruba," 40.

15 *Pan Aruban*, vol. 14, Nr. 7. Archivo Nacionale, Oranjestad, Aruba.

16 War Diary (KTB), *U-156*, 2. Unternehmung, PG 30,143/2, BA-MA.

17 Ibid.

18 "The War Years at Lago," special insert, November 29, 1946.

19 War Diary (KTB), *U-502*, 2. Feindfahrt, PG 30,539/2, BA-MA.

20 *"U-156/U-502 Roundtable Newsletter #6,"* http://www.lago-colony.com.

21 "The Attack on Aruba," 42.

22 Ibid., Appendix B, "List of Ships Sunk by Enemy Submarines in the Caribbean Area."

23 *Miami Herald*, February 17 and 18, 1942.

4: MARTINIQUE

1 Local Caribbean time; 0155 in the KTB (German General Time [DGZ]).

2 War Diary (KTB), *U-156*, 2. Unternehmung, PG 30,143/2, BA-MA.

3 French doctors amputated more of Borne's leg. In May 1944, he was taken to New York, put on the Swedish liner *Gripsholm*, and landed at Barcelona. He was eventually exchanged for Allied prisoners of war and returned to Germany. After the war, Borne joined the navy of the Federal Republic.

4 Lee A. Dew, "The Day Hitler Lost the War," http://www.lago-colony.com.

5 C. Alphonso Smith, "Martinique in World War II," *United States Naval Institute Proceedings* 81 (February 1955): 169ff.

6 Johnston and Shoultz, "History · of the Trinidad Sector and Base Command," 195.

7 Michael C. Desch, *When the Third World Matters: Latin America and United States Grand Strategy* (Baltimore and London: Johns Hopkins University Press, 1993), 65.

8 Reference to Charles de Gaulle, the leader of the Free French government-in-exile in London.

9 See US Chargé in France (Murphy) to Secretary of State Cordell Hull, December 14, 1940. *Foreign Relations of the United States: Diplomatic Papers 1940*, vol. 2, *General and Europe* (Washington: US Government Printing Office, 1957), 490–93.

10 Data taken from *Jane's Fighting Ships of World War II* (New York: McGraw-Hill, 1989), 123–27, 134; *Jane's Fighting Ships 1942* (New York: McGraw-Hill, 1944), 171, 178–79, 182; and Fitzroy André Baptiste, *War, Cooperation, and Conflict: The European Possessions in the Caribbean, 1939–1945* (New York, Westport, CT, and London: Greenwood Press, 1988), 63–64.

11   Entry of June 19, 1940. "Diaries of Prime Minister William Lyon Mackenzie King," MG26-J13, Library and Archives Canada, Ottawa.

12   Entry of June 21, 1940; ibid.

13   Ibid.

14   At one point in these three days, Mackenzie King declared that if force was to be used, the British could pursue with their cruiser once the French ship left. They did not. No doubt Churchill was still mulling over what action he might take regarding the now Vichy-controlled French fleet.

15   Admiral Georges Robert, *La France aux Antilles de 1939 à 1943* (Paris: Plon, 1950), 48.

16   See *The Memoirs of Cordell Hull* (2 vols., New York: MacMillan, 1948), 1:818; and William L. Langer, *Our Vichy Gamble* (New York: A.A. Knopf, 1947), 103. The figure of 12 billion francs is from General Charles-Léon Huntziger's official report to the Germans on August 20, 1940. *Documents on German Foreign Policy 1918–1945*, Series D (1937–1945), vol. 10, *The War Years June 23–August 31, 1940* (London: H.M. Stationery Office, 1957), 516–20. Conversion to 2007 US dollars is from http://eh.net/hmit/compare/result.

17   Johnston and Shoultz, "History of the Trinidad Sector," 192.

18   Ibid., 198.

19   Vice Admiral Kurt von dem Borne was a former chief of staff and now head of the Kriegsmarine's economics section.

20   *Foreign Relations of the United States: Diplomatic Papers 1942*, vol. 2, *Europe* (Washington: US Government Printing Office, 1962), 611–12, 616, 619.

21   David J. Bercuson and Holger H. Herwig, *One Christmas in Washington: The Secret Meeting between Roosevelt and Churchill that Changed the World* (New York: Overlook Press, 2005), 145ff.

22   Secretary of the Navy Frank Knox to Cordell Hull, December 26, 1941. Cordell Hull Papers, Library of Congress, Washington, D.C., Reel 21 Correspondence.

23   Johnston and Shoultz, "History of the Trinidad Sector," 207.

24   Smith, "Martinique in World War II," 170–71, 174. The US diplomatic maneuvers are in *Foreign Relations of the United States: Diplomatic Papers*, vol. 2, *General and Europe*, 453ff.

25   Johnston and Shoultz, "History of the Trinidad Sector," 213.

26   Ibid., 220–26.

27   Ibid., "History of the Trinidad Sector," 190.

28   In July 1943, the Free French took control of the French Antilles – and the gold at Fort Desaix. The United States "evacuated" Robert to Puerto Rico, from where he made his way back to Vichy. He was dismissed from the French navy in disgrace after the war but escaped incarceration because his son had been a member of the French resistance.

29   War Diary (KTB), *U-156*, 2. Unternehmung, PG 30,143/2, BA-MA. Hartenstein obviously encoded the Grid Chart locations in case of interception by the Allies: all action had, in fact, taken place in Quadrant ED.

30   Ibid.

31 The German term *Faule Grete* is untranslatable. The following from Erich Glodschey, *U-Boote. Deutschlands Scharfe Waffe* (Stuttgar: Union-Duetsche Verlagsgesellschaft t, 1943), 174–75; and *Das Archiv* 14 (April 2000), 40.

32 War Diary (KTB), *U-156*, 2. Unternehmung, PG 30,143/2, BA-MA; and Just, *Vom Seeflieger zum Uboot-Fahrer*, 79–85.

33 War Diary (KTB), *U-156*, 2. Unternehmung, PG 30,143/2, BA-MA.

34 Ibid.

35 War Diary (KTB), *U-156*, 2. Unternehmung, PG 30,143/2, BA-MA. See also Glodschey, *U-Boote*, 174–75; and *Das Archiv* 14, 40.

36 War Diary (KTB), *U-156*, 2. Unternehmung, PG 30,143/2, BA-MA.

37 This was the standard German measurement for fuel. One cubic meter (cbm) is equal to 1,000 liters or 220 imperial gallons.

38 War Diary (KTB), *U-156*, 2. Unternehmung, PG 30,143/2, BA-MA.

5: "THE FERRET OF PORT OF SPAIN"

1 See Hans Goebeler, *Steel Boat, Iron Hearts: A U-Boat Crewman's Life aboard* U-505 (New York: Savas Beatie, 2005), 37–38.

2 "Ständige Befehle für U-Boots-Besatzungen," Heft 1, RM 91/18, BA-MA.

3 See Hirschfeld, *The Secret Diary of a U-Boat*, 130, 137, 152.

4 This painful procedure, named after Arthur Kollmann, a 19th-century German urologist, consisted of dilating the urethra,

"a canal about 20 cm in length that opens at the extremity of the glans penis." The first successful treatment of a patient using penicillin took place in Britain in March 1942.

5 War Diary (KTB), *U-156*, 2. Unternehmung, PG 30,143/2, BA-MA.

6 Dönitz's evaluation, RM 98/360 KTB "U156," BA-MA.

7 Peillard, *U-Boats to the Rescue*, 14, 101.

8 Local time, six hours behind German General Time (DGZ).

9 War Diary (KTB), *U-161*, 2. Unternehmung, PG 30,148/2, BA-MA.

10 Haussmann, "Latin American Oil in War and Peace," 355ff.

11 *Dictionary of American Naval Fighting Ships* (Washington: Navy Department, 1959), I:130.

12 Kelshall, *U-Boat War in the Caribbean*, 22.

13 War Diary (KTB), *U-161*, 2. Unternehmung, PG 30,148/2, BA-MA.

14 Ibid.

15 Ibid. Also, "Roth Erinnerungen," 3–4.

16 Kelshall, *U-Boat War in the Caribbean*, 39.

17 Ibid., 40–41.

18 Clay Blair, *Hitler's U-Boat War*, vol. II, *The Hunted, 1942–1945* (New York: Random House, 1998), 506.

19 War Diary (KTB), *U-161*, 2. Unternehmung, PG 30,148/2, BA-MA.

20 Aerial depth charge; *Fliegerbombe* or *Fliebo* for short.

21 A shallow-depth gauge.

22 War Diary (KTB), *U-161*, 2. Unternehmung, PG 30,148/2, BA-MA.

6: WAR COMES TO ST. LUCIA

1   Kelshall, *U-Boat War in the Caribbean*, 60–61.

2   Local time; 0700 DGZ in the war diary (KTB).

3   War Diary (KTB), *U-161*, 2. Unternehmung, PG 30,148/2, BA-MA.

4   "Roth Erinnerungen," 4.

5   St. Lucia National Archives, *The West Indian Crusader*, week of March 2 to 9, 1942.

6   St. Lucia National Archives, *The Voice of St. Lucia*, week of March 2 to 9, 1942.

7   War Diary (KTB), *U-161*, 2. Unternehmung, PG 30,148/2, BA-MA.

8   The following account is from St. Lucia National Archives, The Administrator of St. Lucia (Alban Wright) to the Governor of the Windward Islands, GRENADA, May 28, 1942, including depositions by the Vigie Lighthouse crew.

9   St. Lucia National Archives, *The Voice of St. Lucia*, March 10, 1942, 1.

10 Ibid., Issue of March 17, 1942, 3.

11 Ibid., Issue of March 19, 1942, 1, 3.

12 St. Lucia National Archives, The Administrator of St. Lucia to the Governor of the Windward Islands, GRENADA, May 28, 1942.

13 Ibid.

14 "Roth Erinnerungen," 5; War Diary (KTB), *U-161*, 2. Unternehmung, PG 30,148/2, BA-MA.

15 War Diary (KTB), *U-129*, 4. Unternehmung, PG 30/119/5, BA-MA.

16 Ibid.

17 Ibid.

18 Ibid.

19 Ibid.

20 Clay Blair, *Hitler's U-Boat War*, vol. I, *The Hunters 1939–1942* (New York: Random House, 1996), 507; Paterson, *Second U-Boat Flotilla*, 136; and http://www.ubootwaffe. net.

21 Glodschey, *U-Boote*, 163.

22 Dönitz evaluation, RM 98/365 KTB "U-161," BA-MA.

23 *Miami Herald*, February 24, 1942.

24 Winston S. Churchill, *The Second World War*, vol. IV, *The Hinge of Fate* (Boston: Houghton-Mifflin, 1950), 119.

25 The "tragedy of 20 April 1942" has been largely ignored in British and American works on the war in the Caribbean. The story was unearthed in the Centraal Historisch Archief in Curaçao by Junnes Sint Jago, *De Tragedie van 20 April 1942. Arbeidsconflict Chinese zeelieden en CSM mondt uit in bloedbad* (Curaçao: Imprenta Atiempo, 2000). Marjan Eggermont at the University of Calgary kindly helped with the Dutch translation.

26 Currency conversion is at best an approximation. In the period under discussion, it took 1.55 florin or guilder to buy one US dollar. Thus, the Chinese were paid $32 per month in 1942, which http:// www.eh.net translates into 428 US 2010 dollars. I am indebted to my colleague Herb Emery of the Department of Economics, University of Calgary, for his assistance in this.

27 Max Domarus, ed., *Hitler. Reden und Proklamationen*, vol. II/2,

Untergang (Munich: Süddeutscher Verlag, 1965), 1862; Henry Picker, ed., *Hitlers Tischgespräche im Führerhauptquartier 1941–1942* (Stuttgart: Seewald, 1963), 281–83.

28   The aircraft is frequently referred to as the Me-110, but it had been designed and built by the Bayerische Flugzeugwerke (hence the designation Bf) before Messerschmitt acquired the firm in 1938.

29   The strategic failure to target the lake tankers is discussed in Karl M. Hasslinger, "The U-Boat War in the Caribbean: Opportunities Lost," a paper submitted to the Department of Operations, US Naval War College, Newport, March 1996.

7: TORPEDO JUNCTION

1   Kelshall, *U-Boat War in the Caribbean*, 72; Blair, *Hitler's U-Boat War*, I:537–38.

2   Memorandum of March 15, 1942. I SKL, Teil C IV, KTB U-Bootskriegsführung 1942, RM 7/846, BA-MA.

3   Raeder to Dönitz, March 26, 1942; ibid.

4   Dönitz to Raeder, March 28, 1942; ibid.

5   Staff telegram of April 2, 1942; ibid.

6   Dönitz's war diary, April 14, 1942. Kriegstagebuch, KTB 2.1—30.4 1942, RM 87/5 BdU, BA-MA.

7   Dönitz's war diary, April 30, 1942; ibid.

8   Data from Rössler, *Geschichte des deutschen Ubootbaus*, 195, 229; see also John F. White, *U Boat Tankers*

1941–1945 (Annapolis: Naval Institute Press, 1998).

9   Blair, *Hitler's U-Boat War*, I: 576.

10   Kelshall, *U-Boat War in the Caribbean*, 71.

11   War Diary (KTB), *U-156*, 3. Unternehmung, PG 30,143/3, BA-MA. Entry for April 23, 1942.

12   Kelshall, *U-Boat War in the Caribbean*, 75.

13   Also known as "Cuprex," it was a blue-green copper-sulfate ointment produced by Merck Pharmaceutical.

14   Local time. 2005 hours in the KTB (German General Time).

15   War Diary (KTB), *U-156*, 3. Unternehmung, PG 30,143/3, BA-MA. Entry for April 23, 1942. See also Just, *Vom Seeflieger zum Uboot-Fahrer*, 86ff; and Busch and Röll, "*Schatten voraus!*," 27ff.

16   Theodore Taylor, *Fire on the Beaches* (New York: Norton, 1958), 158.

17   Cited in Max Paul Friedman, *Nazis and Good Neighbors: The United States Campaign against the Germans of Latin America in World War II* (Cambridge: Cambridge University Press, 2003), 62

18   Just, *Vom Seeflieger zum Uboot-Fahrer*, 99.

19   "War Damage Report, USS Blakeley," June 4, 1942. RG 38 Serial 028, Loc. 370 45/1/3 Box 853, National Archives, College Park, Maryland. *Blakeley* was towed to Philadelphia, where it received a new bow. It returned to convoy escort duty in the Caribbean until February 1945. *Dictionary of American Fighting Ships*, I: 130.

20   Just, *Vom Seeflieger zum Uboot-Fahrer*, 103.

21 Wilhelm Polchau, Engineer Report, 3rd War Patrol, 22.4–7.7. 1942. *"U156"* KTB Ing., RM 98/525, BA-MA.

22 Dönitz's evaluation, *U-156.* War Diary (KTB) *U-156*, 3. Unternehmung, PG 30, 143/3, BA-MA.

23 Topp, *Fackeln über dem Atlantik*, 92; in English, *The Odyssey of a U-Boat Commander: Recollections of Erich Topp* (Westport and London: Praeger, 1992), 85.

24 Freiherr Karl Friedrich Hieronymus von Münchhausen (1720–1797) was noted for his outrageous tall tales about service with the Russian army in Turkey in the 1740s.

8: HUNTING OFF THE ORINOCO

1 There is no equivalent rank in the British or United States navies as it falls between those of commander and captain.

2 See "Jürgen Wattenberg, U-162," in Melanie Wiggins, *U-Boat Adventures: Firsthand Accounts from World War II* (Annapolis: Naval Institute Press, 1999), 1–12.

3 Dönitz's evaluation, *U-162.* War Diary (KTB) *U-162*, 2. Unternehmung, RM 98/366, BA-MA.

4 Ibid.

5 Ibid. See also Hans Kreis, "Schweinejagd auf dem Atlantik," Kriegsmarine press release, Summer 1942, File *U-162*, DU-B-M; and Machinist Walter Hartmann memoir, ibid., 46.

6 Stephen W. Roskill, *The War at Sea, 1939–1945*, vol. II, *The Period of Balance* (London: H.M. Stationery Office, 1956), 103.

7 Ibid., 104; Blair, *Hitler's U-Boat War*, I:532–33.

8 Kelshall, *U-Boat War in the Caribbean*, 85–86.

9 War Diary (KTB), *U-162*, 2. Unternehmung, PG 30, 149/2, BA-MA.

10 Whitsuntide (or Pentecost) is observed on the seventh Sunday after Easter.

11 Dönitz's war diary, May 14, 1942. BdU, Kriegstagebuch, RM 87/22, BA-MA.

12 Entry for June 17, 1942. Gerhard Wagner, ed., *Lagevorträge des Oberbefehlshabers der Kriegsmarine vor Hitler 1939–1945* (Munich: Lehmann, 1972), 396.

13 Dönitz's war diary, June 1, 1942. BdU, Kriegstagebuch, RM 87/6, BA-MA.

14 War Diary (KTB), *U-161*, 3. Unternehmung, PG 30,148/3, BA-MA.

15 Ibid. Also, "Roth Erinnerungen," 6–7.

16 Ibid., 7.

17 Two books claim that Achilles had dived deep after launching his torpedoes and was then rammed when he had resurfaced to press the attack: Kelshall, *U-Boat War in the Caribbean*, 102–3; Paterson, *Second U-Boat Flotilla*, 161–62. Neither Achilles' KTB nor Roth's memoirs support such a scenario.

18 The usually reliable U-Boat Museum credits Achilles with sinking the *Scottsburg*; other sources merely credit him with an unidentified 4,000-ton freighter. Both steam freighters were sunk in Grid Quadrant ED 94; Achilles at the time listed his position as Grid Quadrant ED 73.

9: WAR BENEATH THE SOUTHERN CROSS

1 Information concerning *U-157*'s departure, route, and attack on SS *Hagan* is from Blair, *Hitler's U-Boat War*, I: 611; and Homer H. Hickam, *Torpedo Junction: U- Boat War Off America's East Coast, 1942* (Annapolis: Naval Institute Press, 1989), 288.

2 Blair, *Hitler's U-Boat War*, I 611; Hickam, *Torpedo Junction*, 289; Wiggins, *Torpedoes in the Gulf*, 106; and "U-157," at http://www.uscg.mil.history/uscghist/U157.asp.

3 Wiggins, *Torpedoes in the Gulf*, 91–95.

4 Kelshall, *U-Boat War in the Caribbean*, 92–110.

5 Ibid., 110–11.

6 The official US Atlantic Fleet report of the sinking accompanied by a series of sketches and photographs; http://www.uboatarchive.net/U-158Analysis.htm.

7 US Office of Chief of Naval Operations, Intelligence Report, Curaçao, 21 June 1942; File *U-161*, DU-B-M.

8 Kelshall, *U-Boat War in the Caribbean*, 85.

9 "Roth Erinnerungen," 8.

10 War Diary (KTB), *U-161*, 3. Unternehmung, PG 30,148/3, BA-MA.

11 "Roth Erinnerungen," 8–9.

12 War Diary (KTB), *U-161*, 3. Unternehmung, PG 30,148/3, BA-MA.

13 Ibid.

14 War Diary (KTB), *U-162*, 2. Unternehmung, BA-MA. Brazil postponed its formal declaration of war against Germany until August 22, 1942.

15 Frank D. McCann, Jr., "Brazil and World War II: The Forgotten Ally. What Did You Do in the War Zé Carioca?," *Estudios Interdisciplinarios de América Latina y el Caribe* 6 (July 1995): 9–10.

16 Werner Rahn and Gerhard Schreiber, eds., *Kriegstagebuch der Seekriegsleitung 1939–1945* (Herford and Bonn: Mittler, 1992), 34:295.

17 Ibid., 497.

18 Paterson, *Second U-Boat Flotilla*, 155–56.

19 Rahn and Schreiber, eds., *Kriegstagebuch*, 36:411, 451, 463.

20 Roskill, *The War at Sea, 1939–1945*, II:203.

21 McCann, "Brazil and World War II," 11–13.

22 Cited in Wagner, ed., *Lagevorträge*, 420–24.

23 See especially F. H. Hinsley, *British Intelligence in the Second World War* (5 vols., London: H.M. Stationery Office, 1979–90); and David Kahn, *Seizing the Enigma: The Race to Break the German U-Boat Codes, 1939–1943* (Boston: Houghton-Mifflin, 1991).

24 U-Boat Command diary entry for September 28, 1942. ISKL, Teil CIV, KTB U-Bootskriegsführung 1942, RM 7/846, BA-MA.

10: THE ALLIES REGROUP

1 Conn et al., *The Western Hemisphere: Guarding the United States and its Outposts*, 409–11.

2 Ibid., 412–13.

3 Ibid., 417–19.

4 Monica Rankin, "Industrialization through Unity," in Thomas M.

Leonard and John F. Bratzel, eds., *Latin America During World War II* (Plymouth: Rowan and Littlefield, 2007), 20–22.

5   "Anti-Submarine Activities in the Caribbean Defense Command," 45–46.

6   For information on Cuban armed forces in World War II, see "Cuban Aviation," http://www.geocities.com/urrib2000/Miu11-4-e.html?200918; and "Submarine Warfare around Cuba during WW II," http://archiver.rootsweb.ancestry.com/th/read/Mariners/2008-01/1199209641.

7   George M. Lauderbaugh, "Bolivarian Nations: Securing the Northern Frontier," in Leonard and Bratzel, eds., *Latin America*, 115–19.

8   "Anti-Submarine Activities in the Caribbean Defense Command," 49–50.

9   Lauderbaugh, "Bolivarian Nations," 120.

10 Johnston and Shoultz, "History of the Trinidad Sector and Base Command," 33–34.

11 Ibid., 40–44.

12 Ibid., 229–40.

13 Kelshall, *U-Boat War in the Caribbean*, 73.

14 "Anti-Submarine Activities in the Caribbean Defense Command," 14–15.

15 Charles H. Bogart, "From the Coast to the Field," *Field Artillery Journal* 15 (Sept.–Oct. 1986): 28–30.

16 "Anti-Submarine Activities in the Caribbean Defense Command," 73–74.

17 Kelshall, *U-Boat War in the Caribbean*, 166, 218–19.

18 Ibid., 320–21.

19 Caribbean Commission, U.S. Sector, *The Caribbean Islands and the War: A Record of Progress in Facing Stern Realities* (Washington: US Government Printing Office, 1943), 43.

20 Ibid., 70–71.

21 See Ralph de Boissière, *Rum and Coca-Cola* (London: Allison and Busby, 1984), 89.

22 Ibid., 105, 121.

23 Samuel Selvon, *Ways of Sunlight* (London: St. Martin's Press, 1957), 90–91.

24 Annette Palmer, *World War II in the Caribbean: A Study of Anglo-American Partnership and Rivalry* (Randallstown, MD: Block Academy Press, 1998), 67–68.

25 "The American Soldier's Guide Book to Trinidad," n.p., n.d.; US Army Military History Institute, U113.3.T7 A44 1940.

26 Palmer, *World War II in the Caribbean*, 88.

27 Stephen J. Randall and Graeme S. Mount, *The Caribbean Basin: An International History* (London and New York: Routledge, 1998), 81.

28 Gomes, *Through a Maze of Colour*, 5, 8, 13, 87–88, 135.

29 Robert Antoni, *My Grandmother's Erotic Folktales* (New York: Grove Press, 2000), 7–8.

30 See Conn et al., *The Western Hemisphere: Guarding the United States and its Outposts*, 404–5.

31 Antoni, *My Grandmother's Erotic Folktales*, 159.

32 In the late 1940s, the entertainer Morey Amsterdam claimed authorship of the song and recorded his version with the Andrews Sisters; it became the best-selling record of the 1940s, with 4 million singles. In the 1950s

"Lord Invader" (Rupert Grant) filed a lawsuit in US court against Amsterdam and was awarded $150,000 in back royalties. http://www.lightyearsip.net/ip_trinidad_song.shtmal. "Koomahnah" was a bastardized form of Cumana, which lies between Port of Spain and Chaguaramas.

33 Cited in Palmer, *World War II in the Caribbean*, 83.

34 Cited in ibid., 70.

35 Ibid., 71.

36 De Boissière, *Rum and Coca-Cola*, 70.

37 Cited in Palmer, *World War II in the Caribbean*, 125.

11: WHITE CHRISTMAS

1 http://www.lanepl.org/BlountJBCOLSS/documents; Searchable Columns by Jim Blount of the *Hamilton Journal-News*, #226, December 21, 1992—"Christmas first real holiday in wartime 1942."

2 James F. de Rose, *Unrestricted Warfare* (Edison, NJ: Wiley, 2006), 38.

3 Michael Gannon, *Operation Drumbeat* (New York: Harper & Row, 1991), various pages.

4 "March a Month of Drastic Shrinkages," *The Oil Weekly* 105 (30 March 1942), 9.

5 Robert C. Fisher, "'We'll Get Our Own': Canada and the Oil Shipping Crisis of 1942," in Naval History.CA, http://www.familyheritage.ca?Articles/oilarticle.html.

6 Haussmann, "Latin American Oil in War and Peace," 355–56.

7 "East Coast Trouble Upsetting Industry," *The Oil Weekly* 105 (March 9, 1942), 10.

8 Ibid.

9 Ibid.

10 United Kingdom, Department for Business Enterprise & Regulatory Reform, "Crude Oil and Petroleum products; Imports by product; 1920–2006."

11 "Tanker losses ahead of new launchings," *The Oil Weekly* 105 (March 9, 1942), 10.

12 Robert Goralski and Russel W. Freeburg, *Oil and War: How the Deadly Struggle for Fuel in WWII Meant Victory or Defeat* (New York: Morrow, 1987), 111.

13 Ibid., 103.

14 Ibid., 117.

15 Ibid.; "East Coast Shortage Brings Long Expected Rationing Announcement," *The Oil Weekly* 105 (April 27, 1942).

16 Arthur M. Johnson, *Petroleum Pipelines and Public Policy, 1906–1959* (Cambridge, MA: Harvard University Press, 1967), 313.

17 Nash, *United States Oil Policy*, 162–63.

18 "Tank Trucks Relieve Tank Cars" and "East Coast Shipments at Record Levels," in *The Oil Weekly* 105 (March 16, 1942).

19 "Offer Partial Solution to East Coast Oil Shortage," in ibid., March 23, 1942.

20 Igor I. Kavass and Adolf Sprudzs, eds., *A History of the Petroleum Administration for War, 1941–1945* (Buffalo, NY: W.S. Hein, 1974), 83.

21 Johnson, *Petroleum Pipelines*, 307–8.

22 *The Big Inch and Little Big Inch Pipelines: The Most Amazing*

Government–Industry Cooperation
Ever Achieved, The Louis Berger
Group Inc. (East Orange, NJ,
n.d.), 4.

23 Ibid., 13–14.

24 Kavass and Sprudzs, eds., History
of the Petroleum Administration,
102.

25 "Rejects East Coast Pile Line
Again," The Oil Weekly 105 (March
9, 1942), 11.

26 Johnson, Petroleum Pipelines,
314–16.

27 "Washington Sees Big Pipe Line
Only Solution," The Oil Weekly 105
(April 27, 1942), 12.

28 Kavass and Sprudzs, eds., History
of the Petroleum Administration,
105–6.

29 Johnson, Petroleum Pipelines, 324.

30 A good précis of the entire
project, based on a wide variety of
secondary sources, is The Big Inch
and Little Big Inch Pipelines.

31 Harold F. Williamson et al., The
American Petroleum Industry:
The Age of Energy, 1899–1959
(Westport, CT: Northwestern
University Press, 1981), 764.

32 Kavass and Sprudzs, eds., History
of the Petroleum Administration,
110.

33 All statistics from Williamson et
al., American Petroleum Industry,
766–67.

12: THE ALLIES STRIKE BACK

1   Peter Kemp, Decision at Sea:
The Convoy Escorts (New York:
Elsevier-Dutton, 1978), 53.

2   W.A.B. Douglas et al., No Higher
Purpose: The Official Operational
History of the Royal Canadian Navy
in the Second World War, 1939–1943,

vol. II, part 1 (St. Catharines, ON:
Vanwell, 2002), 396–97.

3   Samuel Eliot Morison, History of
United States Naval Operations in
World War II, vol. I, The Battle of the
Atlantic: September 1939–May 1943
(Boston: Little, Brown, 1962),
146–47.

4   "Q-Ships," http://www.
destroyerhistory.org/desron12/
index.htm

5   Ibid.

6   "Destroyer Squadron 12," http://
www.destroyerhistory.org/
desron12/index.htm

7   Paterson, Second U-Boat Flotilla,
157.

8   Kenneth Wynn, U-Boat Operations
of the Second World War, vol.
I, Career Histories, U1–U510
(Annapolis: Naval Institute Press,
1998), 117.

9   Thomas F. Wright, ed., "Short
History of the 'Lucky L': USS
Lansdowne DD 486, 1942–1945,"
USS Lansdowne, DD 486
Association, 1973, 6.

10 John D. Alden, Flush Decks and
Four Pipes (Annapolis: Naval
Institute Press, 1965), 33.

11 Hasslinger, "The U-Boat War in
the Caribbean: Opportunities
Lost."

12 For HMS Churchill and HMS
Havelock, see http://www.naval-
history.net and http://www.uboat.
net/allies/warships.

13 Kelshall, U-Boat War in the
Caribbean, 147–49.

14 Douglas et al., No Higher Purpose,
vol. II, part 1, 407–10.

15 "Analysis of Aircraft Action
Report," September 22, 1941. RG
38 Serial 00768, Loc. 370 45/7/2-
3 Box 1141, National Archives,

College Park, Maryland, U.S.A. (hereafter cited as NA).

16 Canadian action from The Commanding Officer, H.M.C.S. OAKVILLE to Captain (D) Halifax, August 9, 1941, "REPORT OF ATTACK ON 'U-94'."

17 Ibid.

18 Hal Lawrence, *A Bloody War: One Man's Memories of the Royal Canadian Navy, 1939–1945* (Toronto: Macmillan of Canada, 1979), 95–104. See also "Confidential," Navy Department, Office of the Chief of Naval Operations, Serial No. 5, "Report on the Interrogation of Survivors from U-94 sunk on August 27, 1942," which seems to be the basis for most subsequent treatments of this encounter; RG 38, 10th Fleet ASW Files, File No. 1529, No. 1574 Box 78 and 79, 370/47/1/7, NA. Finally, http://www.uboatarchive.net/U-94VP-92ActionReport.htm.

19 From H.G.D. de Chair, "Sinking of U-162 – H.M.S. 'Vimy' – Narrative," September 15, 1942, RG 38, 10th Fleet ASW Files, File No. 1529, No. 1574, Box 78 and 79, 370/47/1/7, NA; and Commander E. A. Gibbs, Secret Report, H.M.S. "Pathfinder," 5th September 1931, to Senior British Naval Officer, Trinidad.

20 U.S. Navy Department, Office of the Chief of Naval Operations, Washington, Serial No. 6, "Report on the Interrogation of Survivors From U-162 Sunk on September 2, 1941," ibid.; and "Monthly Submarine Report, November 1942," CB 4050/42, December 15, 1942, 26–27, courtesy of The

Royal Navy Submarine Museum, Gosport, UK.

21 Paterson, *Second U-Boat Flotilla*, 172–73; Henry Graham de Chair, *Let Go Aft: The Indiscretions of a Salt Horse Commander* (Tunbridge Wells: Parapress, 1993), 143–46; http://www.uboatarchive.net/U-162PathfinderReport.htm.

22 de Chair, "Sinking of U-162."

23 Conn et al., *The Western Hemisphere: Guarding the United States and its Outposts*, 429.

24 Wesley Frank Craven and James Lea Cate, eds., *The Army Air Forces in World War II*, vol. I, *Plans and Early Operations, January 1939 to August 1942* (Chicago: University of Chicago Press, 1948), 302.

25 Ibid., 531.

26 "Anti-Submarine Activities in the Caribbean Defense Command," 63.

27 Ibid., 63–65.

28 Lawrence Paterson, *U-Boat War Patrol: The Hidden Photographic Diary of U 564* (London: Greenhill Books, 2004), 135–36. The photos were taken by a war correspondent from the Navy's Propaganda Service.

29 Conn et al., *The Western Hemisphere: Guarding the United States and its Outposts*, 433.

13: A HARD WAR: HARTENSTEIN AND *U-156*

1 Blair, *Hitler's U-Boat War*, I:685–97.

2 Josef W. Konvitz, "Bombs, Cities, and Submarines: Allied Bombing of the French Ports, 1942–1943," *International History Review* 14 (February 1992): 28–29; and Randolph Bradham, *Hitler's*

U-Boat Fortresses (Westport and London: Praeger, 2003), 8ff.

3   ULTRA was the name British code-breakers used to decrypt Enigma machine messages. German operators deployed "form letters" for daily weather reports, using the same settings on the Enigma machine almost every day.

4   Diary entry for September 9, 1942. RM 7/846 ISKL. Teil CIV, KTB U-Bootskriegsführung 1942, BA-MA.

5   Note of August 10, 1942. RM 87/7 KTB Grand Admiral Dönitz, BA-MA.

6   Diary entry for September 30, 1942; RM 87/23, BA-MA.

7   Diary entry for December 19, 1942; RM 87/7, BA-MA.

8   Diary entry, May 1943; RM 87/27, BA-MA. This undated message was attached to the entry for May 15, 1943.

9   Diary entry for May 24, 1943; ibid.

10 Entries for December 30, 1942 and January 1, 1943. RM 7/259 ISKL. Teil CA, Grundlegende Fragen der Kriegführung 1942–43, BA-MA.

11 Notes of the meeting in Wagner, ed., *Lagevorträge*, 452–54.

12 Ibid., 475–78.

13 Diary entries for spring 1943. Kriegstagebuch des B.d.U., RM 87/8 1.1.—30.6.1943, BA-MA. See also Dönitz, *Memoirs*, 340–41.

14 From Bercuson and Herwig, *One Christmas in Washington*, 227, 229.

15 Diary entry for January 8, 1943. Elke Fröhlich, ed., *Die Tagebücher von Joseph Goebbels*, part 2, *Diktate 1941–1945* (Munich: K.G. Saur, 1994), III:69–70.

16 John Ellis, *World War II: A Statistical Survey* (New York: Facts on File, 1993), 280.

17 War Diary, *U-156*, 4. Unternehmung, PG 30,143/4, BA-MA.

18 Michael L. Hadley, "Grand Admiral Karl Dönitz (1891–1980): A Dramatic Key to the Man behind the Mask," *The Northern Mariner* 10 (2000), 12.

19 Dönitz, *Memoirs*, 259–63.

20 Dan van der Vat, *The Atlantic Campaign: World War II's Great Struggle at Sea* (New York: Harper and Row, 1988), 353–54.

21 Assistant Chief of Staff, Intelligence, Historical Division, "The AAF Command" April 1945, 139; obtained from http://www.uboatarchive.net/AAFHistory.htm.

22 Ibid., 139–40.

23 Van der Vat, *The Atlantic Campaign*, 336; "US Navy – Tenth Fleet Fights the U-boats," on http://www.uboat.net/allies/ships/.us_10thfleet.htm.

14: HIGH NOON

1   *Flieger-Abwehr-Kanone*, or anti-aircraft gun.

2   Radio messages intercepted by Bletchley Park. RG 38, Intercepted Radio Traffic. U-156. Box 106, 370/1/4/6. OP20-6 *Ultra* intercepts/decrypts, NA. At the time, it was still taking four weeks to get ULTRA decrypts to the front in the Caribbean.

3   Ibid.

4   The B-18 encounters from Kelshall, *U-Boat War in the Caribbean*, 269–71.

5   War Diary (KTB), *U-156*, PG 30,143/5, BA-MA. This is the brief, reconstituted war diary of

the fifth war patrol on the basis of signals received from Hartenstein.

6   Following from Dryden's "Report of Antisubmarine Action by Aircraft" of March 8, 1943, as well as official evaluation reports of the action. RG 38, ASW Assessment Files, No. 2646, Box 94, 370/47/2/3, "Tenth Fleet ASW Files, VP 53 and the destruction of U-156," NA.

7   For a graphic account, see Lee A. Dew, "The Sinking of U-156," *Red River Valley Historical Journal of World History* 4 (Autumn 1979): 64–76.

8   Nickname for a Type B-4 inflatable lifejacket based on the buxom figure of Hollywood actress Mae West (1893–1980).

9   Author's tour of the Villa Kerillon on July 22, 2006, courtesy of Admiral Pierre Martinez, Commandant la Marine à Lorient.

10 Van der Vat, *The Atlantic Campaign*, 333, suggests this. He also claims that Peter-Erich Cremer was in command at the time, but Cremer did not again take command of *U-333* until May 18, 1943. See also http://www.uboat.net/boats/u333.htm.

11 Chris Bishop, *The Essential Submarine Identification Guide: Kriegsmarine U-Boats, 1939–45* (London: Amber Books, 2006), 86.

12 "Anti-Submarine Activities in the Caribbean Defense Command," 103–5.

13 Ibid.

14 Ibid., 105–8.

15 "Analysis of Anti-Submarine Action by Aircraft," May 17, 1943, in http://www.uboatarchive.net/U-128-11-Anl.htm

16 "Anti-Submarine Activities in the Caribbean Defense Command," 113–14; also, "Appendix B, List of Ships Sunk in the Caribbean Area," 20–21.

17 http://www.uboat.net/boats/u590.htm.

18 http://www.uboat.net/boats/u759.htm.

19 Anita Lesko, "Mariner's Victory at Sea," *Naval Aviation News* (Jan.–Feb. 2000): 22–23.

20 "Report of Antisubmarine Action by Aircraft, July 21, 1943," in http://www.uboatarchive.net/U-662VP94ASW6_5_43.htm.

21 http://www.uboat.net/boats/u359.htm.

22 "Report of Antisubmarine Action by Aircraft, July 29, 1943," in http://www.uboatarchive.net/U-159ASW-6.htm.

23 http://www.uboat.net/boats/u572.htm.

15: GUNDOWN: *U-615* AND *U-161*

1   See Kelshall, *U-Boat War in the Caribbean*, Chapter 19, "The greatest battle."

2   Gaylord T. M. Kelshall, "Ralph Kapitzky: Battle in the Caribbean and the Death of U-615," in Theodore P. Savas, ed., *Silent Hunters: German U-Boat Commanders of World War II* (Annapolis: Naval Institute Press, 2003), 43–44. This account in many details differs from the one in *U-Boat War in the Caribbean*.

3   War Diary (KTB), *U-615*, 1. Unternehmung, PG 30,646/1, BA-MA.

4   Taffrail [Taprell Dorling], *Blue Star Line at War 1939–1945*

(London: Foulsham, 1973), 109–12.

5   War Diary (KTB), *U-615*, 1. Unternehmung, PG 30,646/1, BA-MA.

6   Following from War Diary (KTB), *U-615*, 2. Unternehmung, PG 30,646/2, BA-MA.

7   Ibid.

8   Ibid.

9   Cited in Blair, *Hitler's U-Boat War*, II:363.

10 From Wolfgang Ott, *Sharks and Little Fish: A Novel of German Submarine Warfare* (Guilford, CT: Pantheon, 2003), 299–301.

11 Recollections of Machinist Mate Reinhold Abel, June 14, 1985. File *U 615*, DU-B-M. Hereafter cited as "Abel Recollections."

12 A "top secret" electric night telescope, nicknamed *Seehund Drei* (seal three), that allowed a broader field of vision and greater brightness.

13 RG 38, Translations of German Intercepts U-615, Box 141, 370/1/5/7, NA.

14 Blair, *Hitler's U-Boat War*, II:363.

15 The ensuing hunt for *U-615* taken from "Enemy Action Summary," U.S. Naval Base Trinidad, Reports from July 31 to August 11, 1943. RG 38, 10th Fleet, Trinidad Daily Summaries, August 1–11, 1943, Box 55, 370/47/1/4, NA.

16 Local time; GMT –5.

17 "Abel Recollections."

18 Kelshall, *U-Boat War in the Caribbean*, 387.

19 Three separate searches by three separate researchers at the National Archives (NA) failed to unearth Crockett's (mandatory) after-action report.

20 "Abel Recollections."

21 Kelshall, *U-Boat War in the Caribbean*, 399.

22 Kelshall, "Ralph Kapitzky," 69. The "Kapitzky Diary" has never been found. Schlipper made no deposition with the Deutsches U-Boot Museum in Cuxhaven-Altenbruch.

23 Herbert Skora, "Letzte Feindfahrt von U 615," September 17, 1992. File *U 615*, DU-B-M.

24 War Diary, USS *Walker*, RG 38, Records of the Office of the Chief of Naval Operations USS *Walker*, August 7, 1943; NA.

25 War Diary (KTB), *U-615*, 4. Unternehmung, PG 30,646/4, BA-MA. Reconstituted on the basis of Enigma signals traffic.

26 RG 38, Translations of German Intercepts U-615, Box 141, 370/1/5/7, NA.

27 War Diary (KTB), *U-161*, 4. Unternehmung, PG 30,148/5, BA-MA.

28 Paterson, *Second U-Boat Flotilla*, 215.

29 Local time; GMT –2.

30 Confidential "Report of an Interview with the Chief Officer— Mr. E.C. Martyn," presented by the Royal Navy Submarine Museum, Gosport, to the then U-Boot-Archiv, Cuxhaven-Altenbruch, May 13, 1988. File *U 161*, DU-B-M.

31 File *U 161*, DU-B-M; and http://www.ubootwaffe.net.

32 See ibid. for a suggestion that the ship actually collided with the motor-ship *Aracati*, and that its owners sued for damages in the Tribunal Marítimo Rio de Janeiro. Virtually every other sources credit Achilles with the "kill."

33 Following action from Patterson's "Report of Antisubmarine Action by Aircraft," September 27, 1943. RG 38, ASW Assessment Files, No. 4619, Box 120, 370/47/2/6, Tenth Fleet ASW Files, VP 74 and destruction of U-161, NA.

34 Kelshall, *U-Boat War in the Caribbean*, 420.

35 Reconstituted War Diary (KTB), *U-161*, 5. Unternehmung, PG 30,148/6, BA-MA.

36 Lieutenant-Commander.

37 Cited in Bercuson and Herwig, *Deadly Seas*, 295.

CONCLUSION

1 Since writing this chapter, Holger Herwig has summarized some of the conclusions in "Slaughter in Paradise," *Naval History* 24 (February 2010): 56–63.

2 Committed to paper two days later as "Operationsbefehl 'West Indien' No 51, Secret. For Commanders Only!" RM 7/2336 Chefsache, vol. 3, U-Boote. Allgemein, BA-MA. Signed "Dönitz."

3 See Sint Jago, *De Tragedie van 20 April 1942*.

4 See Frey and Ide, eds., *A History of the Petroleum Administration for War 1941–1945*.

5 War Cabinet and Cabinet Office: Historical Section: War Histories, "Statistics of petroleum supplies, disposal and stocks in the UK 1938 and 1940–50," Civil CAB 102/588, NA, Kew.

6 Wiggins, *Torpedoes in the Gulf*, 148–49.

7 The Times, *British War Production, 1939–1945*, 135–36; and Baptiste, "The Exploitation of Caribbean Bauxite and Petroleum," 110–13.

8 Statistics from Desch, *When the Third World Matters*, 68–72.

9 Look Magazine, *Oil for Victory*, 14–15.

10 British Library of Information, http://www.ibiblio.org.

11 Compiled from the website http://www.uboat.net.

12 From Kelshall, *U-Boat War in the Caribbean*, 467–68.

13 Diary entry, May 1943. RM 87/27, Kriegstagebuch (KTB) des BdU, BA-MA. The undated message was attached to the diary entry for May 15, 1943.

14 Diary entry for May 24, 1943. Ibid.

15 Compiled from Kelshall, *U-Boat War in the Caribbean*, 461–71.

16 See White, *U Boat Tankers 1941–1945*.

17 Raeder to Dönitz, February 11 and 16, 1942. RM 7/2336 Chefsache Bd. 3: U-Boote. Allgemein, BA-MA.

18 Note of August 10, 1942. RM 87/7 KTB des BdU, BA-MA.

19 Raeder to Dönitz, March 26, 1942. RM 7/846 I SKL, Teil C IV, KTB U-Bootskriegsführung 1942, BA-MA.

20 Diary entry of April 14,1941. RM 87/5, KTB des BdU, BA-MA.

21 See Holger H. Herwig, *Politics of Frustration: The United States in German Naval Planning, 1889–1941* (Boston and Toronto: Little, Brown, 1976), 243ff.

22 Entry for June 17, 1942. Wagner, ed., *Lagevorträge*, 396.

23 Cited in Paterson, *Second U-Boat Flotilla*, 180.

24 Hasslinger, "The U-Boat War," 9–10.

25 War Diary (KTB), *U-161*, 2. Unternehmung, PG 30,148/2. BA-MA.

26 War Diary (KTB), *U-156*, 2. Unternehmung, PG 30, 143/2, BA-MA.

27 Wilhelm Polchau, Engineer Report, 3rd War Patrol, 22.4–7.7.1942. RM 98/525 "U156" KTB Ing., BA-MA.

28 War Diary (KTB), *U-161*, 3. Unternehmung, PG 30, 148/3, BA-MA.

29 Hasslinger, "The U-Boat War," 14.

30 Entry for April 14, 1942. RM 87/5, KTB des BdU, BA-MA.

31 Blair, *Hitler's U-Boat War*, II:705, 707; Alex Niestlé, *German U-Boat Losses During World War II* (Annapolis: Naval Institute Press, 1998), *passim*.

32 For a recent tally, see Timothy Mulligan, *Neither Sharks nor Wolves: The Men of Nazi Germany's U-Boat Arm* (Annapolis: Naval Institute Press, 1999).

33 Blair, *Hitler's U-Boat War*, vol. II, 705; Michael Salewski, "The Submarine War: A Historical Essay," in Lothar-Günther Buchheim, *U-Boat War* (New York: Bonanza Books, 1978), n.p.; Holger H. Herwig, "Germany and the Battle of the Atlantic," in *A World at Total War: Global Conflict and the Politics of Destruction, 1937–1945*, eds. Roger Chickering, Stig Förster, and Bernd Greiner (Cambridge: Cambridge University Press, 2005), 71–88.

BIBLIOGRAPHY

The documentary record for Operation Neuland is at the German Federal Military Archive (Bundesarchiv-Militärarchiv, or BA-MA) at Freiburg. At the operational level, Admiral Karl Dönitz's war diary is of critical importance: RM 87 Kriegstagebuch des Befehlshabers der U-Boote, vols. 19–31; RM 87/58 BdU, Operationsbefehle; and N 236/13 Nachlaß Dönitz, Mat. Sammlung Großadmiral Dönitz. Also relevant are several position papers by the First Supreme Command of the Navy (ISKL) that deal with both the overall maritime situation and specific U-boat operations: RM 7/258 Grundlegende Fragen der Kriegführung August–Dezember 1941, Betrachtung der allgemeinen strategischen Lage nach Kriegseintritt Japan/USA; RM 7/259 ISKL, Teil CA, Grundlegende Fragen der Kriegführung 1942–43; RM 7/815 ISKL, Weisungen zur Führung des Handelskrieges; RM 7/841 ISKL, Teil BIV, Ergänzung zur Lage U-Boote 1. Januar 1942 – 31. Mai 1942; RM 7/846 ISKL, Teil CIV, KTB U-Bootskriegsführung 1942; RM 7/2319 ISKL, Iu, U-Boote, Allgemein; and RM 7/2336, ISKL, Iu, Chefsache Allgemein, U-Boote. The medical side of service on the U-boats is detailed in RM 87/88 "German Undersea Medical Research" (Dr. K. W. Essen, Kriegsmarine).

The Deutsches U-Boot Museum (formerly Traditionsarchiv Unterseeboote, U-Boot-Archiv), Cuxhaven-Altenbruch, Germany, constitutes a treasure trove of U-boat materials. Records researched include the technical data of the Neuland boats used in this study (*U-94*, *U-156*, *U-161*, *U-162*, *U-615*), the personal files for the skippers who commanded those boats (Albrecht Achilles, Werner Hartenstein, Otto Ites, Ralph Kapitzky, Jürgen Wattenberg), and the photographic collections pertaining to these boats and men. Of course, the most important records are the war diaries (*Kriegstagebuch*, or KTB) of the boats: *"U 94"* August 16, 1941 – August 27, 1942; *"U 156"* September 4, 1941 – February

28, 1943; *"U 161"* July 8, 1941 – September 29, 1943; *"U 162"* September 9, 1941 – June 8, 1942; and *"U 615"* March 26, 1941 – July 28, 1943.

The intercepted signals from, and the ends of, the U-boat raiders in the Caribbean (and elsewhere) were reconstructed from Allied records, mainly at the National Archives at College Park, Maryland. For *U-94*, RG 38 Translations of German intercepts U-94, Box 101, 370/1/4/4; and RG 38 10th Fleet ASW files, File No. 1529, No. 1574, Box 78 and 79, 370/47/1/7. For *U-156*, RG 38 Intercepted Radio Traffic U-156, Box 106, 370/1/4/6, and OP20-6 *Ultra* intercepts/decrypts; and RG 38 Tenth ASW Assessment Files No. 2646, Box 94, 370/47/2/3. For *U-161*, RG 38 Intercepted Radio Traffic U-161, Box 106, 370/1/4/6, and OP20-6 *Ultra* intercepts/decrypts; and RG 38 Tenth Fleet ASW Files, No. 4619, Box 120, 370/47/2/6. For *U-162*, RG 38 Tenth Fleet ASW Files, No. 1529 and 1574, Box 78 and 79, 370/47/1/7. And for *U-615*, RG 38 Translations of German Intercepts U-615, Box 141, 370/1/5/7; and RG 38 10th Fleet, Trinidad Daily Summaries, August 1–11, 1943, Box 55, 370/47/1/4. *U-615* survivors' interrogation records are in RG 38 Op-16-Z, "U-615" Box 28, 370/15/9/6; and RG 165 MIS-Y, "U-615" Box 734, 390/35/14/7. The end of *U-162* has also benefited from: Royal Navy, Submarine Museum, Gosport, CB 4050/42 (11), Monthly Anti-Submarine Report (November 1942), "Narratives."

The service records of U-boat commanders are in Manfred Dörr, *Die Ritterkreuzträger der U-Boot-Waffe* (2 vols., Osnabrück, 1988); and Rainer Busch and Hans-Joachim Röll, eds., *German U-Boat Commanders of World War II: A Biographical Dictionary* (London, 1999). Memoirs and biographies are few and far between: three of the five commanders did not survive the war; the two who did left no memoirs. Otto Ites receives but scant attention from Theodore Taylor, *Fire on the Beaches* (New York, 1958); and Albrecht Achilles from Erich Glodschey, *U-Boote. Deutschlands scharfe Waffe* (Stuttgart, 1943). Very little biographical information exists on Ralph Kapitzky. Gaylord T. M. Kelshall, "Ralph Kapitzky: Battle in the Caribbean and the Death of U-615," in Theodore P. Savas, ed., *Silent Hunters: German U-Boat Commanders of World War II* (Annapolis, 2003), has provided a Wagnerian interpretation of Kapitzky's final hours. Jürgen Wattenberg drew the attention of Melanie Wiggins, *U-Boat Adventurers: Firsthand Accounts from World War II* (Annapolis, 1999). Werner Hartenstein's torpedoing of the troop transport *Laconia* in September 1942 naturally assured him due attention: Léonce Peillard, *U-Boats to the Rescue: The* Laconia *Incident* (London, 1961); and Frederick Grossmith, *The Sinking of the* Laconia: *A Tragedy in the Battle of the Atlantic* (Stamford, Lincolnshire, 1994). Hartenstein's actions off Aruba were heroically portrayed in Glodschey, *U-Boote*; and more calmly by Executive Officer Paul Just, *Vom Seeflieger zum Uboot-Fahrer. Feindflüge und Feindfahrten 1939–1945* (Stuttgart, 1979). His relations with Dönitz are alluded to by fellow U-boat commander Erich Topp, *Fackeln über dem Atlantik. Lebensbericht eines U-Boot-Kommandanten* (Herford and Bonn, 1990); in English, *The Odyssey of a U-Boat-Commander: Recollections of Erich Topp* (Westport, CT, and London, 1992). The end of *U-156* has been described by Lee A. Dew, "The Sinking of U-156," *Red River Valley Historical Journal of World History* 4 (Autumn 1979): 64–76.

After the war Grand Admiral Dönitz left his recollections of the U-boat war: *Zehn Jahre und zwanzig Tage* (Bonn, 1958); in English, *Memoirs: Ten Years and Twenty Days* (Annapolis, 1990). An assessment of his famous "wolf-pack" tactics is by Bernard Edwards, *Dönitz and the Wolf Packs* (London, 1996). Grand Admiral Erich Raeder barely touches the U-boat war in his memoirs: *Mein Leben* (2 vols., Tübingen, 1956–57); and *My Life* (Annapolis, 1960), an abridged English-language edition.

There are now countless recollections of life on the U-boats by former officers and crew members. Some of the most useful include Hans Goebeler, *Steel Boat, Iron Hearts: A U-Boat Crewman's Life aboard* U-505 (New York, 2005); Wolfgang Hirschfeld, *Hirschfeld: The Secret Diary of a U-Boat* (London, 2000); and Werner Hirschmann, *Another Place, Another Time: A U-Boat Officer's Wartime Album* (Toronto and London, 2004). Lothar-Günther Buchheim, *The Boat* (New York, 1975), remains a best-selling (and controversial) classic; while Wolfgang Otto's novel, *Sharks and Little Fish: A Novel of German Submarine Warfare* (Guilford, CT, 2003), is superb. David J. Bercuson and Holger H. Herwig, *Deadly Seas: The Duel between the* St. Croix *and the* U-305 *in the Battle of the Atlantic* (Toronto, 1997), offers a work of historical recreation of the U-boat war in the Atlantic. Michael L. Hadley, *Count not the Dead: The Popular Image of the German Submarine* (Montreal, 1995), is without rival.

Germany's Atlantic U-boat bases have been treated by Jak P. Mallmann Showell, *Hitler's U-Boat Bases* (Annapolis, 2002); Randolph Bradham, *Hitler's U-Boat Fortresses* (Westport, CT, and London, 2003); Lawrence Paterson, *First U-Boat Flotilla* (Barnsley, 2002), as well as *Second U-Boat Flotilla* (Barnsley, 2003); Christophe Cérino and Yann Lukas, *Keroman. Base de sous-marins, 1940–2003* (Plomelin, 2003); Gordon Williamson and Ian Palmer, *U-Boat Bases and Bunkers 1941–45* (Botley, Oxford, 2003); and Luc Braeur, *U-Boote! Saint-Nazaire* (Le Pouliguen, 2006).

Technical data on the U-boats is readily available in Bodo Herzog, *Die deutschen Uboote 1906 bis 1945* (Munich, 1959); and Eberhard Rössler, *Geschichte des deutschen Ubootbaus* (Munich, 1975). U-boat development during World War II has been analyzed by Werner Rahn, "Die Entstehung neuer deutscher U-Boot-Typen im Zweiten Weltkrieg. Bau, Erprobung und erste operative Erfahrungen," *Militärgeschichte* 2 (1993): 13–20. Overviews for the entire war period are provided by J. Rohwer and G. Hümmelchen, *Chronik des Seekrieges 1939–1945* (Herrsching, n.d.); Kenneth G. Wynn, *U-Boat Operations of the Second World War* (2 vols., Annapolis, 1997–98); Rainer Busch and Hans-Joachim Röll, eds., *Der U-Boot-Krieg, 1939–1945* (5 vols., Hamburg: 1996–2003); and Clay Blair, *Hitler's U-Boat War* (2 vols., New York, 2000). The so-called "milk cow" supply ships are in John F. White, *U Boat Tankers 1941–45: Submarine Suppliers to Atlantic Wolf Packs* (Annapolis, 1998). Two websites were also useful: in English, http://uboat.net; and in German, http://www.ubootwaffe.net. The German Navy's grid charts (*Quadratkarten*) of the Caribbean Sea are at BA-MA, Files 1909G and 1912G.

Research sites in the Caribbean included the newspaper, manuscript, and photography archives of the Biblioteca Nacional as well as the Lago Colony and Refinery on Aruba; and the Bibliotheek van de Universiteit van de Nederlandse Antillen, the Maritime Museum, and the Centraal Historisch Archief on Curaçao. The St. Lucia National Archives provided a good deal of "local color" for the attack by *U-161*. Unfortunately, a fire in 1948 destroyed much of the capital, Castries, as well as many irreplaceable official documents. But two local newspapers survived for 1942: *The Voice of Saint Lucia* and *The West Indian Crusader*. The Archive's director, Margo Thomas, unearthed reports of eyewitness testimonies to the attack in March 1942 as well as the official after-action report by the Administrator of St. Lucia, May 1942. Finally, the National Archives of the Bahamas at Nassau provided two excellent newspapers for 1942 and 1943, *The Nassau Daily Tribune* and *The Nassau Guardian*, as well as a collection of documents: Archives of the Government, The Bahamas in the World Wars, 1914, 1918, 1939–45.

The war in the Caribbean basin has been portrayed by Gaylord T. M. Kelshall, *The U-Boat War in the Caribbean* (Annapolis, 1994); and Fitzroy André Baptiste, *War, Cooperation, and Conflict: The European Possessions in the Caribbean,*

1939–1945 (New York, Westport, London, 1988). A brief first overview was provided by C. Alphonso Smith, "Battle of the Caribbean," *United States Naval Institute Proceedings* 80 (September 1954): 976–82. The British Ministry of Defence used Kriegsmarine records to reconstruct the various battles in the Caribbean as part of: *German Naval History: The U-Boat War in the Atlantic 1939–1945* (London, 1989). The United States defense of the Caribbean is in *United States Army in World War II: The Western Hemisphere. Guarding the United States and its Outposts* (Washington, D.C., 1964); *The Army Air Forces in World War II*, vol. 1, *Plans and Early Operations January 1939 to August 1942* (Chicago, 1948); and Michael C. Desch, *When the Third World Matters: Latin America and United States Grand Strategy* (Baltimore and London, 1993). The role of Allied intelligence in the war at sea was first detailed in: United States, Naval Security Group 1979, "Intelligence Reports on the War in the Atlantic 1942–1945"; and later by David Syrett, *The Battle of the Atlantic and Signals Intelligence: U-Boat Situations and Trends, 1941–1945* (Aldershot, 1998). Central to the debate remains F. H. Hinsley, *British Intelligence in the Second World War: Its Influence on Strategy and Operations* (4 vols., London, 1979–90).

The prewar history of British colonial Trinidad has been examined by Arthur Calder-Marshall, *Glory Dead* (London, 1939). How the arrival of American forces affected the socio-ethnic-economical structure of Trinidad has been gleaned from a number of works: R. R. Kuczynski, *Demographic Survey of the British Colonial Empire* (3 vols., London, 1948–53); M. G. Smith, *The Plural Society in the British West Indies* (Berkeley and Los Angeles, 1965); and Kelvin Singh, *Race and Class Struggles in a Colonial State: Trinidad 1917–1945* (Calgary: University of Calgary Press, 1994). United States assessments of the Caribbean were taken from Caribbean Commission, U.S. Section, *The Caribbean Islands and the War* (Washington, DC, 1943); and US policy with regard to sending black troops into the Caribbean from *United States Army in World War II*, Special Studies, *The Employment of Negro Troops* (Washington, DC, 1966). Anglo-American relations in the Caribbean have been sketched by Annette Palmer, *World War II in the Caribbean: A Study of Anglo-American Partnership and Rivalry* (Randallstown, MD, 1998).

Popular novels dealing with the war and the arrival of the Americans include Robert Antoni, *My Grandmother's Erotic Folktales* (New York, 2000); Ralph de Boissière, *Rum and Coca-Cola* (London, 1984); and Samuel Selvon,

Ways of Sunlight (London, 1957). A popular journalist's account is Albert Gomes, *Through a Maze of Colour* (Port of Spain, 1974).

The specific issue of Caribbean oil and the outcome of World War II have been addressed in: Frederick Haussmann, "Latin American Oil in War and Peace," *Foreign Affairs* 21 (January 1943): 354–61; *Look Magazine, Oil for Victory: The Story of Petroleum in War and Peace* (New York, 1946); John W. Frey and H. Chandler Ide, eds., *A History of the Petroleum Administration for War 1941–1945* (Washington, D.C., 1946); Henry Longhurst, *Adventure in Oil: The Story of British Petroleum* (London, 1959); A. J. Payton-Smith, *Oil: A Study of War-Time Policy and Administration* (London, 1971); and Robert Goralski, *Oil and War: How the Deadly Struggle for Fuel in WWII Meant Victory or Defeat* (New York, 1987). Both bauxite and oil are surveyed by Fitzroy Baptiste, "The Exploitation of Caribbean Bauxite and Petroleum, 1914–1945," *Social and Economic Studies* 37 (1988): 107–42. The "what-might-have-been" for the Germans is in Karl M. Hasslinger, "The U-Boat War in the Caribbean: Opportunities Lost," US Naval War College, Newport, RI, Department of Operations paper, March 1996.

The Esso Lago refinery at San Nicolas on Aruba has a dedicated website maintained by Don D. Gray: http://www.lago-colony.com. The Biblioteca Nacionale Aruba has a full run of the newspapers *Aruba Esso News* and *Pan Aruban*. A special supplement of the *Aruba Esso News* has encapsulated the events of February 1942 as "The War Years at Lago: 1939–A Summing up–1945." With regard to secondary works, the German attack on San Nicolas on February 16, 1942, has been sketched in pamphlet form by William C. Hochstuhl, *German U-Boat 156 Brought War to Aruba February 16, 1942* (Oranjestad, 2001). Tales of the reactions to the attack by *U-156* have been collected in James L. Lopez, *The Lago Colony Legend: Our Stories* (Conroe, TX, 2003). The revolt of the Chinese stokers at Willemstad, Curaçao, in April 1942, has been researched by Junnes Sint Jago, *De Tragedie van 20 April 1942: Arbeidsconflict Chinese zeelieden en CSM mondt uit in bloedbad* (Curaçao, 2000).

The testy issue of French aircraft, gold, and ships controlled by the Vichy regime at Martinique was gleaned from several sources. C. Alphonso Smith provided an early overview: "Martinique in World War II," *United States Naval Institute Proceedings* 81 (February 1955): 169–74. A more detailed account is in Baptiste, *War, Cooperation, and Conflict*; and William L. Langer, *Our Vichy Gamble* (New York, 1947). Admiral Georges Robert penned his recollections after the war: *La France aux Antilles de 1939 à 1943* (Paris, 1950). The position of the

United States Department of State is in *Foreign Relations of the United States: Diplomatic Papers 1940*, vol. 2, *General and Europe* (Washington, D.C., 1957), and ibid., *Diplomatic Papers 1942*, vol. 2, *Europe* (Washington, D.C., 1962); *The Memoirs of Cordell Hull* (2 vols., New York, 1948); and *Cordell Hull: A Registry of His Papers in the Library of Congress*, Manuscript Division, 2000. Official German views on French gold are in *Documents on German Foreign Policy 1918–1945, Series D (1937–1945)*, vol. 10, *The War Years June 23–August 31, 1940* (London, 1957); while the Canadian transshipment of gold from Halifax to Martinique is in C. P. Stacey, *Canada and the Age of Conflict: A History of Canadian External Policies*, vol. 2, *1921–1948, The Mackenzie King Era* (Toronto, Buffalo, London, 1981).

Allied warships in the Caribbean involved in the antisubmarine war 1942–43 were researched in several primary and secondary sources. For US Navy ships, at the National Archives at College Park, Maryland: USS *Blakeley*, RG 38 Serial 028, Loc. 370 45/1/3 Box 853; USS *Lea*, RG 38 Serial 00768, Loc. 370 45/7/2-3 Box 1141; USS *Walker*, 10th Fleet, War Diary 799–800, and RG 38 Records of the Office of the Chief of Naval Operations. And for VP-92, PBY-5 Flying Boat, RG 38, Records of the Office of the Chief of Naval Operations VP-92, August 13, 1942.

The antisubmarine activities of the Canadian frigate HMCS *Oakville* are at Library and Archives Canada, Ottawa: RG 24, National Defence, Series D-1, D-4, D-10, and D-13; and General Information HMCS OAKVILLE, Movements HMCS OAKVILLE, Sub attacks HMCS OAKVILLE, and Captain (D) Halifax HMCS OAKVILLE. There are also two stirring books by a participant of the encounter between *U-94* and HMCS *Oakville*: Hal Lawrence, *A Bloody War: One Man's Memories of the Canadian Navy 1939–1945* (Toronto, 1979); and *Tales of the North Atlantic* (Toronto, 1985).

Technical data on Allied warships, merchantmen, and tankers were researched in *Jane's Fighting Ships 1942* (London, 1943); *Jane's Fighting Ships of World War II* (New York, 1989); M. J. Whitley, *Destroyers of World War II: An International Encyclopedia* (London, 1988); *Dictionary of American Naval Fighting Ships* (8 vols., Washington, D.C., 1959); and Ken Macpherson and Marc Milner, *Corvettes of the Royal Canadian Navy 1939–1945* (St. Catharines, ON, 1993). Aircraft data came from *Jane's Fighting Aircraft of World War II* (New York, 1989); the website http://www.warbirdalley.com; and the various websites of aircraft producers such as Boeing, Consolidated, Martin Mariner, Lockheed, Short, and

Douglas. Last but not least, information on Allied shipping sunk by the U-boats was taken from the same sources that the commanders used in 1942–43: *Lloyds Register of British and Foreign Shipping* (London, 1886–); Erich Gröner, *Taschenbuch der Handelsflotten* (Munich, 1940); and *Weyers Flottentaschenbuch* (Munich, n.d.). All claims by the Kaleus for tonnage sunk were checked in *Lloyd's Register of Shipping 1940–1941* (London, 1940).

BOOKS AND ARTICLES

Achong, Tito P., *The Mayor's Annual Report: A Review of the Activities of the Port-of- Spain City Council, with Discourses on Social Problems Affecting the Trinidad Community, for the Municipal Year 1942–43* (Boston: US Government Printing Office, 1943)

Alden, John D., *Flush Decks and Four Pipes* (Annapolis: Naval Institute Press, 1965)

Antoni, Robert, *My Grandmother's Erotic Folktales* (New York: Grove Press, 2000)

Baptiste, Fitz A., *The United States and West Indian Unrest, 1918- 1939* (Mona, Jamaica: University of the West Indies Press, 1978)

Baptiste, Fitzroy André, *War, Cooperation, and Conflict: The European Possessions in the Caribbean, 1939–1945* (New York, Westport, CT, and London: Greenwood Press, 1988)

Bercuson, David J., and Holger H. Herwig, *One Christmas in Washington: The Secret Meeting between Roosevelt and Churchill that Changed the World* (New York: Overlook Press, 2005)

Bercuson, David J., and Holger H. Herwig, *Deadly Seas: The Duel between the* St Croix *and the* U305 *in the Battle of the Atlantic* (Toronto: Random House Canada, 1997)

Bishop, Chris, *The Essential Submarine Identification Guide: Kriegsmarine U-Boats, 1939–45* (London: Amber Books, 2006)

Blair, Clay, *Hitler's U-Boat War*, vol. I, *The Hunters 1939–1942* (New York: Random House, 1996)

Blair, Clay, *Hitler's U-Boat War*, vol. II, *The Hunted, 1942–1945* (New York: Random House, 1998)

Boissière, Ralph de, *Rum and Coca-Cola* (London: Allison and Busby, 1984)

Bradham, Randolph, *Hitler's U-Boat Fortresses* (Westport and London: Praeger, 2003)

Buchheim, Lothar-Günther, *U-Boat War* (New York: Bonanza Books, 1978)

Busch, R., and H. J. Röll, *"Shatten voraus!" Feindfahrten von U-156 unter Werner Hartenstein* as a special edition of *Der Landser Grossband* (Rastatt: Pabel-Moewig, 1996)

Calder-Marshall, Arthur, *Glory Dead* (London: M. Joseph Ltd., 1939)

Caribbean Commission, U.S. Sector, *The Caribbean Islands and the War: A Record of Progress in Facing Stern Realities* (Washington: US Government Printing Office, 1943)

Cérino, Christophe, and Yann Lukas, *Keroman. Base de sous-marins, 1940–2003* (Plomelin: Palantines, 2003)

Chair, Henry Graham de, *Let Go Aft: The Indiscretions of a Salt Horse Commander* (Tunbridge Wells: Parapress, 1993)

Chickering, Roger, Stig Förster, and Bernd Greine, eds., *A World at Total War: Global Conflict and the Politics of Destruction, 1937–1945* (Cambridge: Cambridge University Press, 2005)

Churchill, Winston S., *The Second World War*; vol. IV, *The Hinge of Fate* (Boston: Houghton-Mifflin, 1950)

Conn, Stetson, Rose C. Engelman, and Byron Fairchild, eds., *United States Army in World War II: The Western Hemisphere, Guarding the United States and its Outposts* (Washington: US Government Printing Office, 1964)

Craven, Wesley Frank, and James Lea Cate, eds., *The Army Air Forces in World War II*, vol. I, *Plans and Early Operations, January 1939 to August 1942* (Chicago: University of Chicago Press, 1948)

Desch, Michael C., *When the Third World Matters: Latin America and United States Grand Strategy* (Baltimore and London: Johns Hopkins University Press, 1993)

Dictionary of American Naval Fighting Ships, vol. I (Washington: Navy Department, 1959)

Documents on German Foreign Policy 1918–1945, Series D (1937–1945), vol. 10, *The War Years June 23–August 31, 1940* (London: H.M. Stationery Office, 1957)

Domarus, Max, ed., *Hitler. Reden und Proklamationen*, vol. II/2, *Untergang* (Munich: Süddeutscher Verlag, 1965)

Dönitz, Karl, *Memoirs: Ten Years and Twenty Days* (London: Weidenfeld and Nicolson, 1959)

Dörr, Manfred, *Die Ritterkreuzträger der U-Boot-Waffe* (2 vols., Osnabrück: Biblio, 1988)

Douglas, W.A.B., et al., *No Higher Purpose: The Official Operational History of the Royal Canadian Navy in the Second World War, 1939–1943*, vol. II, part 1 (St. Catharines, ON: Vanwell, 2002)

Eckes, Jr., Alfred E., *The United States and the Global Struggle for Minerals* (Austin and London: University of Texas Press, 1979)

Ellis, John, *World War II: A Statistical Survey* (New York: Facts on File, 1993)

Fanning, Leonard M., *American Oil Operations Abroad* (New York: McGraw-Hill, 1947)

Foreign Relations of the United States: Diplomatic Papers 1940, vol. 2, *General and Europe* (Washington: US Government Printing Office, 1957)

Foreign Relations of the United States: Diplomatic Papers 1942, vol. 2, *Europe* (Washington: US Government Printing Office, 1962)

Frey, John W., and H. Chandler Ide, eds., *A History of the Petroleum Administration for War 1941–1945* (Washington: US Government Printing Office, 1946)

Friedman, Max Paul, *Nazis and Good Neighbors: The United States Campaign against the Germans of Latin America in World War II* (Cambridge: Cambridge University Press, 2003)

Fröhlich, Elke, ed., *Die Tagebücher von Joseph Goebbels*, part 2, *Diktate 1941–1945* (Munich: K.G. Saur, 1994)

Gannon, Michael, *Operation Drumbeat* (New York: Harper & Row, 1991)

Gannon, Michael, *Operation Drumbeat: The Dramatic True Story of Germany's First U-boat Attacks Along the American Coast in World War II* (Annapolis: Naval Institute Press, 2009)

Gericke, Bernd, *Die Inhaber des Deutschen Kreuzes in Gold, des Deutschen Kreuzes in Silber der Kriegsmarine und die Inhaber der Ehrentafelspange der Kriegsmarine* (Osnabrück: Bilbio, 1993)

Gilbert, Martin, ed., *The Churchill War Papers*, vol. 3, *The Ever Widening War, 1941* (New York and London: W.W. Norton, 2000)

Glodschey, Erich, *U-Boote. Deutschlands Scharfe Waffe* (Stuttgart: Union-Deutsche Verlagsgesellschaft, 1943)

Goebeler, Hans, *Steel Boat, Iron Hearts: A U-Boat Crewman's Life aboard* U-505 (New York: Savas Beatie, 2005)

Gomes, Albert, *Through a Maze of Colour* (Port of Spain: Key Caribbean Publications, 1974)

Goodhart, Philip, *Fifty Ships that Saved the World* (Garden City, NY: Doubleday, 1965)

Goralski, Robert, and Russel W. Freeburg, *Oil and War: How the Deadly Struggle for Fuel in WWII Meant Victory or Defeat* (New York: Morrow, 1987)

Herwig, Holger H., *Germany's Vision of Empire in Venezuela 1871–1914* (Princeton: Princeton University Press, 1986)

Herwig, Holger H., *Politics of Frustration: The United States in German Naval Planning, 1889–1941* (Boston and Toronto: Little, Brown, 1976)

Herzog, Bodo, *Die deutschen Uboote 1906 bis 1945* (Munich: J.F. Lehmann, 1959)

Hessler, Günter, "The U-Boat War in the Atlantic 1939–1945" (2 vols., London: Navy Department, 1989)

Hickam, Homer H., *Torpedo Junction: U-Boat War off America's East Coast, 1942* (Annapolis: Naval Institute Press, 1989)

Hinsley, F. H., *British Intelligence in the Second World War* (5 vols., London: H.M. Stationery Office, 1979–90)

Hochstuhl, William C., *German U- Boat 156 Brought War to Aruba February 16, 1942* (Oranjestad: Aruba Scholarship Foundation, 2001)

Hull, Cordell, *The Memoirs of Cordell Hull* (2 vols., New York: MacMillan, 1948)

Jago, Sint, *De Tragedie van 20 April 1942 Arbeidsconflict Chinese zeelieden en CSM mondt uit in bloedbad* (Curaçao: Imprenta Atiempo, 2000)

Jane's Fighting Ships 1942 (New York: McGraw-Hill, 1944)

Jane's Fighting Ships of World War II (New York: McGraw-Hill, 1989)

Johnson, Arthur M., *Petroleum Pipelines and Public Policy, 1906–1959* (Cambridge, MA: Harvard University Press, 1967)

Just, Paul, *Von Seeflieger zum Uboot-Fahrer: Feindflüge und Feindfahrten 1939–1945* (Stuttgart: Motorbush-Verlag, 1979)

Kahn, David, *Seizing the Enigma: The Race to Break the German U-Boat Codes, 1939–1943* (Boston: Houghton-Mifflin, 1991)

Kavass, Igor I., and Adolf Sprudzs, eds., *A History of the Petroleum Administration for War, 1941–1945* (Buffalo, NY: W.S. Hein, 1974)

Kelshall, Gaylord T.M., *The U-Boat War in the Caribbean* (Annapolis: Naval Institute Press, 1994)

Kemp, Peter, *Decision at Sea: The Convoy Escorts* (New York: Elsevier-Dutton, 1978)

Kuczynski, R. R., *Demographic Survey of the British Colonial Empire*, vol. III, West Indian and American Territories (London, New York, Toronto: A.M. Kelley, 1948–53)

Langer, William L., *Our Vichy Gamble* (New York: A.A. Knopf, 1947).

Larson, Henrietta M., Evelyn H. Knowlton and Charles Popple, *New Horizons: 1927–1950 (History of Standard Oil Company (New Jersey)* (New York: Harper, 1971)

Lawrence, Hal, *A Bloody War: One Man's Memories of the Royal Canadian Navy, 1939–1945* (Toronto: Macmillan of Canada, 1979)

Leonard, Thomas M., and John F. Bratzel, eds., *Latin America During World War II* (Plymouth: Rowan and Littlefield, 2007)

Mallmann Showell, Jak P., *Hitler's U-Boat Bases* (Annapolis: Naval Institute Press, 2002)

Milner, Marc, *Battle of the Atlantic* (Stroud, UK: Vanwell, 2003)

Morison, Samuel Eliot, *History of United States Naval Operations in World War II*, vol. I, *The Battle of the Atlantic: September 1939-May 1943* (Boston: Little, Brown, 1962)

Mulligan, Timothy, *Neither Sharks nor Wolves: The Men of Nazi Germany's U-Boat Arm* (Annapolis: Naval Institute Press, 1999)

Nash, Gerald D., *United States Oil Policy: 1890–1964* (Westport, CT: Greenwood Press, 1968)

Niestlé, Alex, *German U-Boat Losses During World War II* (Annapolis: Naval Institute Press, 1998)

O'Connor, P.E.T., *Some Trinidad Yesterdays* (Port of Spain: Imprint Caribbean, 1978)

Ott, Wolfgang, *Sharks and Little Fish: A Novel of German Submarine Warfare* (New York: Pantheon, 1957)

Overy, Richard, *Why the Allies Won* (New York and London: Norton, 1995)

Palmer, Annette, *World War II in the Caribbean: A Study of Anglo-American Partnership and Rivalry* (Randallstown, MD: Block Academy Press, 1998)

Paterson, Lawrence, *Second U-Boat Flotilla* (Barnsley: L. Cooper, 2003)

Paterson, Lawrence, *U-Boat War Patrol: The Hidden Photographic Diary of U 564* (London: Greenhill Books, 2004)

Payton-Smith, D.J., *Oil: A Study of War-Time Policy and Administration* (London: H.M. Stationery Office, 1971)

Peillard, Léonce, *U-Boats to the Rescue: The* Laconia *Incident* (London: Coronet, 1961)

Picker, Henry, ed., *Hitlers Tischgespräche im Führerhauptquartier 1941–1942* (Stuttgart: Seewald, 1963)

Popple, Charles Sterling, *Standard Oil Company (New Jersey) in World War II* (New York: Standard Oil Co., 1952)

Rahn, Werner, and Gerhard Schreiber, eds., *Kriegstagebuch der Seekriegsleitung 1939–1945* (Herford and Bonn: Mittler, 1992)

Randall, Stephen J., and Graeme S. Mount, *The Caribbean Basin: An international history* (London and New York: Routledge, 1998)

Robert, Georges, *La France aux Antilles de 1939 à 1943* (Paris: Plon, 1950)

Rose, James F. de, *Unrestricted Warfare* (Edison, NJ: Wiley, 2006)

Roskill, Stephen W., *The War at Sea, 1939–1945*, vol. II, *The Period of Balance* (London: H.M. Stationery Office, 1956)

Rössler, Eberhard, *Geschichte des deutschen Ubootbaus* (Munich: Lehmanns, 1975)

Savas, Theodore P., ed., *Silent Hunters: German U-Boat Commanders of World War II* (Annapolis: Naval Institute Press, 2003)

Selvon, Samuel, *Ways of Sunlight* (New York: St. Martin's Press, 1957)

Shaffer, Ed, *Canada's Oil and the American Empire* (Edmonton: Hurtig, 1983)

Smith, M.G., *The Plural Society in the British West Indies* (Berkeley and Los Angeles: University of California Press, 1965)

Spector, Ronald H., *At War at Sea: Sailors and Naval Combat in the Twentieth Century* (New York: Viking, 2001)

Taffrail, [Taprell Dorling], *Blue Star Line at War 1939–1945* (London: Foulsham, 1973)

Taylor, Theodore, *Fire on the Beaches* (New York: Norton, 1958)

Texas Eastern Transmission Corporation, *The Big Inch and Little Big Inch Pipelines: The Most Amazing Government-Industry Cooperation Ever Achieved*, The Louis Berger Group Inc. (Shreveport, LA: 1957 , n.p.)

Topp, Erich, *The Odyssey of a U-Boat Commander: Recollections of Erich Topp* (Westport and London: Praeger, 1992)

Topp, Erich, *Fackeln über dem Atlantik. Lebensbericht eines U-Boot-Kommandanten* (Herford and Bonn: Mittler, 1990)

United States Army in World War II. The Technical Services, The Transportation Corps: Operations Overseas (Washington: U.S. Government Printing Office, 1957)

Vat, Dan van der, *The Atlantic Campaign: World War II's Great Struggle at Sea* (New York: Harper and Row, 1988)

Wagner, Gerhard, ed., *Lagevorträge des Oberbefehlshabers der Kriegsmarine vor Hitler 1939–1945* (Munich: Lehmann, 1972)

White, John F., *U Boat Tankers 1941–1945* (Annapolis: Naval Institute Press, 1998)

Wiggins, Melanie, *U-Boat Adventures: Firsthand Accounts from World War II* (Annapolis: Naval Institute Press, 1999)

Williams, Eric, *History of the People of Trinidad and Tobago* (London: Praeger, 1962)

Williamson, Gordon, and Darko Pavlovic, *U-Boat Crews 1914–45* (London: Osprey, 1996)

Williamson, Gordon, and Ian Palmer, *U-Boat Bases and Bunkers 1941–45* (Botley, Oxford: Osprey, 2003)

Williamson, Harold F., et al., *The American Petroleum Industry: The Age of Energy, 1899–1959* (Evanston, IL: Northwestern University Press, 1963)

Wynn, Kenneth, *U-Boat Operations of the Second World War*, vol. I, *Career Histories, U1–U510* (Annapolis: Naval Institute Press, 1998)

INDEX

German attacks on, 57–71 passim, 112,
181, 282–83
oil production, xv, xvi
oil refineries, xv, 7, 9, 281;
Operation Neuland, 45–49 passim,
57–71, 82–95, 104–5, 112,
123–25, 135, 150, 152, 181,
184–87, 214–18, 228, 244–45,
250–51, 259, 275–83
Operations Order No. 51 "West Indies,"
xvi
U-boats, xvii, 46, 48, 51, 53–54, 70, 82,
87, 92, 95, 104–5, 135, 184,
228, 244–45, 250, 275–57,
282–83
UK, defense of, 15,
UK, supply to, 7–9,
US aircraft, 35,
US troops on, 9, 31, 35, 187

B

Battle of the Atlantic xiv, xix. *See also*
Dönitz; Hitler (zone of destiny);
tonnage war
ASW, use of, 274
escorts, shortage of, 213
Operation Neuland and, xviii
peak of, 252
US forces, buildup in Caribbean, 35
US, role in, 236
U-boats, role of, 176, 232
bauxite
airplane production, role in, 11
Allied supply, 275, 276
British Guiana, production of, 11, 281
importance of, 11
Operation Neuland, 278
shipments, US and Canada annual, 278
sources of, 118,
tankers/freighters, 145, 148; German
sinking of, 147, 149, 151, 153,
177, 195, 276
transport, stop of, 151
UK and, 278

Bay of Biscay
attacks against, 238, 250, 258, 279
U-boats, xi, 6, 123, 126, 129, 130, 237,
265, 269; congestion, 126;
sinking of, 171, 216
Bender, Werner, xi, xx, xviii, 36, 93, 95–96,
98, 104–5, 108, 110–15
Bermuda. *See also* Lend-Lease
convoys, 151
US troops, 31, 35
U-boats, 127, 163, 216
USS Blakeley, 46, 94, 137–38
Brazil
air force, 174, 254, 274
allies, as potential partners, 147; shift
to allies, 148; declaration of
war against Axis, 172 pro-
American basing policy and
Hitler, 173
German attacks on ships, 148, 252, 271;
on ports, 173
navy, 174
neutrality of, 34, 141, 147
Pot of Gold, operation, 172
Rio Conference, 173
US, cooperation with military,
173; economic and trade
concessions, 172
Brest, U-boat base at, 128
British Colonial Administration
taxation, 24
Trinidad, 24
unrest, 24
British Empire Workers
Citizens Home Rule Party, 23
British Guiana
Trinidad Department, Caribbean
Defense Command, 16
US airbases, 31; forces, 31
British Petroleum, 2
Bungalow Plan, 75

LONG NIGHT OF THE TANKERS

C

Calder-Marshall, Arthur, 24, 28
Canada. *See also* King, William Lyon
    Mackenzie; Royal Canadian Navy;
    tankers
  Aluminum Company of Canada, 11
  ASW, role in, 213, 240
  Atlantic Convoy Conference, 240
  Canadian Trade, 8
  convoys, 7, 151
  Emile Bertin (Fr), and Vichy France, 76
  escorts, groups, 151, 187, 218–19, 221
  Newfoundland, relations with, 15
  oil, resources, 11; domestic production,
    200–201; exports to US, 201;
    shortage, 198, 200–201, 218
  shipping, construction of, 231
  tankers, 8, 104; losses, 111, 114, 117, 149,
    163, 216; routes to, 15, 210
  troops based in Central and South
    America, 94
Caribbean Defense Command, 16, 26, 32,
    34, 54, 55, 68, 184, 194, 239
Caribbean Sea Frontier, 16, 17, 79, 214, 239
Castries (St. Lucia), 105, 107–13, 119, 139,
    167, 269
Charguaramas (Trinidad)
  US base on, 27–29, 94–95, 98, 105, 187,
    189, 227, 245, 247, 249, 261–63
Chinese tanker crews, rebellion of, 120–22,
    277
choke points, 152, 186, 278
Churchill, Winston, 1, 2, 6, 7, 16, 22, 27, 35,
    75, 120, 128, 191, 194, 199, 217,
    239, 279
  Soviet Union, German invasion of, 1
coal, 1, 2, 86, 197, 200, 257
Cole Pipeline Bill, 204
Colombia, 280
convoys, xi, xiv, xviii, xix, 7, 137, 146, 154,
    176, 198, 214, 217
  Atlantic Convoy Conference, 240
  Canadian ports and, 7; RCN, 213;
    escorts, 218

Caribbean and North Atlantic,
    differences between, 152, 235,
    283
  cycles of, 151
  escort aircraft, 129, 226, 233, 244, 253
    escorts, 173 (see ASW)
  fast or slow, 7
  German attacks on, 6, 146, 152, 156,
    157, 158, 168, 169, 170, 215,
    217, 218, 220, 234, 244,
    245, 246, 247, 255, 257, 259,
    261–62, 274
  labor unrest, 122
  routes of, 213, 218, 245
  temporary stoppage of, 128
  U-boats, effects, 152
  US protection of, 120–21, 227, 238 (*see*
    ASW)
  US, 215, 238, 240
Cuba
  Caribbean Sea Frontier, 16–17
  Guantánamo, US base at, 14, 19, 35,
    214, 245,
  Gulf of Mexico, defense of, 183, 280
  U-boats, attacks near, 129, 134, 161,
    162, 218, 228, 252
  US intervention in, 181
Cumuto Reserve, 26, 27
Curaçao, 48, 92, 104, 105, 120, 123, 152,
    155, 215, 228, 275
  Britain's domestic oil supply, 7
  convoy routes, 152
  defense of, 15, 31, 35, 95, 184, 186–87,
    218, 251
  German attacks on, 48, 55, 65, 92, 125,
    259, 275, 277, 282, 283
  harbors, 43, 63, 186
  Operation Neuland, 45
  refineries, 9
  Royal Dutch Shell Refinery, labor
    unrest, 121–22, 277;
    Willemstad, xvi, 108; Santa
    Anna, 108; Shottegat plant,
    276
  U-boat stationing near, xv
Curaçaose Shipping, 120

D

Davies, Ralph K., 13
destroyers for bases deal, 15, 25, 94, 107, 137, 213, 279, 281
diesel fuel
 Aruba production, 282
 engines, 37, 39, 42, 47, 70, 71, 83, 132
 French purchases on Aruba, 71
 production per day, Aruba, xv
 stocks, xix
Dönitz, Karl ("the Great Lion"), xi–xix passim, 80, 89, 129, 137, 140, 179, 194, 228, 244. *See also* Enigma
 Allied air cover over Aruba, 125
 Allied ASW, thoughts on, 127, 231, 127, 155
 Battle of Atlantic, and Caribbean offensive effect on, 151; and reassessment of, 232–33, 236
 civilian tanker crews, treatment of, 175–76, 271
 German command, relationship with, 282–84
 Operation Neuland Caribbean offensive: first wave, 43, 51, 62–65, 92, 114–19 passim, 123, 232–37, 243–50, 275–76, 281, 282, 285; second wave, 125–26, 140, 141, 142, 143, 145, 146, 147, 154; third wave, 154–56
 post war oil supply, 210
 technical developments, and reaction to, 176–77, 258, 269
 tonnage war, strategy of, 233, 236, 281, 284
 Triton-Null Order (Laconia Order), 237
 U-boat, effects on Allied war effort, 123, 124, 278
 U-boat losses, reaction to, 161–63, 172–77, 228, 235–37, 249–54, 260, 268, 272–74
 U-boat tankers, strategy of, 127, 281
 war of memoranda with Raeder, 126

Dragon's Mouth, 27, 93, 95, 98, 152, 157, 186
Drum Beat, Operation (*Paukenschlag*), xvi, xviii, 120, 275, 277
Dutch West Indies, 8–9

E

Eagle Oil and Shipping Company, 135
economic war
 German preparations for, 278
Edinburgh, air base at, 188, 227, 245, 247, 261, 266
Embargo
 French West Indies, British, 78, 80
 Brazilian threat, 173
Enigma, 41, 42, 45, 51, 64, 83, 85, 130, 134, 135, 176, 232, 244, 246, 247, 259, 261, 269, 274, 280

F

"Ferret." *See* Achilles
Fisher, Sir John, First Sea Lord, 1, 183
Florida Strait, xvi
 fog of uncertainty and, 285
 U-boats in, 119, 152, 161, 162, 163
Four Freedoms (of speech and worship, from want and fear), 25
 Trinidad, 194
Fort-de-France (Martinique) 75, 77
 U-boats, 69, 73, 74, 74, 78, 79, 136, 137, 138, 139, 141; US, 78, 79, 134
France, 10, 14, 64, 76, 128
 entry into WWII, 4
 Gaulism, 75
 Gold reserves, 75
 oil imports from US, 3
 surrender of, 4, and German advantages 6, 35
 Vichy, 74, 75, 119, 134, and U-boat ports, 123

G

garbage tour, 39, 88, 129, 154
gasoline, 279, 282
    100-octane gas, 9, 13, 200
    civilian use, 197, 198, 200
    hybrid, 9–10
    pipelines, 208
    shortages, 203, 277
    supplies, xv, xvi, xvii, xix
    tankers, 87, 104
    UK, and imports, 199; and shortages
        of, 278
    US, 13
Germany
    aircraft carriers, lack of, 16
    ASW, countermeasures for, 281
    Brazil, attacks on, 172–73
    Britain, battle of, 176
    British attacks on, 127
    Central and South America, relations
        with, 172
    economic warfare, preparation for, 278
    France's surrender, advantages of to
        Germany, 6
    Hess, Rudolf (Deputy Führer), 123
    lack of oil reserves, 2, 279
    military machine, myth of, 1
    naval buildup, 14
    Operation Blue, 232
    Operation Neuland, strategy of, 282,
        285
    spheres of influence, 172
    spy ring in Panama Canal Zone, 195
    suppliers, 2–3
    USSR, 2–3
Goebbels, Joseph, 45, 112, 118, 236
Golden West, 37, 40, 155, 261, 274, 276
Gomes, Albert, activist, 20, 22, 23, 192, 195
Greenslade, Rear Admiral J. W., 26, 27,
    29, 78

H

Hartenstein, Werner (*U-156*), xiv–xvii,
    43–103 passim, 118, 124, 129–53
    passim, 186, 231, 236–50 passim,
    259, 274, 279, 283, 285
Hitler, Adolf, xv, 25, 121, 123, 128, 155,
    234, 282–85
    Brazil, relations with, 156, 172–73;
        spheres of influence, 175
    new technology, push for, 176–77
    North Africa, Allied invasion of, 175,
        281; effect on Operation
        Neuland, 175
    oil, 2, Allied production of, 175, 236;
        attacks on Caribbean refineries,
        xv; German imports, cost of, 2
    Operation Neuland, interference in,
        282, 285
    Raeder, relations with, 122–23, 155,
        282, 284
    Russia, attack on, 3; effect on Operation
        Neuland, 283
    surface fleet, disenchantment with,
        234–35
    U-boat losses, 235–36
    war in the east, 232, 283
Hull, Cordell (US Secretary of State), 10
    and destroyers for bases deal, 16
    and Vichy, 77

I

Ickes, Harold (Secretary of the Interior), 12,
    201, 203–4, 211. *See also* Petroleum
        Administrator for War; Petroleum
        Coordinator for War; War
        Production Board
    domestic oil companies, 13, 203
    pipelines, 202–4
    railway, use of for oil transfer, 201
    sea going tankers, obstacles to use of,
        201
    tanker losses, impact of, 203
I.G. Farben, 2

R

radar, 14, 31, 61, 73, 123, 259, 260, 280
  air to surface, 103, 161–64, 175–76,
       227–28, 231, 245–46, 261, 265,
       274, 280
  centimetric, 175, 224, 247, 280
  decoys, 177
  detectors, 42, 176–77, 233, 244, 245,
       247, 252, 281
Raeder, Grand Admiral Erich (Commander
  in Chief, *Kriegsmarine*), xvii, 122,
  234
  Dönitz, relationship with, 128, 282
  Operation Neuland, decisions in, 126,
      129, 155, 175
  retirement of, 235
  tactics, 282, 284
  tonnage warfare, 281
  U-boats, 233
Rainbow 5, 180
Robert, Admiral Georges, 134
Roosevelt, Franklin D.
  and Anglo-American Caribbean
      Commission, 194
  Caribbean bases, 26–27
  Casablanca Conference and, 239
  Congress, and, 16, 27
  defense of Caribbean, 94, 172, 179–81
  Four Freedoms, 194
  labor unrest in Caribbean, 188
  Lend-Lease, 16
  military production plans, 11–13, 151,
      175, 214–15, 236
  oil shortages, 203
  Operation Pot of Gold, 172
  Petroleum Coordinator for National
      Defense, 12
  pipelines, 202
  tankers, attacks on, 72, 120, 199
  UK, 168
  Vichy France, 77
Rosenstiel, Jürgen von (*U-502*), xiv, xv, 55,
    57–60, 63, 68, 70–71, 93, 124, 158,
    165, 171, 276, 279
Royal Air Force. *See also* United Kingdom

bombing attacks on German forces, 231
Brest, raids on, 128
coastal command, 235
fuel supply, 24, 278, 284, 286
German attacks on, 5
Royal Canadian Navy, 240. *See also* Canada
  convoy system, 7
  escort vessels, 187, 213, 221
  U-boats and, 223, 281
Royal Commission on Conditions in the
  Caribbean, 24
Royal Commission on Oil Supplies, 1
Royal Dutch Shell refinery Arend (Eagle),
  Santa Anna, xvii, 9, 61, 121, 135
Royal Dutch Shell, Oranjestad, Willemstad,
  xvi, 9, 63
Royal Navy, 1, 3, 23, 174. *See also* Oil
  Control Board; United Kingdom;
  ASW
  ASW patrols, 31
  coastal Command, 238
  commercial traffic, control of, 7
  convoys, development of, 7
  escorts, 151, 184, 187, 213, 217–18, 240,
    245
  fleet tankers, 210
  oil, dependence on, 1
  oil reserves, 1, 5, 278
  port of calls, 5
  protective measures, 6
  Royal Canadian Navy, relationship with,
    7, 187
  Trinidad, bases in, 31
  US Navy, relationship with, 31
  U-boats, attacks on, 98, 99, 223

S

Sherwood Foresters, 23
shipping, attacks on coastal, 5
St. Nazaire, U-boat base at, 128
Standard Oil of New Jersey–Lago, San
  Nicolas, xvi, 8
  Lago Oil and Transport Co., 9
  production, 8
  refineries, 49, 276

submarines ("gray sharks"). *See also* Dönitz;
Hitler
Operation Neuland, U-boats use in:
  U-124, 244; *U-129*, 244;
  U-156, xiv, xv, 36, 42–103
    passim, 112, 121, 129–56
    passim, 186, 231, 236–85
    passim; *U-157*, 163–63, 171,
    217; *U-158*, 163–65, 171;
  U-159, 167, 254; *U-161*, xi,
    xii, xiii, xv, 36–37, 42–43, 46,
    93–121 passim, 155–70 passim,
    187, 236, 255–276 passim,
    283; *U-203*, 244; *U-333*,
    250; *U-359*, 254; *U-406*, 263;
  U-459, 127–28, 143, 236, 274,
    281; *U-502*, xiv, xv, 36, 48,
    55–60, 63, 70–71, 93, 121, 156,
    158, 171, 218, 276; *U-510*, 244;
  U-572, 254; *U-615*, 255–76
    passim; *U-66*, 43, 125, 130,
    253; *U-662*, 253; *U-67*, xiv,
    xv, 36, 48, 63–65, 90, 93, 120,
    275; *U-68*, xv; *U-759*, 253;
  U-94, 218–23, 231, 274
U-boat tanker fleet milk cows
    (*Milchkühe*, type XIV), 127,
    154, 155, 174, 276, 281; *U-487*,
    258, 274, 281; *U-459*, 127–28,
    143, 236, 274, 281; *U-461*, 179,
    236, 274, 281; *U-462*, 274,
    281; *U-489*, 274, 281
targeting, German strategy of, 281–82
Suriname (Dutch Guiana), 11, 31, 35, 116

T

tankers, Allied reliance on, 3, 5, 6, 12, 123,
    125, 278, 285. *See also* convoys;
    Ickes; tonnage war
Axis capacity, 2
capacity: Italy, 2; Germany, 2;
    Norwegian, 8; UK, 2–7; US, 8
construction of, 210, 236, 277
fuel shortages, 198–200, 277; and
    solutions to, 200; Canada,

200–201; overland routes in
    US, 208–9, 277; US, 200–201;
    UK, 200
halt in traffic, 128
losses to U-boats, 3, 4, 6, 12, 88, 198,
    203, 205; and effects on US,
    12, 120
routes, xvii, 6–7, 15, 28, 48, 218, 276
surplus, at beginning of war, 4
tankers and freighters sunk by U-boats:
    Anglo-Canadian (Cdn),
    216; *Arkansas* (*Aryan*), 62;
    Athelempress, 146; *Barrdale*
    (US), 133; *Bayou*, 11, 72 ; *Beth*,
    153; *British Colony* (UK), 150;
    British Consul (UK), 96, 121,
    218; *British Governor* (UK),
    101–2; *Cabadello*, 72; *Circle
    Shell* (UK), 72, 99–100; *Cold
    Harbor* (US), 159; *Darina*
    (Cdn), 163; *Delmundo* (US),
    218; *Delplata*, 72, 83; *Eastern
    Sword* (US), 148; *Empire Cloud*,
    218; *Empress Gold*, 118; *Esso
    Copenhagen*, 72; *Esso Houston*,
    150; *Everasma*, 72; *Edward
    B. Dudley* (US), 257; *Fairport*
    (US), 170; *Faja de Oro*, 182;
    Frank B. Bair (Cdn), 163;
    Franklin K. Lane (US), 218;
    George L. Torain (Cdn), 72,
    149; *Gobeo*, 249; *Hagan* (US),
    161; *Hooiberg* (UK), 49, 51,
    55, 61, 70, 80; *J.N. Pew*, 72;
    Kennebec (UK), 3; *Kennox*, 72;
    Kongsgaärd, 72; *La Carriere*,
    72, 84–85; *Lady Nelson* (Cdn),
    111; *Lihue*, 72, 100–102;
    Macgregor, 72, 86; *Mary*
    (US), 118; *Mokihana*, 121;
    Monagas, 55–57, 121; *Mont
    Louis* (Cdn), 149; *Nordvangen*,
    72, 116; *Oranjestad*, 72, 121;
    Oregon, 72, 87; *Pedernales*
    (UK), 52–66 passim, 72, 121;
    Potlatch (US), 216; *Potrero*

War Production Board, 201, 204–5. *See also*
    Supply Priorities and Allocations
    Board
Wattenberg, Jurgen (*U-162*), 145–50,
    153–54, 223–26, 231, 274
wildcat strike, Trinidad, 23
Winant, John G. (US Ambassador to the
    UK), 16

wolf packs, xviii, 283
    tactics of, xviii, 283

Y

Young, Sir Hubert (UK Governor of
    Trinidad), 21–22, 25–29, 43, 94,
    181, 190, 193